Lecture Notes in Computer Science 6387

Rainer Böhme
Philip W.L. Fong
Reihaneh Safavi-Naini (Eds.)

Information Hiding

12th International Conference, IH 2010
Calgary, AB, Canada, June 28-30, 2010
Revised Selected Papers

Springer

Volume Editors

Rainer Böhme
International Computer Science Institute
1947 Center Street, Suite 600
Berkeley, CA 94704, USA
E-mail: rainer.boehme@icsi.berkeley.edu

Philip W.L. Fong
University of Calgary
Department of Computer Science
2500 University Drive NW
Calgary, AB, T2N 1N4, Canada
E-mail: pwlfong@ucalgary.ca

Reihaneh Safavi-Naini
University of Calgary
Department of Computer Science
2500 University Drive NW
Calgary, AB, T2N 1N4, Canada
E-mail: rei@ucalgary.ca

Library of Congress Control Number: 2010936196

CR Subject Classification (1998): E.3, K.6.5, D.4.6, E.4, H.5.1, I.4

LNCS Sublibrary: SL 4 – Security and Cryptology

ISSN 0302-9743
ISBN-10 3-642-16434-X Springer Berlin Heidelberg New York
ISBN-13 978-3-642-16434-7 Springer Berlin Heidelberg New York

springer.com

Printed in Germany

Typesetting: Camera-ready by author, data conversion by Scientific Publishing Services, Chennai, India
Printed on acid-free paper 06/3180

Preface

IH 2010 was the 12th Information Hiding Conference, held in Calgary, Canada, June 28–30, 2010. This series of conferences started with the First Workshop on Information Hiding, held in Cambridge, UK in May 1996. Since then, the conference locations have alternated between Europe and North America. The conference has been held annually since 2005.

For many years, information hiding has captured the imagination of researchers. This conference series aims to bring together a number of closely related research areas, including digital watermarking, steganography and steganalysis, anonymity and privacy, covert and subliminal channels, fingerprinting and embedding codes, multimedia forensics and counter-forensics, as well as theoretical aspects of information hiding and detection. Since its inception, the conference series has been a premier forum for publishing research in these areas. This volume contains the revised versions of 18 accepted papers (incorporating the comments from members of the Program Committee), and extended abstracts of two (out of three) invited talks.

The conference received 39 anonymous submissions for full papers. The task of selecting 18 of them for presentation was not easy. Each submission was reviewed by at least three members of the Program Committee or external reviewers reporting to a member of the Program Committee. In the case of co-authorship by a Program Committee member, five reviews were sought. There is no need to say that no member of the Program Committee reviewed his or her own work. Each paper was carefully discussed until consensus was reached. The contributions of invited speakers were not formally reviewed.

The invited speakers of IH 2010 were:

Gábor Tardos ..... Capacity of collusion-secure fingerprinting—a trade-off between rate and efficiency

Pim Tuyls .... Hardware instrinsic security

Boris Škorić ...... Security with noisy data

We would like to thank all those who helped with the organization of the conference and in particular the members of the local organizing team whose unrelenting effort ensured a smooth running of the conference. We would like to thank Kris Narayan for his continued effort in maintaining the Web pages and the iChair submission system, and for lending us a hand whenever it was needed. The conference benefitted from the generous financial support of Alberta Innovates, the European Office of Aerospace Research and Development, Technicolor, and the University of Calgary's Department of Computer Science, Faculty of Science, and Institute for Security, Privacy & Information Assurance (ISPIA).

We gratefully appreciate the work of the 24 external reviewers and 4 shepherds who lent us their experience in shepherding four conditionally accepted papers.

Finally, we would like to thank the authors of all submitted papers for their hard work, and also all presenters and attendees of the conference whose support ensured the success of this conference.

August 2010

Rainer Böhme
Philip W. L. Fong
Reihaneh Safavi-Naini

Organization

Information Hiding 2010 was hosted by the Department of Computer Science, University of Calgary, Alberta, Canada.

Executive Committee

General Chair	Philip W. L. Fong (University of Calgary, Canada)
Program Chairs	Rainer Böhme (ICSI Berkeley, USA) Rei Safavi-Naini (University of Calgary, Canada)

Program Committee

Ross Anderson	University of Cambridge, UK
Mauro Barni	University of Siena, Italy
Patrick Bas	CNRS, France
Francois Cayre	GIPSA-lab Grenoble, France
Ee-Chien Chang	University of Singapore
Christian Collberg	University of Arizona, USA
Ingemar J. Cox	University College London, UK
Gwenaël Doërr	Technicolor, France
Hany Farid	Darthmouth College, USA
Jessica Fridrich	SUNY Binghamton, USA
Teddy Furon	INRIA, France
Neil F. Johnson	Booz Allen Hamilton & JJTC, USA
Stefan Katzenbeisser	TU Darmstadt, Germany
Darko Kirovski	Microsoft Research, USA
John McHugh	Univ. North Carolina & RedJack LLC, USA
Ira S. Moskowitz	Naval Research Lab, USA
Andreas Pfitzmann	TU Dresden, Germany
Ahmad-Reza Sadeghi	Ruhr-University Bochum, Germany
Phil Sallee	Booz Allen Hamilton, USA
Berry Schoenmakers	TU Eindhoven, The Netherlands
Kaushal Solanki	Mayachitra Inc., USA
Kenneth Sullivan	Mayachitra Inc., USA

Organizing Team

Local Organizers	Mohd Anwar, Mina Askari, Martin Gagné, Kris Narayan, Camille Sinanan
External Advice	Stefan Katzenbeisser (TU Darmstadt)
Volunteers	Mohsen Alimomeni, Hoi Le, Mohammed Tuhin

External Reviewers

André Adelsbach
Hadi Ahmadi
Gerard Allwein
Manuela Berg
Stefan Berthold
Tomas Filler
Elke Franz
Martin Gagne
Steven J. Greenwald
Christian Grothoff
Stefan Köpsell
Matthias Kirchner
Yali Liu
Hans Löhr
Cathy Meadows
Bartolomeo Montrucchio
Kris Narayan
Richard E. Newman
Dagmar Schönfeld
Haya Shulman
Boris Škorić
Robin Sommer
Andreas Westfeld
Antje Winkler

Shepherds

Rainer Böhme	ICSI Berkeley, USA
Christian Grothoff	TU München, Germany
Teddy Furon	INRIA, France
Matthias Kirchner	TU Dresden, Germany

Sponsoring Institutions

Alberta Innovates — Technology Future, Edmonton, Canada
European Office of Aerospace Research and Development, London, UK
Technicolor S. A., France
University of Calgary (Department of CS, Faculty of Science, ISPIA)

Table of Contents

FPGA Time-Bounded Unclonable Authentication

Mehrdad Majzoobi, Ahmed Elnably, and Farinaz Koushanfar

Rice University, Electrical and Computer Engineering, Houston TX 77005, USA
{mehrdad.majzoobi,ahmed.elbanby,farinaz}@rice.edu

Abstract. This paper introduces a novel technique for extracting the unique timing signatures of the FPGA configurable logic blocks in a digital form over the space of possible challenges. A new class of physical unclonable functions that enables inputs challenges such as timing, digital, and placement challenges can be built upon the delay signatures. We introduce a suite of new authentication protocols that take into account non-triviality of bitstream reverse-engineering in addition to the FPGA's unprecedented speed in responding to challenges. Our technique is secure against various attacks and robust to fluctuations in operational conditions. Proof of concept implementation of the signature extraction and evaluations of the proposed methods are demonstrated on Xilinx Virtex 5 FPGAs. Experimental results demonstrate practicality of the proposed techniques.

1 Introduction

Any security mechanism is centered on the concept of a secret. Classic cryptography protocols relay on a secret key for reversing trapdoor functions. While such protocols are often secure against algorithmic attacks, it is well-known that the digitally stored secret keys can be attacked and cloned. Furthermore, conventional SRAM-based FPGAs suffer from major limitations in secret key storage since their fabrication technology cannot easily integrate non-volatile memory (NVM). Thus, the keys need to be stored on off-chip memory where communicating the keys to- and from- the off-chip components demands secure channels and additional protocols. On-chip NVM requires the overhead of a constant power source. Such a reliance not only incurs additional overhead, but increases the vulnerability to attacks.

Physical unclonable functions (PUFs) are an efficient enabling mechanism for many security applications [1,2]. PUFs exploit the secret information that inherently exists in the unclonable physical variations of the silicon devices. Moreover, PUFs provide a unique chip-dependent mapping from a set of digital inputs (challenges) to digital outputs (responses) based on the unique properties of the underlying physical device. PUFs can be employed to provide security at multiple levels and for addressing a range of problems in securing processors [3], software protection [4], IP protection [5], and IC authentication [6]. Even though a number of methods for realizing PUFs on FPGAs were proposed and implemented to date [5,7], the scope of the existing FPGA PUF methods is limited. Either they have a limited number of challenge-response pairs, or they are limited by the routing constraints on FPGAs, or they are vulnerable to learning and reverse engineering attacks.

R. Böhme, P.W.L. Fong, and R. Safavi-Naini (Eds.): IH 2010, LNCS 6387, pp. 1–16, 2010.

In this paper, we introduce a novel approach for implementing FPGA PUFs. In this proposed approach, first, the timings of each configurable logic block is accurately characterized for different combinations of inputs. Then a PUF is implemented such that a challenge to the PUF queries relationship of the clock pulses with respect to the delays for a subset of configurable blocks. Reproducing the responses requires the knowledge of the extracted delays as well as the structure and placement of the PUF circuit. We introduce a dynamic authentication protocol, where the placement of the PUF is randomly updated in each round of authentication, forcing the adversary to constantly reverse-engineer the configuration bit stream to discover the PUF structure and placement. The protocol is based on the fact that it is infeasible to reverse-engineer the configuration bit stream and generate the challenge-response signatures through simulation in a short time duration compared to evaluation on hardware. Our contributions are as follows:

- Introduction of structures for efficiently extracting the unclonable analog delay signatures of each of the FPGA configurable logic blocks over the possible inputs (challenges) using only commodity test equipments.
- A suite of new authentication protocols are built upon the extracted timing signatures.A new time-bounded protocol further takes advantage of the gap of between PUF simulation time and evaluation time on the authentic physical hardware to enhance the security of authentication against learning attacks.
- A linear calibration method is developed to compensate for the effects of variations in operating temperature and supply voltage to assure signature consistency.
- Experimental results and proof-of-concept implementation are demonstrated on Xilinx Virtex 5 FPGAs. The results show the applicability of the proposed methods, uniqueness of the signatures, and their robustness against the fluctuations in operational and environmental conditions.

The remainder of the paper is organized as follows. Section 2 summarizes the related literature. In Section 3 we present our method for timing signature extraction from a circuit for each possible challenges. Section 4 introduces time-bounded authentication protocol and its variants. Attacks and countermeasures are discussed in Section 4.3. Calibration for signature and response robustness against fluctuations in operational conditions is discussed in Section 5. Experimental evaluations are demonstrated in Section 6. Finally, we conclude in Section 7.

2 Related Work

The idea of using the complex unclonable features of a physical system as an underlying security mechanism was initially proposed by Pappu et al. [1]. The concept was demonstrated by mesoscopic physics of coherent light transport through a disordered medium. Another group of researchers observed that the manufacturing process variability present in modern silicon technology can be utilized for building a PUF. They proposed the arbiter-based PUF architecture which was based on the variations in CMOS logic delays, and demonstrated its implementation in the ASIC technology, and discussed a number of attacks and countermeasures [2,8,9,6,10]. In particular, they made the observations that the linear arbiter-based PUF is vulnerable to modeling attacks and proposed using nonlinear feed forward arbiters and hashing to safeguard against this attack [2]. Furthermore, they made the important observation that the PUF responses may not all be stable, and to alleviate the issue, proposed utilizing error correcting codes [11]. Further efforts were made to address the PUF vulnerability issues by adding input/output networks, adding nonlinearities to hinder machine learning and enforcing an upper bound on the PUF evaluation time [7,12,13,14,15]. The recent work in [7] demonstrated that even though successful ASIC implementation of arbiter PUF was shown, FPGA implementation of this PUF is not possible in the state-of-the-art technology. This is because of the routing constraints for implementing the similar parallel paths enforced by the regularities in the underlying FPGA fabric. For implementing PUFs on FPGA, Ring oscillator (RO) PUFs were proposed [6]. The major drawback of the RO PUFs is having only a quadratic number of challenges with respect to the number of ROs [12]. Furthermore, the ROs (while in use) consume significant dynamic power due to frequent transitions during oscillations. SRAM PUFs suffer from the same limitation in terms of the number of possible challenge combinations [12].

This paper presents the first practical method and proof of concept FPGA implementation of a PUF with exponential number of possible challenges of different kind including placement challenges. The new proposed PUF uses the unique cell-by-cell characteristics of the FPGA array. Note that time-bounded properties of FPGA PUFs was briefly mentioned by [7]. The authors in [13] exploited the time-boundedness characteristics of generic public key protocol by PUFs but no practical implementation options were discussed.

Besides the ongoing research on PUFs, several other relevant work on delay characterization serve as the enabling thrust for realization of our novel PUF structures. To perform delay characterization, Wong et al. in [16] proposed a built-in self-test mechanism for fast chip level delay characterization. The system utilizes the on chip PLL and DCM modules for clock generation at discrete frequencies.

3 Delay Signature Extraction

To measure the delays of components inside FPGA, we exploit the device reconfigurability to implement a delay signature extraction circuit. A high level view of the delay extraction circuitry is shown in Figure 1. The target circuit/path delay to be extracted is called the *Circuit Under Test (CUT)*. Three flip flops (FFs) are used in this delay extraction circuit: *launch FF*, *sample FF*, and *Capture FF*. The clock signal is routed to all three FFs as shown on the Figure. Assume for now that the binary challenge input to the CUT is held constant and thus, the CUT delay is fixed.

Assuming the FFs in Figure 1 were originally initialized to zero, a low-to-high signal is sent through the CUT by the launch flip flop at the rising edge of the clock. The output is sampled T seconds later on the falling edge of the clock (T is half the clock period). Notice that the sampling register is clocked at the falling edge of the clock. If the signal arrives at the sample flip flop before sampling takes place, the correct signal value would be sampled, otherwise the sampled value would be different producing an error. The actual signal value and the sampled value are compared by an XOR logic and the result will be held for one clock cycle by the capture FF.

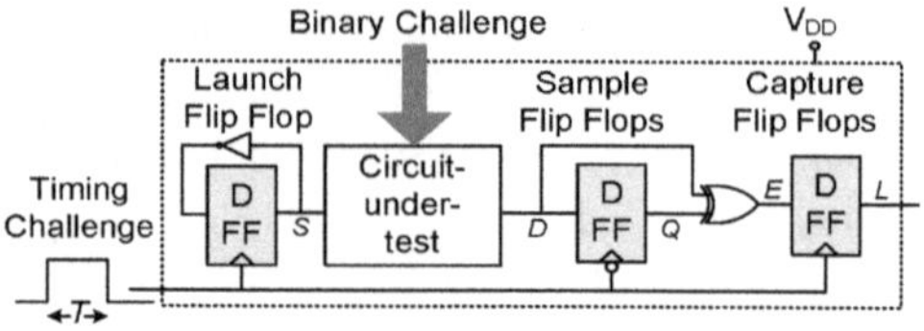

Fig. 1. The timing signature extraction circuit

A more careful timing analysis of the circuit reveals the relationship between the delay of the CUT (t_{CUT}), the clock pulse width (T), the clock-to-Q delay at the launch FF t_{clk2Q}, and the clock skew between the launch and sample FFs t_{skew}. The setup/hold of the sampling register and the setup/hold time of the capture register are denoted by t_{setS}, t_{holdS}, t_{setC}, and t_{holdC} respectively. The propagation delay of the XOR gate is denoted by t_{XOR}. The time it takes for the signal to propagate through CUT and reach the sample flip flop from the moment the launch flip flop is clocked is represented by t_P. Based on the circuit functionality in Figure 1, $t_P = t_{CUT} + t_{clk2Q} - t_{skew}$.

As T approaches t_P, the sample flip flop enters a metastable operation because of the setup and hold time violations and its output becomes nondeterministic. The probability that the metastable state resolves to a 0 or 1 is a function of how close T is to t_P. For instance, if T and t_{CUT} are equal, the signal and the clock simultaneously arrive at the sample flip flop and metastable state resolves to a 1 with a probability of 0.5. If there are no timing errors in the circuit, the following relationships must hold:

$$t_{holdC} < t_P < T - t_{setS} \tag{1}$$

$$t_P < 2T - (t_{setC} + t_{XOR}) \tag{2}$$

The errors start to appear if t_p enters the following interval:

$$T - t_{setS} < t_P < T + t_{holdS} \tag{3}$$

The rate (probability) of observing timing error increases as t_p gets closer to the upper limit of Equation 3. If the following condition holds, then timing error happens every clock cycle:

$$T + t_{holdS} < t_P < 2T - (t_{setC} + t_{XOR}) \tag{4}$$

Notice that in the circuit in Figure 1, high-to-low and low-to-high transitions travel through the CUT every other clock cycles. The propagation delay of the two differ in practice. Suppose that the low-to-high transition propagation delay ($t_p^{l\rightarrow h}$) is smaller than high-to-low transition propagation delay ($t_p^{h\rightarrow l}$). Then, for example if for low-to-high transitions, $t_p^{l\rightarrow h}$ satisfies Inequalities 1, 2 and for high-to-low transitions, $t_p^{h\rightarrow l}$ satisfies Inequality 4, timing errors happen only for high-to-low transitions and as a result timing error can only be observed 50% of the times. Figure 2 (a) illustrates this scenario.

Observability of timing errors follow a periodic behavior. In other words, if t_p goes beyond $2T - (t_{setC} + t_{XOR})$ in Inequality 4, the rate of timing errors begin to decrease again. However this time the decrease in the error rate is not a result of proper operation yet it is because the errors can not be observed and captured by the capture flip flip. Inequalities 5 and 6 correspond to the transition from the case where timing error happens every clock cycle Inequality 4) to the case where no errors can be detected Inequality 7.

$$2T - (t_{setC} + t_{XOR}) < t_P < 2T + (t_{holdC} - t_{XOR}) \quad (5)$$
$$t_P < 3T - t_{setS} \quad (6)$$

$$2T + (t_{holdC} - t_{XOR}) < t_P < 3T - t_{setS} \quad (7)$$

Timing errors no longer stay undetected if t_p is greater than $3T - t_{setS}$. Timing errors begin to appear and be captured if t_p falls into the following intervals:

$$3T - t_{setupS} < t_p < 3T + t_{holdS} \quad (8)$$
$$t_p < 4T - (t_{setupC} + t_{XOR}) \quad (9)$$

If the following condition holds, then timing error gets detected every clock cycle.

$$3T + t_{holdS} < t_p < 4T - (t_{setupC} + t_{XOR}) \quad (10)$$

This periodic behavior continues the same way for integer multiples of T, however it is physically upper bounded by the maximum clock frequency of the FPGA device. In general, if T is much larger than the XOR and flip flop delays, the intervals can be simplified to $n \times T < t_p < (n+1) \times T$ and timing errors can only be detected for odd values of n where n=0,1,2,3,...

In the rest of the paper, we refer to the characterization circuit that includes the as a *characterization cell* or simply a *cell*. Each cell in our implementation on FPGA is pushed into one configurable logic block (CLB). The circuit under test consists of four cascaded look-up tables (LUT) each implementing a variable delay inverter. We explain in Section 3.3 how the delay of the inverters can be changed.

3.1 Signature Extraction System

In this subsection, we present the system that efficiently extracts the probability of observing timing failure as a function of clock pulse width for a group of components on FPGA. The circuit shown in Figure 1 only produces a single bit flag of whether errors happen or not. We also need a mechanism to measure the rate or probability at which errors appear at the output of the circuit in Figure 1.

To measure the probability of observing error at a given clock frequency, an error histogram accumulator is implemented by using two counters. The first counter is the error counter whose value increments by unity every time an error takes place. The second counter counts the clock cycles and resets (clears) the error counter every 2^N clock cycles, where N is the size of the counters. The value of the error counter is stored in the memory exactly one clock cycle before it is cleared. The stored number of errors normalized to 2^N yields the error probability value. The clock frequency to the system is swept linearly and continuously in T_{sweep} seconds from $f_i = \frac{1}{2T_i}$ to $f_t = \frac{1}{2T_t}$, where $T_t < t_p < T_i$. A separate counter counts the number of clock pulses in each frequency sweep. This counter acts as an accurate timer that bookmarks when a timing error happened. The value of this counter is retrieved every time the number of counted errors are recorded (i.e. every 2^N cycles). A unique time stamp (T_m) can be calculated which indicates at what point during the sweep time a certain value appears in the clock counter. By knowing T_m, the frequency associated to the m-th error value

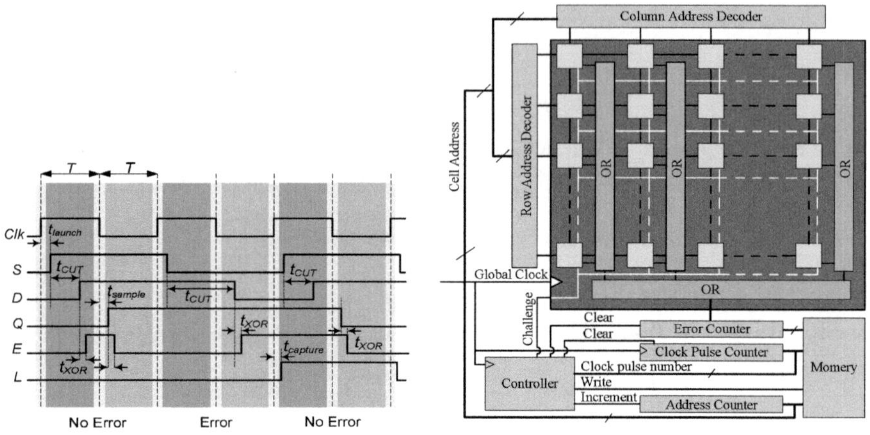

(a) (b)

Fig. 2. (a)The timing diagram showing occurrence of timing error. (b) The architecture for chip level delay extraction of logic components.

can be easily calculated using the linear interpolation $f_m = (f_t - f_i) \times \frac{T_m}{T_{sweep}} + f_i$. The system shown in Figure 2 (b) is used for extracting the delays of an array of CUTs on the FPGA. Each square in the array represents the characterization circuit shown in Figure 1 and is referred to as a *cell* in the remainder of the text. Any logic configuration can be utilized within the CUT in the characterization circuit. In particular, the logic inside the CUT can be made a function of binary challenges, such that its delay varies by the given inputs. The system in Figure 2 (b) characterizes each cell by sweeping the clock frequency once. Then, it increments the cell address and moves to the next cell. The cells are characterized in serial. The row and column decoders activate the given cell while the rest of the cells are deactivated. Therefore, the output of the deactivated cells remain zero. As a result, the output of the OR function solely reflect the timing errors captured in the activated cell. Each time the data is written to the memory, three values would be stored: the cell address, the accumulated error value, and the clock pulse number at which the error has occurred. The clock counter is reset at each new sweep. The whole operation iterates over different binary challenges to the cells. Please note that the scanning can also be performed in parallel to save time.

3.2 Parameter Extraction

So far we have described the system that measure the probability of observing timing error for different clock pulse widths. The error probability can be represented compactly by a set of few parameters. These parameters are directly related to the circuit component delays and flip flop setup and hold time. It can be shown that the probability of timing error can be expressed as the sum of shifted Gaussian CDFs [7]. The Gaussian nature of the error probabilities can be explained by the central limit theorem. Equation 11 shows the parameterized error probability function.

$$f_{\mathbf{D},\Sigma}(t) = 1 + 0.5 \sum_{i=1}^{|\Sigma|-1} -1^{\lceil i/2 \rceil} Q(\frac{t - d_i}{\sigma_i}) \tag{11}$$

where $Q(x) = \frac{1}{\sqrt{2\pi}} \int_x^\infty \exp\left(-\frac{u^2}{2}\right)$ and $d_{i+1} > d_i$. To estimate the timing parameters, f is fit to the set of measured data points (t_i, e_i), where e_i is the error value recorded when the pulse width equals t_i.

3.3 Challenge Configuration

To enable authentication, we require a mechanism for devising challenge inputs to the device and observing the device invoked responses. Fortunately, the capture flip flop yields a binary response. Assuming that the flip flop characteristics are known and constant, the response is a function of the clock pulse width T, and the t_{CUT}. Thus, one way to challenge the circuit and read its response out is to change the clock pulse width. The use clock pulse width has a number of implications. The response from the PUF will be deterministic if the T are either too high and too low. Predictability of responses makes it easy for the attacker to impersonate the PUF.

Another way to challenge the PUF is to alter the t_{CUT}. So far, we assumed that the delay of CUT is not changing which means the CUT have a specific configuration and specific input vector. Changing the input vector can alter the signal path delay, and hence the response. CUT is implemented by a set of LUT Figure 3 shows the internal circuit structure of an example 3-input LUT. In general, a Q-input LUT consists of 2^Q-1 2-input MUXs which allow selection of 2^Q SRAM cells. The SRAM cell values are configured to implement a pre-specific functionality. In this example, the SRAM cell values are configured to build a negated parity generation function of the three inputs to the LUT. If the number of ones in $A_1A_2A_3$ are even, then the output will be equal to 1. If the inputs A_1A_2 are held constant, the function $O = f(A_1)$ implements an inverter regardless of A_1A_2. However changing the input A_1A_2 can alter the delay of the inverter due to the changes in the signal propagation path inside the LUT and process-dependant variations in delay of paths with the same length. The LUTs in Xilinx Virtex 5 FPGAs, consist of 6 inputs. Five inputs of the LUT can be used to alter the inverter delay yielding 2^5 = 32 distinct delays for each LUTs.

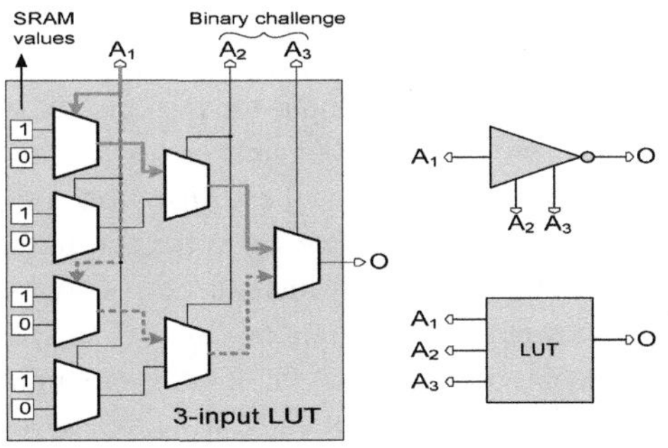

Fig. 3. The internal structure of LUTs. The signal propagation path inside the LUTs change as the inputs change.

4 Authentication

In this section, we show how the extracted cell characteristics in Section 3 can be utilized for FPGA authentication. The following terminology is used in the rest of the paper. The *verifier* (V) authenticates the *prover* (P) who owns the FPGA device. The verifier authenticates the device by verifying the unique timing properties of the device. The challenge vectors are denoted by c_i, $i = 1, \ldots, N$, and the corresponding responses are denoted by r_i, $i = 1, \ldots, N$. The PUF that performs the challenge to response transformation is denoted by $\mathcal{T} : \mathcal{T}(c_i) := r_i$, $i = 1, \ldots, N$.

4.1 Classic Authentication

The registration and authentication processes for the classic authentication case are demonstrated in the diagram in Figure 4 (a) and (b) (disregard the darker boxes for now). The minimum required assumptions for this case are (i) the verifier is not constrained in power (ii) it is physically impossible to clone the FPGA (iii) the characteristics of the FPGA owned by the prover is a secret only known to the prover and verifier.

As shown in Figure 4(a), during the registration phase, the verifier extracts and securely stores the cell delay parameters by performing characterization as explained in Sections 3. By knowing the FPGA-specific features in addition to the structure and placement of the configured PUF circuit, the verifier is able to predict the responses to any challenges to the PUF circuit. After registrations, the FPGA along with the pertinent PUF configuration bitstream is passed to the end-user.

At the authentication, end-user (prover) is queried by the verifier to make sure he is the true owner of the FPGA. Classic authentication is shown in Figure 4(b). To authenticate the ownership, the verifier utilizes a random seed and generates a set of pseudorandom challenge vectors for querying the prover. The prover responds to the challenges she receives from the verifier by applying them to the configured FPGA hardware. The verifier then compares the received responses from the prover with the predicted ones, and authenticates the chip if the responses are similar.

To ensure robustness against errors in measuring the delays and the change in delay measurement conditions, the registration entity may also compute the error correction information for the responses to the given challenges. To prevent information leakage via the error correction bits, secure sketch techniques can be used. A secure sketch produces public information about its input that does not reveal the input, and still permits exact recovery of the input given another value that is close to it [17].

The device is authenticated if the response after error correction would be mapped to the the verifier-computed hash of responses. Otherwise, the authentication would fail. Alternatively, the verifier can allow for some level of errors in the collected responses and remove the error correction and hashing from the protocol. However, accepting some errors in the responses, the verifier would be more suspectable to emulation/impersonating attacks [2,15].

4.2 Time-Bounded FPGA Authentication Using Reconfigurability

After the FPGA registration, the verifier is able to compute and predict the responses to any set of challenges by knowing (i) the cell-level features of the pertinent FPGA, (ii)

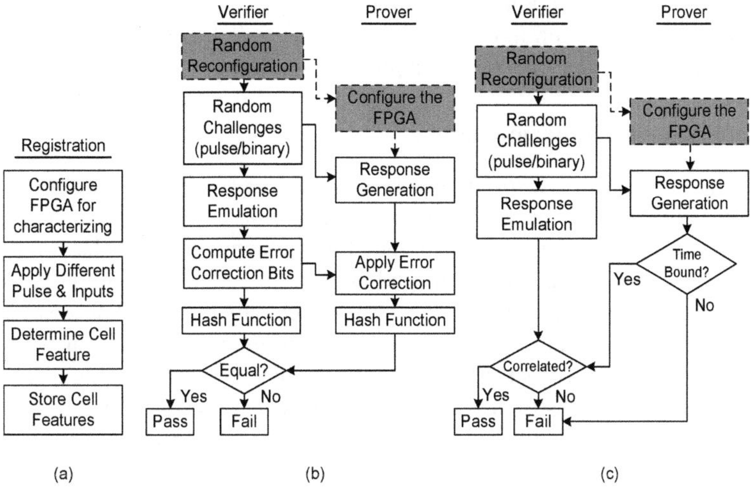

Fig. 4. (a) FPGA registration (b) Classic authentication flow (c) Time-bound authentication flow

the circuit structure and (iii) placement of the PUF circuit. The information on the PUF circuit structure and placement is embedded into the configuration bitstream. In the classic authentic method, the bitstream is never changed. A dishonest prover, off-line and given enough time and resources can (i) extract the cell-level delays of the FPGA and (ii) reverse engineer the bitstream to discover the PUF structure and its placement on the FPGA. During the authentication, he can compute the responses to the given challenges online by simulating the behavior of the PUF on the fly and produce the responses that pass the authentication.

A stronger set of security protocols can be built upon the fact that the prover is the only entity who can compute the correct response to a random challenge within a specific time bound since he has access to the actual hardware. In this protocol, prior to the beginning of the authentication session, the FPGA is blank. The verifier then sends a bitstream to the device in which a random subset of LUTs are configured for authentication. After the device is configured, the verifier starts querying the FPGA with random challenges. The verifier accepts the responses that are returned back only if $\Delta t \leq \Delta t_{\max}$ where Δt is the time lapsed on the prover device to compute the responses after receiving the configuration bitstream, and $\Delta t_{\max}$ is the upper bound delay estimated computation of responses by the authentic FPGA prover device, which is composed of device configuration, response generation, error correction, and hashing time all performed in hardware. The verifier would authenticate the device, only if the time the device takes to generate the response is less than $\Delta t_{\max}$. We denote the minimum emulation time by $t^{emu}_{\min}$, where $t^{emu}_{\min} >> \Delta t_{\max}$. Time-bounded authentication protocol can be added to the authentication flow, as demonstrated in Figure 4(c). Compared to the classic authentication flow, a time bound check is added after the hash function. While performing the above authentication, we emphasize on the assumption that the time gap between the hardware response generation and the simulation (or emulations) of the prover must be larger than the variation in channel latency. The time-bound assumption would be enough for providing the authentication proof [7,13,18].

4.3 Attacks and Countermeasures

Perhaps the most dangerous of all attacks is the impersonation attack. Impersonation attack aims at deceiving the verifier into authentication by cloning the same physical device, reverse-engineering and simulation of the authentic device behavior, or storing and replaying the communication, or random guessing. Among these threats only the reverse engineering and simulation attack may stand any chance of success. To break the time-bound protocol, an adversary needs to find the response to a new challenge, he has to reverse engineering the bitstream and simulate (or emulate) the PUF behavior within the given time constraint. Even after many years of research in rapid simulation technologies for hardware design and validation, fast and accurate simulation or emulation of a hardware architecture is extremely slow compared to real device. In addition, even though bitstream reverse-engineering have partially been performed on some FPGAs [19], performing it would require a lot of simulations and pattern matching. Thus, it would take many more cycles than the authentic hardware where the verifying time is dominated by bitstream configuration time (order of $100\mu s$). Simulating bitstream on software models would also take many more cycles than hardware and cannot be done within the limited time-bound. A great advantage of this authentication method is the large degree of freedom in selecting the LUTs that would be queried.

5 Robustness

The extracted delay signatures at characterization phase are subject to changes due to aging of silicon devices, variations in the operating temperature and supply voltage of the FPGA. Such variations can undermine the reliability of the authentication process. In this paper, we take a pragmatic approach to the problem which in conjunction with existing error correction methods [11] can significantly leverage performance and reliability of key generation and authentication. The proposed method performs calibration on clock pulse width according to the operating conditions.

Fortunately, many modern FPGAs are equipped with built-in temperature and core voltage sensors. Before authentication begins, the prover is required to send to the verifier the readings from the temperature and core voltage sensors. The prover then based on the current operating conditions calibrate the clock frequency. The presented calibration method linearly adjusts the pulse width using Equation 12.

$$T_{calib} = \alpha_{temp} \times (temp_{current} - temp_{ref}) + T_{ref} \tag{12}$$

$$T_{calib} = \alpha_{vdd} \times (vdd_{current} - vdd_{ref}) + T_{ref} \tag{13}$$

$temp_{ref}$ and vdd_{ref} are the temperature and FPGA core voltage measured at the characterization time. $temp_{current}$ and $vdd_{current}$ represent the current operating conditions. The responses from the PUF to the clock pulse width T_{calib} are then treated as if T_{ref} were sent to the PUF at reference operating condition. The calibration coefficients

α_{temp} and α_{vdd} are device specific. These coefficients can be determined by testing and characterizing each single FPGA at different temperatures and supply voltages. For example, if $D_i^{temp_1}$ and $D_i^{temp_2}$ are i-th extracted delay parameter under operating temperatures $temp_1$ and $temp_2$, then $\alpha_{t,i} = \frac{D_i^{t_1} - D_i^{t_2}}{t_1 - t_2}$.

6 Experimental Evaluations

In this section, the implementation details of the signature extraction system are presented. We demonstrate results obtained by measurements performed on Xilinx FPGAs and further use the platform to carry out authentication on available population of FPGAs. For delay signature extraction, the system shown in Figure 2 (b) is implemented on Xilinx Virtex 5 FPGAs. The systems contains a 32$\times$ 32 array of signature extraction circuits as shown in Figure 1. The CUT inside the characterization circuit consists of 4 inverters each being implemented using one 6-input LUT. The first LUT input (A_1) is used as the input of the inverter and the rest of the LUT inputs (A_2,...,A_6) serve as the binary challenges to alter the effective delay of the inverter. The characterization circuit is pushed into 2 slices (one CLB) on the FPGA. In fact, this is lower limit on number of slices that can be used to implement each characterization circuit. This is because interconnections inside the FPGA forces all the flip flops inside the same slice use either rising edge or falling edge clocks. Since the launch and sample flip flop must operate on different clock edges, they cannot be placed inside the same slices. In total, 8 LUTs and 4 flip flops are used (within two slices) to implement the characterization circuit. The error counter size (N) is set to 8, and the accumulated error values are stored if they are between 7 and 248.

We use an ordinary desktop function generator to sweep the clock frequency from 8MHz to 20MHz and afterwards shift the frequency 34 times up using the PLL inside the FPGA. The sweeping time is set to 1 milliseconds (due to the limitations of the function generator, the lower sweeping time could not be reached). The measured accumulated error values are stored on an external memory and the data are transferred to computer for further processing. Notice that the storage operation can easily be performed without the logic analyzer by using any off-chip memory.

The system is implemented on twelve Xilinx Virtex 5 XC5VLX110 chips and the measurements are taken under different input challenges and operating conditions. The characterization system in total uses 2048 slices for the characterization circuit array and 100 slices for the control circuit out of 17,280 slices.

The measured samples for each cell and the input challenge is processed and the twelve parameters as defined in Section 3.2 are extracted. Figure 5 shows the measured probability of timing error versus the clock pulse width for a single cell and a fixed challenge. The (red) circles represent original measured sample points and the (green) dots show the reconstructed samples. As explained earlier, to reduce the stored data size, error samples with values of 0 and 1 (after normalization) are not written to the memory and later are reconstructed from the rest of the sample points. The solid line shows the Gaussian fit on the data expressed in equation 11. Parameter extraction procedure is repeated for all cells and challenges. Figure 6 shows the extracted parameters d_1 and σ_1 for all PUF cells on chips #9 while the binary challenge is fixed. The pixels in the

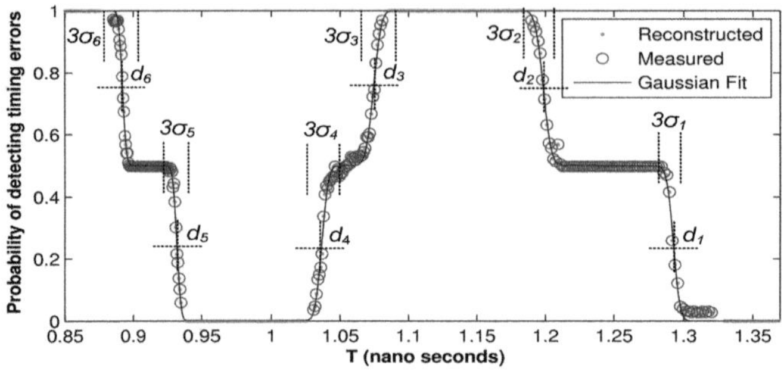

Fig. 5. The probability of detecting timing errors versus the input clock pulse width T. The solid line shows the Gaussian fit to the measurement data.

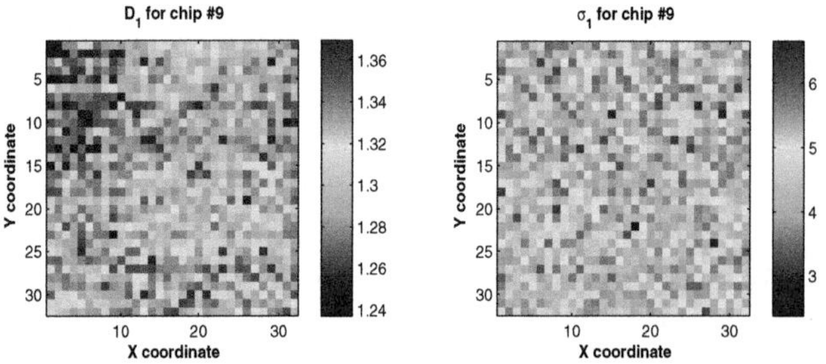

(a) (b)

Fig. 6. The extracted delay parameters d_1 (a) and σ_1 (b) for chips 9

images correspond to the cells of the 32×32 array on FPGA, and their value represent the corresponding extracted parameters. Some levels of spatial correlation among d_1 parameters can be observed on the FPGA fabric. The boxplot in Figure 7 (a) shows the distribution of the delay parameters d_i for i=1,2,...,6 over all 12 chips and 1024 cells and 2 challenges. The central mark on the boxplot denotes the median, the edges of the boxes correspond to the 25th and 75th percentiles, the whiskers extent to the most extreme data points and the red plus signs show the outlier points.

Now using the measured data from the twelve chips, we investigate different authentication scenarios. The existing authentication parameters within defined framework substantially increases the degree of freedom under which authentication may take place. These parameters include the number of clock pulses (denoted by N_p), the number of tried challenges (denoted by N_c), the clock pulse width (denoted by T), and the number of PUF cells (N_{cell}) being queried. In other words, in each round of authentication, N_c challenges are tried during which N_p pulses of width T are sent to N_{cell} cells on the chip. The response for each challenge can be regarded as the percentage of ones in the N_p response bits.

In the first experiment, we study the effect of the number of cells and the width of clock pulse on the probability of detection (p_d) and false alarm (p_f). Detection error occur in cases where the test and target chip are the same, but due instability and noise

Table 1. Probability of false alarm (a) and probability of detection (b)

(a)

N_{cell}	Challenge Pulse Width					
	1.23	1.15	1.06	1.03	0.9	0.87
64	0.96	0	0	0	0	1.52
128	2.04	0	0	0	0	1.52
256	4.55	0	0	0	0	1.52

(b)

N_{cell}	Challenge Pulse Width					
	1.23	1.15	1.06	1.03	0.9	0.87
64	93.3	96.2	100	100	100	100
128	94.2	98.8	100	100	100	100
256	99.85	100	100	100	100	100

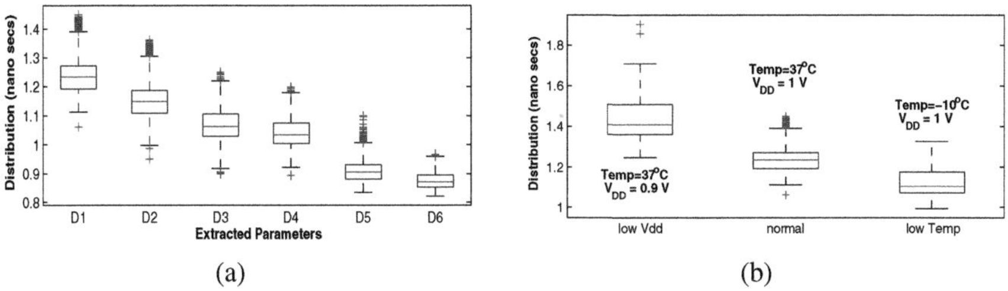

Fig. 7. (a) Distribution of delay parameters d_i. (b) The distribution of d_1 for normal, low operating temperature, and low core voltage.

in responses to fail to be authenticated as the same. On the other hand, false alarm corresponds to the cases where the test and target chip are the different, but they are identified at the same chips. During the experiment, the binary challenges to PUF cells are fixed and the number clock pulses is set to $N_p = 8$. Next, we study different cases where the clock width (T) is set to each of the medians of the values shown in Figure 7 (a). Setting the clock pulse width to the median values result in least predictability in responses. The same experiment is repeated for 10 times to obtain 10 response vectors for each chip. After that, the distance between the responses from the same chips (intra-chip) over repeated evaluations are measured using the normalized L_1 distance metric. The same procedure is performed on responses from different chips over difference evaluations in order to find the inter-chip responses.

If the distance between the test chip and the target chip responses is smaller than a pre-specified detection threshold, then the chip is successfully authenticated. In the experiments the detection threshold is set at 0.15.

Tables 1 shows the probability of detection and false alarm for different clock pulse widths and number of queried PUF cells. As it can be observed the information extracted from even the smallest set of cells is sufficient to reliably authenticate the FPGA chip if the pulse width is correctly set. In the next experiments, we study the effect of fluctuations in the operating conditions (temperature and core supply voltage) on the probabilities of detection and false alarm. Moreover, we demonstrate how linear calibration on the challenge clock pulse width can improve the reliability of detection.

To determine the calibration coefficient defined in Equation 12, we repeat the delay extraction process and find the delay parameters for all twelve chips at temperature -10^oC and core voltage 0.9 Volts. The chip operates at the temperature 37^oC and core

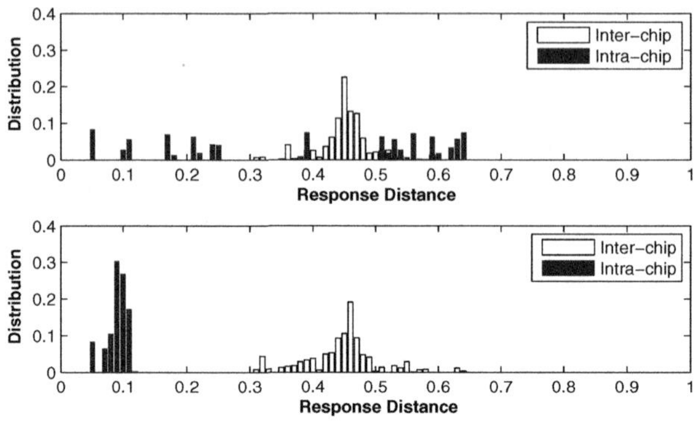

Fig. 8. The inter-chip and intra-chip response distances for $T = 0.95\ ns$ and $N_c = 2$ before (top) and after (bottom) calibration against changes in temperature

Table 2. The probability of detection and false alarm before and after performing calibration on the challenge pulse width in presence of variations in temperature and core voltage

		No Calibration								Calibrated							
		N_C=1				N_C=2				N_C=1				N_C=2			
		v_{low}		t_{low}		v_{low}		t_{low}		v_{low}		t_{low}		v_{low}		t_{low}	
		p_d	p_f	p_d	p_f	p_d	p_f	p_d	p_f	p_d	p_f	p_d	p_f	p_d	p_f	p_d	p_f
T	1.23	18.4	0	33.3	16.7	18.4	0	33.3	22.29	100	0	75	0	100	0	75	0
	1.06	18.4	0	18.4	0	18.4	0	18.4	0	50	0	50	0	57.3	0	50	0
	1.01	18.4	0	16.7	0	18.4	0	16.7	0	66.6	0	75	0	68.2	0	75	0
	0.95	18.4	0	16.7	0	18.4	0	16.7	0	66.7	0	100	0	84.9	0	100	0
	0.9	16.7	0	25	0	16.7	0	25	0	83.3	0	91.7	0	83.4	0	100	0
	0.87	25	0	25	0	25	1.5	25	0	100	0	100	0	100	0	100	0

voltage of 1 volts in the normal (reference) condition. We use the built-in sensors and the Xilinx Chip Scope Pro package to monitor the operating temperature and core voltage. To cool down the FPGAs, liquid compressed air is consistently sprayed over the FPGA surface. Figure 7 (b) depicts the changes in the distribution of the first delay parameter d_1 at the three different operating conditions.

The probabilities of detection and false alarm are derived before and after performing calibration on the challenge pulse width for different clock pulse widths and number of binary challenges to the cells. In this experiment, all 1024 PUF cells on the FPGA are queried for the response. The number of pulses sent for each binary challenge is set N_p =8 as before. As it can be seen in Table 2, the detection probabilities are significantly improved after performing linear calibration based on the coefficients extracted for each chip. The variables v_{low} and t_{low} correspond to -10^oC temperature and 0.9 supply voltages respectively. The reported probabilities of Table 2 are all in percentage. Also note that for the challenge pulse width of $T = 0.87\ ns$, the probability of detection reaches 100% and probability of alarm falls to zero after calibration. The same holds true for

$N_c = 2$ and $T = 0.87, 0.9, 0.95$. Thus, increased level of reliability can be achieved during authentication with proper choice of pulse width and number of challenges.

Figure 8 shows how performing calibration decreases the intra-chip response distances in presence of temperature changes. The histogram correspond to $T = 0.95\ ns$ and $N_c = 2$ in Table 2 before and after calibration.

7 Conclusions

We presented a technique for FPGA authentication that takes advantage of the unclonable timings variability present in FPGAs, the FPGA reconfigurability, and its unprecedented speed. Authentication comprises of two phases; namely registration and authentication. During registration, cell level timing features are extracted and stored in a database. Later at the authentication phase, the verifier generates a random configuration bitstream and sends to the prover. A unique aspect of the new method is its high degree of freedom in placing the PUF cells and selection of challenges. The protocol relies on the fact that online reverse-engineering of the bitstream is a non-trivial task. A new calibration method for improving robustness to temperature and voltage fluctuations was demonstrated. Evaluations on Xilinx V5 FPGA show the effectiveness and practicality of the new timing signature extraction and authentication method.

References

1. Pappu, R., Recht, B., Taylor, J., Gershenfeld, N.: Physical one-way functions. Science 297, 2026–2030 (2002)
2. Gassend, B., Clarke, D., van Dijk, M., Devadas, S.: Silicon physical random functions. In: CCS, pp. 148–160 (2002)
3. Suh, G., O'Donnell, C., Devadas, S.: AEGIS: A single-chip secure processor. IEEE Design & Test of Computers 24(6), 570–580 (2007)
4. Atallah, M., Bryant, E., Korb, J., Rice, J.: Binding software to specific native hardware in a VM environment: the PUF challenge and opportunity. In: VMSec, pp. 45–48 (2008)
5. Guajardo, J., Kumar, S., Schrijen, G., Tuyls, P.: FPGA intrinsic PUFs and their use for IP protection. In: Paillier, P., Verbauwhede, I. (eds.) CHES 2007. LNCS, vol. 4727, pp. 63–80. Springer, Heidelberg (2007)
6. Suh, G., Devadas, S.: Physical unclonable functions for device authentication and secret key generation. In: DAC, pp. 9–14 (2007)
7. Majzoobi, M., Koushanfar, F., Potkonjak, M.: Techniques for design and implementation of secure reconfigurable pufs. ACM TRETS 2(1), 1–33 (2009)
8. Gassend, B., Clarke, D., van Dijk, M., Devadas, S.: Controlled physical random functions. In: ACSAC, pp. 149–160 (2002)
9. Lee, J., Daihyun, L., Gassend, B., Suh, G., van Dijk, M., Devadas, S.: A technique to build a secret key in integrated circuits for identification and authentication applications. In: Symp. of VLSI, pp. 176–179 (2004)
10. Holcomb, D., Burleson, W., Fu, K.: Power-up SRAM state as an identifying fingerprint and source of true random numbers. IEEE Trans. Computers (2009)
11. Gassend, B.: Physical random functions. S.m. thesis, MIT (2003)
12. Ruhrmair, U., Solter, J., Sehnke, F.: On the foundations of physical unclonable functions. In: Cryptology ePrint Archive (2009)

13. Ruhrmair, U.: SIMPL system: on a public key variant of physical unclonable function. In: Cryptology ePrint Archive (2009)
14. Majzoobi, M., Koushanfar, F., Potkonjak, M.: Lightweight secure PUFs. In: ICCAD, pp. 670–673 (2008)
15. Majzoobi, M., Koushanfar, F., Potkonjak, M.: Testing techniques for hardware security. In: International Test Conference (ITC), pp. 1–10 (2008)
16. Wong, J.S.J., Sedcole, P., Cheung, P.Y.K.: Self-measurement of combinatorial circuit delays in fpgas. ACM TRETS 2(2), 1–22 (2009)
17. Dodis, Y., Reyzin, L., Smith, A.: Fuzzy extractors: How to generate strong keys from biometrics and other noisy data. In: Cachin, C., Camenisch, J.L. (eds.) EUROCRYPT 2004. LNCS, vol. 3027, pp. 523–540. Springer, Heidelberg (2004)
18. Beckmann, N., Potkonjak, M.: Hardware-based public-key cryptography with public physically unclonable functions. In: Katzenbeisser, S., Sadeghi, A.-R. (eds.) IH 2009. LNCS, vol. 5806, pp. 206–220. Springer, Heidelberg (2009)
19. Note, J., Rannaud, E.: From the bitstream to the netlist. In: FPGA, pp. 264–274 (2008)

A Unified Submodular Framework for Multimodal IC Trojan Detection

Farinaz Koushanfar, Azalia Mirhoseini, and Yousra Alkabani

Rice University, Electrical and Computer Engineering, Houston TX 77005, USA
{farinaz,azalia,yousra}@rice.edu

Abstract. This paper presents a unified formal framework for integrated circuits (IC) Trojan detection that can simultaneously employ multiple noninvasive measurement types. Hardware Trojans refer to modifications, alterations, or insertions to the original IC for adversarial purposes. The new framework formally defines the IC Trojan detection for each measurement type as an optimization problem and discusses the complexity. A formulation of the problem that is applicable to a large class of Trojan detection problems and is *submodular* is devised. Based on the objective function properties, an efficient Trojan detection method with strong approximation and optimality guarantees is introduced. Signal processing methods for calibrating the impact of inter-chip and intra-chip correlations are presented. We propose a number of methods for combining the detections of the different measurement types. Experimental evaluations on benchmark designs reveal the low-overhead and effectiveness of the new Trojan detection framework and provides a comparison of different detection combining methods.

1 Introduction

The prohibitive cost of manufacturing ICs in nano-meter scales has made the use of contract foundries the dominant semiconductor business practice. Unauthorized IP usage, IC overbuilding, and insertion of additional malware circuitry (*Trojans*) are a few of the major threats facing the horizontal IC industry where the IP providers, designers, and foundries are separate entities [8]. The Trojan attacker modifies the original design to enable an adversary to control, monitor, spy contents and communications, or to remotely activate/disable parts of the IC. Trojans are often hidden and are rarely triggered as needed.

A standing challenge for noninvasive IC testing and Trojan detection is dealing with the increasing complexity and scale of the state-of-the-art technology. It is hard to distinguish between the characteristic deviations because of the process variations and the alterations due to the Trojan insertion. What complicates the problem even more is that the space of possible changes by the adversary is large. Very little is known or documented about IC Trojan attacks. The possible adversaries are likely financially and technologically advanced and thus, intelligent attacks are possible. Because of the hidden functional triggering of Trojans, the logic-based testing methods are unlikely to trigger and distinguish

R. Böhme, P.W.L. Fong, and R. Safavi-Naini (Eds.): IH 2010, LNCS 6387, pp. 17–32, 2010.

the malicious alterations. The conventional parametric IC testing methods have a limited effectiveness for addressing Trojan related problems. Destructive tests and IC reverse-engineering are slow and expensive.

This paper formally devises a new unified framework that simultaneously integrates the results of several noninvasive measurement types. Each noninvasive measurement type is called a *modality*: *unimodal* detection employs a single measurement modality for finding the internal characteristics of the chip, while *multimodal* detection combines the measurements from several modalities to reveal the unwanted changes to the original design. We show that the detection objective for each modality is *submodular*. The submodularity property formalizes the intuition that inserting a Trojan would have a higher impact on a small circuit than inserting the same Trojan to a larger circuit that contains the smaller circuit as a subpart. The concept is demonstrated in Figure 1. The design

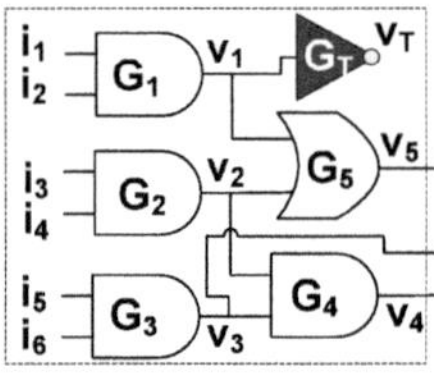

Fig. 1. The submodular property

consists of 9 gates $G_1, \ldots, G_9$, 6 inputs and 2 outputs. A Trojan gate G_T is added. Consider a subcircuit of this design composed of gates $G_1, \ldots, G_5$ in the dotted square that also includes G_T. Now, for one input vector applied to the circuit, the ratio of the current leaked by G_T to the rest of the circuit leakage would be higher for the subcircuit compared to the whole circuit. We exploit the theoretical results known for submodular functions to propose a near-optimal Trojan detection algorithm. Our contributions are as follows: (1) Proposing a unified noninvasive Trojan detection framework, (2) Formulating the optimization problem for simultaneous gate level profiles and Trojan detection for each modality, (3) Exploiting submodularity to achieve a near optimal solution for unimodular detection, and (4) Devising and comparing four methods for combining the results of multiple unimodal detections on benchmark designs.

2 Related Work

Hardware Trojan detection is a new and emerging research area. Agrawal et al. [1] use destructive tests to extract a fingerprint for a group of unaltered chips based on the global transient power signal characteristics. The other chips would be noninvasively tested against the extracted fingerprints by statistical Hypothesis testing. The overhead of destructive testing, sensitivity to noise and process variations, and lack of usage of the logical structure and constraints are the drawbacks of this method.

Banga et al. [4,3] propose a region-based testing that first identifies the problematic regions based on power signatures and then performs more tests on the region. The underlying mathematical and logical circuit structure or the process variations are not considered. Rad et al. [24,23] investigate power supply transient signal analysis methods for detecting Trojans. The focus is on test signatures and not on the lower-level components (e.g., the gate level characteristics). Rad et al. further improved the resolution of power analysis techniques to Trojans by carefully calibrating for process and test environment (PE) variations. The main focus of this research is on the evaluation of four experimental signal calibration techniques, each designed to reduce the adverse impact of PE variations on their detection method. They also investigated the sensitivity of their Trojan detection method in terms of determining the smallest detectable Trojan under conditions such as measurement noise.

Jin and Markis [11] extract the path delay fingerprints by using the well-known principal component analysis that is a statistical dimension reduction technique. They use Hypothesis testing against the delay fingerprints to detect the anomalies. This approach also does not consider the gate level components and would also require exponential path measurements in the worst case. Li and Lach propose adding on chip delay test structures for Trojan detection [16]. Gate level characterization for noninvasive post-silicon IC profiling [12] and for Trojan detection was used in [20,22,2,28]. However, the previous work did not provide a systematic algorithm with any kind of optimality, nor they addressed calibration, sensitivity, or multimodal combining. Our work provides the first rigorous treatment of the multimodal Trojan detection problem, near-optimal solutions, mathematical calibration. Even though a number of authors suggested the potential benefits of combining different measurement types, to the best of our knowledge no systematic approach with evaluation results on combining different test and measurement modalities was reported.

Our method exploits the concept and results of submodular function optimization [21]. The concept has been utilized earlier in a variety of contexts [13], including but not limited to: set cover [9], sensor networks [15], linear regression [7], graph problems [6], and social networks [18]. To the best of our knowledge, our work is the first to use submodularity for IC Trojan detection.

3 Preliminaries

In this section, we provide the necessary background and the measurement setup.

(i) Process variations. As the CMOS dimensions shrink, uncertainty in the device characteristics increases. The variations might be temporal or spatial. In controlled settings, the dominant source of difference between the chips is the spatial variation [27,17]. Spatial variation may be intra-die, or inter-die, and could be systematic or random. We use the widely adopted Gaussian variation models [17]. Timing and dynamic power variations are a linear function of the variations and follow the same Gaussian patterns, while the leakage has an approximately lognormal distribution. Our approach works for the stationary process variation models.

(ii) Trojan threat model. From the conventional testing and inspections point of view, the Trojan IC has exactly the same set of I/O pins, has the same deterministic I/O response as the original plan, and has the same physical form factor. A Trojan causes a change in the statistical distribution of the estimated gate characteristics that otherwise follow the process variation distributions. Our method uses the likelihood of the post-silicon characteristics for detection. The intra-chip correlations are assumed to have a lower amplitude when compared to the impact of the Trojans on the estimated profiles. The nominal values for the gate characteristics are available via the factory provided simulation models needed to ensure design-time power control and timing closure.

(iii) Measurement setup. We use similar measurement set-up as the conventional testing. However, our assumption is that the chip has already passed the standard automatic test pattern generation (ATPG) tests, and does not include any of the standard faults. For timing measurements, we exploit the classic timing test and validation techniques: Given an input vector to a test chip, applying the test input vector with multiple clock frequencies can give us the pass/fail behavior of the chips. The path delay would be the shortest clock period for which the chip does not fail with the intended input vector. We use the testing pattern generation method described in [29].The leakage current can be measured via the commonly known IDDQ test methods, where IDDQ refers to the measurement of the quiescent power-supply current. The IDDQ tests are often done via the off-chip pins by the precision measurement unit (PMU) [26]. The dynamic current tests are referred to as IDDT tests. IDDT tests can be done by averaging methods that do not require high precision or high frequency measurement devices needed for capturing the transient signals [10].

4 Unimodal Trojan Detection

The basis of the unimodal Trojan approach is the gate profiling discussed in Section 4.1. In Section 4.2 we show the detection problem is submodular. We further discuss the complex structure of the general unimodal detection problem which cannot be optimally addressed. We opt to use our prior knowledge about the process variations and submodularity property to address the problem in a hierarchical way. The precursor for our hierarchical method is systematic calibration that is discussed in Section 4.3.

4.1 Gate Profiling

In this subsection, we show how the side-channel measurements can be decomposed to their gate level components post-silicon. One can exploit the linear relationship between the IC's gate level profile and the side-channel measurements (constrained by the logic relations) to estimate the gate level characteristics. We introduce a formal framework for this problem:

Problem. UNIMODAL GATE PROFILING (MODALITY M).

Given. A combinational circuit **C**, with N_I primary inputs $x_1, \ldots, x_{N_I}$, and N_O primary outputs $z_1, \ldots, z_{N_O}$, where the netlist and logic structure is fully available. The circuit consists of interconnections of single-output gates where each gate G_k, $k = 1, \ldots, N_g$ implements an arbitrary logic function. The nominal profile of G_k for the modality M for each possible combination of gate inputs is available from the technology libraries and simulations models.

Measurements. For the modality M, a set of input vectors (V's) that are each an N_I tuple $(v_1, v_2, \ldots, v_{N_I})$, where $v_j \in \{0, 1\}$ for $j = 1, \ldots, N_I$ are available. Component values of V are applied to the primary inputs $x_1, \ldots, x_{N_I}$ which changes the states of the internal gates. For one or more input vectors, the side channel measurement is recorded either from the output pins, from other external pins, or contactless. The side-channel measurement is a linear combination of the gate characteristics in measurement modality M and a measurement error.

Objective. Estimate the post-silicon profile of each gate for the modality M.

A key step in noninvasive profiling of the chips is generation of input vectors that can controllably change the states of the gates such that this modification is observable from the measurement medium. Note that generating the input patterns that can distinctively identify each gate's characteristics is known to be NP-complete and has been a subject of extensive research in circuit testing [10]. We use the best known methods in testing for generation of the input vector patterns that maximally cover all the gates. Although we are limited by the same constraints as testing in terms of gate coverage, the difference here is that we are not detecting a particular fault model or the worst-case behavior (e.g., critical paths or stuck at fault) but we are estimating the gate parameters that may incur a certain error.

In delay testing, the input vectors have to functionally *sensitize* the tested paths such that the output is observable at an output pin. Delay test generation methods for sensitizing paths that achieve a high coverage, i.e., exercising many paths are available. Since there is a linear relationship between the tested paths and the gate delays, the explored paths directly translate to high coverage of the gate delays. We use the path sensitizing method proposed by Murakami et al [19]. Similarly, we use the available high coverage test vector generation methods for IDDQ, and for IDDT testing [25]. The number of generated input vectors is linear with respect to the total number of gates.

(i) Timing modality. The noninvasive timing measurements are taken by changing the inputs and measuring the time propagation of input transition to the output nodes. In this paper we consider the gate delays and ignore the wires. However, we emphasize that since the wire timings are linearly added to the path delays (assuming that crosstalk is bounded by controlling the possible couplings), their inclusion in the linear formulations is straightforward.

One can test J different paths and write a linear system of delay equations. The gate delays ($T(G_{k_j})$) on each chip are the variables (because of the process variation and operation effect) and $T_{meas}(P)$'s are the path delays that are measured on each chip:

$$EQ_j : \Sigma_{k_j=1}^{K_j} T(G_{k_j}) = T_{meas}(P_j), P_j = \{G_{k_j}\}_1^{K_j}, \qquad 1 \leq j \leq J \quad .$$

Noninvasive gate profiling aims at solving the above system of equations in presence of measurement error. If measuring the path delay $T_{meas}(P_j)$ incurs the error $\epsilon_T(P_j)$, the optimization problem objective function (OF) and constraints (C's) can be written as follows:

$$OF : \min_{1 \leq j \leq J} \mathcal{F}(\epsilon_T(P_j)) \tag{1}$$
$$C's : \Sigma_{k_j=1}^{K_j} T(G_{k_j}) = T_{meas}(P_j) + \epsilon(P_j), P_j = \{G_{k_j}\}_1^{K_j}, \qquad 1 \leq j \leq J \quad .$$

where $\mathcal{F}$ is a metric for quantifying the measurement errors; commonly used forms of $\mathcal{F}$ are the maximum likelihood formulation, or the l_p norms of errors defined as: $l_p = (\sum_{j=1}^{J} |\epsilon_T(P_j)|^p)^{1/p}$, *if* $1 \leq p < \infty$, and $l_p = max_{j=1}^{J} |\epsilon_T(P_j)|$, *if* $p = \infty$.

The delay of one gate $T(G_k)$ can be further written in terms of the deviation from the nominal delay of this gate from the value specified in the technology files. If the nominal gate delay value for the gate type G_k is $T^{nom}(G_k)$ and the deviation from nominal for G_k for the chip under measurement is $\theta_T(G_k)$, then $T(G_k) = \theta_T(G_k)T^{nom}(G_k)$ and thus, the unknowns are $\theta_T(G_k)'s$ and $\epsilon(P_j)'s$. The variable $\theta_T(G_k)$ is called the *delay (timing) scaling factor* of G_k. If there were no path measurement errors, the number of equations (J) required to have a full-rank system would be the same as the number of variables (gate delays). In presence of errors, the number of required equations is slightly higher, but the order is still linear in terms of number of gates N_g.

(ii) Leakage power modality. The leakage measurements rely on the fact that leakage is a function of the gate input for each gate type. Since the supply voltage is fixed, the static power is only dependent on the leakage current. For each input vector in quiescent state, the external pin current can be measured and written in terms of the sum of the individual components. For example, for (v_1; v_2; v_3; v_4) =1111, the total measured leakage current can be written as: $\Phi_{meas}(1111) + \epsilon_\Phi(1111) = \Phi_{G_1}(11) + \Phi_{G_2}(11) + \Phi_{G_3}(00) + \Phi_{G_4}(00) + \Phi_{G_5}(11)$ $+ \Phi_{G_6}(11)$, where $\Phi_{G_k}(x_1x_2)$ is the leakage current for gate G_k for its incident input (x_1x_2), and $\epsilon_\Phi(.)$ denotes the measurement error for the incident input. Each gate's leakage can be further decomposed to the nominal leakage value for the gate type and a *leakage scaling factor* denoted by $\theta_\Phi(G_k)$, i.e., $\Phi_{G_k}(x_1x_2) = \theta_\Phi(G_k)\Phi_{G_k}^{nom}(x_1x_2)$. Therefore, the linear optimization can be written over the J leakage measurements:

$$OF : \min_{1 \leq j \leq J} \mathcal{F}(\epsilon_\Phi(X_j)) \tag{2}$$
$$C's : \Sigma_{k=1}^{N_g} \theta_\Phi(G_k)\Phi_{G_k}^{nom}(x_j) = \Phi_{meas}(X_j) + \epsilon_\Phi(X_j) \quad .$$

(iii) Dynamic power modality. The dynamic power is dependent on the input transition. The measured average dynamic current (Section 3(iii)) can be written as the sum of the gate dynamic currents. Thus, the linear optimization for J dynamic current measurements would be:

$$\begin{aligned} OF: &\min_{1\le j\le J} \mathcal{F}(\epsilon_\Psi(X_{j\to j+1})) \\ C's: &\Sigma_{k=1}^{N_g}\theta_\Psi(G_k)\Psi_{G_k}^{nom}(x_{j\to j+1}) = \Psi_{meas}(X_{j\to j+1}) + \epsilon_\Psi(X_{j\to j+1}) \end{aligned} \quad . \tag{3}$$

where $\epsilon_\Psi(.)$ denotes the measurement error for a reading, $\theta_\Psi(G_k)$ is the gate *scaling factor for the dynamic current*, $\Psi_{G_k}^{nom}(.)$ is the nominal dynamic current value for gate G_k for the pertinent transition, $x_{j\to j+1}$ refers to input vector transition from the vector j to j_1, and $\Psi_{meas}(.)$ is the extracted dynamic current measured at the external pin.

We see that each of the modalities can be written in the unified format of a system of linear equations. In the remainder of the paper we use the following generic notations for the gate profiling over the different modalities.

(i) **OF:** $\min\mathcal{F}(\epsilon)$, **Constraints:** $A\theta = B+\epsilon$; where $A_{[J\times N_g]}$ and $B_{[J]}$ are given by the technology values and J measurements, $\theta_{[N_g]}$ is a vector of unknown scaling factors, and $\epsilon_{[N_g]}$ is a vector of measurement errors.

(ii) Alternatively, the optimization problem can be written as **OF:** $\max\mathcal{L}(\epsilon)$, **Constraints:** $A\theta = B+\epsilon$, where $\mathcal{L}$ is the likelihood that the variations are coming from a certain distribution, e.g., normal distribution. Under the assumption of normal error distribution, maximizing the likelihood corresponds to minimizing the $\mathcal{F} = l_2$ error norm.

4.2 Unimodal Detection

Let us assume that the gates are positioned at the locations $\mathcal{D}$ in the 2D layout space. For a single modality we can find an estimation of each gate's profile. As we described in Section 3(i), the profile of a benign gate can be modeled as the sum of its inter-chip and intra-chip systematic process variations, the random process variations, and measurement noise. The global objective of Trojan detection is to maximize the probability of Trojan detection (P_D) and to minimize the probability of false alarm (P_{FA}). However, explicit formulation of the two objectives is not plausible, since probability of detection/false alarm can only be determined for cases where we know the exact Trojan attack and the ground truth. Instead, our Trojan detection attempts at removing the impact of the anomalous gates by reweighing. The reweighing is done such that the likelihood of the remaining benign gate profiles being drawn from the process variation and noise distribution after mapping to the benign space is maximized. Based on this criteria, the objective here is to select a subset of gates $\Gamma \subseteq \mathcal{D}$ for the linear program and reweigh them such that the likelihood $\mathcal{L}(\mathcal{D}\backslash\Gamma,\epsilon)$ is maximized (i.e., the gate profiles fall into the benign space for maximizing the P_D), subject to the cost constraint $\mathcal{Q}(\Gamma)$ for selecting the O gates as Trojan (for minimizing the P_{FA}). Reweighing is done by setting the gate scaling factor to its nominal value of unity, assuming the systematic variations are calibrated. Let q_u denote the budget for the cost $\mathcal{Q}(\Gamma)$. Thus, one can write the objective function and the constraints of the problem as follows:

$$OF: \max_{\Gamma \subseteq \mathcal{D}} \mathcal{L}(\mathcal{D} \backslash \Gamma, \epsilon) \tag{4}$$
$$C's: \mathcal{Q}(\Gamma) \leq q_u; A\theta_{\{\theta \in [(\Gamma \subseteq \mathcal{D}) \cup (\Gamma = 1)]\}} = B_{\{\theta \in [(\Gamma \subseteq \mathcal{D}) \cup (\Gamma = 1)]\}} + \epsilon \quad .$$

The first constraint corresponds to the cost budget for the number of Trojans. The second constraint set corresponds to the gate level profiling discussed in the previous section (after reweighing the anomalies). Assuming the distribution of random process variations is Gaussian and the systematic process variations follows a 2D Gaussian in the spatial domain, the above likelihood function will always be lowered if we select to reweigh the maximum number of anomalies q_u. This is because the reweighing can make the noisier observations more consistent and therefore improve the likelihood results. But this is not usually desirable since it would unnecessarily increase the P_{FA}. Notice that the OF in Equation 4 has two simultaneous goals, one is to find the location of the gates that maximize the likelihood, and the other is to maximize the likelihood of the estimation error ϵ. Generally speaking, detecting guaranteed anomalies in problems like ours where there is an uncertainty about the value and interval of the variables (dependent on the other variables values) was demonstrated to be NP-hard [14]. Thus, we can only hope for heuristics and approximations to address the problem.

Iterative hierarchical detection and profiling. To simultaneously address the two goals embedded in Equation 4, we take a hierarchical approach for solving the problem by separation of concerns paradigm and iterative evaluations. We first present the high level view of this algorithm and then propose a class of formulations for which we can derive tight bounds on the solutions obtained by our approach. Our method is presented in Algorithm 1.

The procedure iteratively increases the maximum allowed number of anomalies q_u, starting from zero (Step 1). The stopping criteria of the iterative algorithm is improvement above a certain threshold (Step 2). For each added value of q_u (Step 3), we follow a greedy selection and add the most discrepant gate o to the set Γ (Step 4)). Discrepancy is evaluated as the distance to the projection into the benign gate space (Steps 5-7). A new round of gate level profiling is done after adding the newly reweighed gate o (Step 5). Since the derived gate profiles contain both systematic and random variations, calibration is performed to adjust for the systematic variations (Step 6). Now, the benign gates would only have random variations. The anomaly detection criteria is evaluated for checking the stopping condition for the algorithm (Step 7).

Algorithm 1 - Unimodal anomaly detection

Input. Combinational circuit, noninvasive measurements for J inputs, nominal technology values;
Output. Scaling factors (θ) from gate level profiling (GLP); anomalous gate set (Γ);

Set $\Gamma = \emptyset$, $q_u = 0$; Perform an initial GLP;
While (improvement by anomaly reweighing) do
    q_u++;
    Select the gate o to reweigh ($\Gamma = \Gamma \cup \{o\}$);
    Perform a GLP with the reweighted o;
    Calibrate for systematic variations;
    Evaluate the improvement criteria;

The complexity of Algorithm 1 can be computed as follows. Let N_g denote the total number of gates in the circuit. Assuming that solving the linear system and calibration are at most of polynomial complexity, the worst case complexity of the above algorithm would still be polynomial and is effectively dominated by the time the solver takes to find the gate level profiles. The exact form of the GLP objective function would determine the solver time. For example, maximizing the likelihood for Gaussian distribution corresponds to solving a quadratic optimization problem in each round. The number of iterations is much less than the number of gates N_g since not all gates will be reweighed and the improvement criteria has a diminishing return property that would decrease at each iteration. If the Trojan is so large that many gates need to be reweighed, then the problem becomes trivial: it is well known in statistics that the anomaly detection is only challenging when the outlier characteristics only slightly differ from noise [14].

Greedy anomaly detection. For evaluating the improvement criteria for reweighing the gates in Γ, we propose a formulation of the anomaly detection objective based on likelihood improvement. This objective aims at performing *penalty reduction*. The penalty reduction metric quantifies the expected reward obtained by reweighing a set of gates. The expected penalty reduction due to reweighing the gates in the set Γ is denoted by $\mathcal{R}(\Gamma)$ and is defined as:

$$\mathcal{R}(\Gamma) = \mathcal{L}(\mathcal{D}) - \mathcal{L}(\mathcal{D} \backslash \Gamma) \quad . \tag{5}$$

The above formulation has a number of important properties that we exploit in our framework. A set function $\mathcal{R}$ is called *submodular* if it satisfies the following properties: (i) the penalty will not be reduced if we do not reweigh a new anomalous gate, i.e., $\mathcal{R}(\emptyset) = 0$; (ii) $\mathcal{R}$ is a nondecreasing set function and thus, reweighing a new anomaly could just decrease the associated penalty, i.e., $\mathcal{R}(\Gamma_1) \leq \mathcal{R}(\Gamma_2)$, for $\Gamma_1 \subseteq \Gamma_2 \subseteq \mathcal{D}$; (iii) the set function satisfies the *diminishing return property*: if we reweigh a gate in a smaller set of gates with logic relations (denoted by $\mathcal{D}_s$), we improve the reward by at least as much, as if we reweigh in a larger set of gates (denoted by $\mathcal{D}_l$) with logic relations such that $\mathcal{D}_l \subseteq \mathcal{D}_s$.

Nemhauser et al. [21] have shown that a function $\mathcal{R}$ is submodular if and only if the following theorem holds:

THEOREM 1. For all detected and reweighed Trojans $\Gamma_1 \subseteq \Gamma_2 \subseteq \mathcal{D}$ and a new candidate point $o \in \mathcal{D} \backslash \Gamma_2$ the following holds:

$$\mathcal{R}(\Gamma_1 \cup \{o\}) - \mathcal{R}(\Gamma_1) \geq \mathcal{R}(\Gamma_2 \cup \{o\}) - \mathcal{R}(\Gamma_2) \quad . \tag{6}$$

It can be shown that the reward function $\mathcal{R}$ satisfies the above theorem [15]. Now our optimization problem over Γ can be expressed as:

$$OF : \max_{\Gamma \subseteq \mathcal{D}} \mathcal{R}(\Gamma) \qquad\qquad C's : \mathcal{Q}(\Gamma) \leq q_u \quad .$$

Since solving the above problem has been shown to be NP-hard for the most interesting instances [9], we address the above optimization problem by the greedy procedure described in Algorithm 1. This is because a key result states that for submodular functions, the greedy algorithm achieves a constant factor approximation:

THEOREM 2 [21]. For any submodular function $\mathcal{R}$ that satisfies the above three properties, the set Γ_G obtained by the greedy algorithm achieves at least a constant fraction $(1 - 1/e)$ of the objective value obtained by the optimal solution, or,$\mathcal{R}(\Gamma_G) \geq (1 - 1/e) \max_{|\Gamma \leq q_u|} \mathcal{R}(\Gamma)$. Perhaps more surprisingly, Feige has shown that no polynomial time algorithm can provide a better approximation guarantees unless P=NP [9]. Thus, for any class of submodular objective functions, the proposed greedy selection algorithm results in the best achievable solution.

4.3 Calibration

To perform the anomaly detection, it is required that we calibrate for the systematic variations after profiling the gates. As mentioned in Section 3(i), the systematic variations consist of inter-chip and intra-chip variations. The inter-chip variations are simply affecting the mean of the variations and can be adjusted for by shifting the mean extracted profile values to have a mean of unity. The intra-chip variations are in form of a spatial distribution, e.g., 2D Gaussian in our model. The key observation is that the spatial rate of change of the neighboring gate level profiles due to the systematic intra-chip variations (spatial correlations) is slower than the rate of change because of the Trojan insertion. The larger Trojans that would affect many gates in a larger area are trivial to detect and would not be a challenge to address. This suggests using a high-pass filter over the 2D discrete space of the gate layouts for the identification of the sharp edges that have high frequency components in their frequency transformation. In this paper, we use the 2D Discrete Cosine Transform (DCT).

5 Multimodal Trojan Detection

The next step of our approach is to combine the results for anomalous gate detection over the M modalities. While there are a number of possible methods to accomplish this task, our goal is to combine the unimodal methods to optimize the P_D and P_{FA} results. Assume that $\mathcal{C}_m(G_k)$ is the anomaly vote for gate G_k in modality m:

$$\mathcal{C}_m(G_k) = \begin{cases} 1 \text{ for } G_k \text{ anomalous in modality } m; \\ 0 \text{ otherwise.} \end{cases}$$

We propose four methods for combining the results of different modalities.

(i) Unanimous voting: In this voting approach, the Trojan gates are those that have been marked anomalous by all the M modalities. For example, for the three modalities the following constraint should hold for marking a gate as Trojan: $\mathcal{C}_T(G_k)+\mathcal{C}_\Phi(G_k)+\mathcal{C}_\Psi(G_k) = 3$ where the subscripts T, Φ, and Ψ denote the timing, quiescent current, and dynamic current measurement modalities respectively. This voting method is likely to decrease P_D but improves P_{FA}. It would also give the minimum achievable P_{FA} (lower bound) by any linear combination of the unimodal detection methods.

(ii) Conservative voting: A gate that has been marked anomalous by any of the modalities is marked as a Trojan by the conservative voting method. In our case, the following constraint is necessary and sufficient for marking a gate G_k as Trojan by conservative voting: $\mathcal{C}_T(G_k) + \mathcal{C}_\Phi(G_k) + \mathcal{C}_\Psi(G_k) \geq 1$. This voting method is likely to increase P_{FA} but also increases P_D. It would also give the maximum achievable P_D (upper bound) by any linear combination of our anomaly detection methods.

(iii) Majority voting: Here, the Trojan gates are those that have been marked anomaly by at least $1 + \lfloor \frac{M}{2} \rfloor$ of the modalities. In our case, the majority voting translates to the following condition: $\mathcal{C}_T(G_k) + \mathcal{C}_\Phi(G_k) + \mathcal{C}_\Psi(G_k) \geq 2$. This method provides a useful trade-off between the P_D and P_{FA} values.

(iv) Weighed voting: The voting methods above assume that all the modalities have the same detection ability. However, this is not true. For example in our experiments we see that there is less controllability/observability for the timing modality. For example, assume that we give the weights S_k^T, S_k^Φ, and S_k^Ψ for gate G_k for timing, leakage, and dynamic current respectively. Now, the votes of the three unimodal detectors over an anomalous gates are combined as follows: $S_k^T\mathcal{C}_T(G_k) + S_k^\Phi\mathcal{C}_\Phi(G_k) + S_k^\Psi\mathcal{C}_\Psi(G_k) \geq$ threshold. If this expression is true, the gate G_k is marked as the Trojan. Changing the detection threshold introduces a tradeoff between P_D and P_{FA} values.

6 Experimental Evaluations

6.1 Evaluation Set-Up

We evaluate the performance of the unimodal detection and the unified multimodal framework on the widely used MCNC benchmark suite. The ABC synthesis tool from UCB was used for mapping the benchmark circuits to NAND2, NAND3, NAND4, NOR2, NOR3, NOR4, and inverter library gates. Placing the gates on the layout placement is done by the Dragon placement tool from UCLA. The gates have different sizes and they are located on irregular grids. The process variation model is as described in Section 3(i). In a number of our experiments, the amount of variations are altered and the detection results are tested against the variation fluctuations. In cases where we do not change the variations, the random variation is 12%, intra-die variation correlation is 60% of the total variation [5]; 20% of the total variation is uncorrelated intra-die variation and the remaining variation is allotted to the inter-die variation. The noninvasive measurement setup was described in Section 3(iii).

To find the values for timing, leakage, and dynamic currents for each of the library gate files over various input states, we performed HSPICE simulations for the 65nm CMOS transistor technology. The linear optimization was performed using the MATLAB optimization toolbox. Several other internal MATLAB functions are used for computing the likelihood and for filtering to calibrate the systematic variations. Each of our reported numbers and statistics are averaged over 100 runs of the random circuit instances.

6.2 Unimodal Trojan Detection

Gate level profiling. Finding the gate level profiles is the essence of the proposed approach. Table 1 shows gate level dynamic power, static power, and timing profiling results on benchmark circuits. For these evaluations the number of measurements is the minimum of double the number of gates and the maximum number of tests that can be done on the circuit. The first column shows the name of the benchmark denoted by *ct*. The second column shows the number of gates in the benchmark denoted by N_g. The number of primary inputs and primary outputs denoted by #*i* and #*o* are shown in the third and fourth columns respectively. The next nine columns show the profile estimation l_2 errors in the presence of 3%, 5%, and 10% measurement error for the three modalities respectively. On the average, the gate level profile estimation error is 3.8%, 4.8%, and 9.6% for the different measurment errors in case of dynamic power. The error in static power profiling is 4%, 6%, and 10% for 3%, 5%, and 10% of the measurement error respectively. For the timing modality the profiling error is 4%, 6%, and 11%.

Table 1. Dynamic power, static power, and timing profile estimation error for MCNC benchmark circuits

				D. Power			S. Power			Timing		
ct	N_g	**#i**	**#o**	**3%**	**5%**	**10%**	**3%**	**5%**	**10%**	**3%**	**5%**	**10%**
C1355	512	41	32	7.8	9.1	11.5	8.5	10	12	4	8	12.3
c8	165	28	18	4.2	6.4	11.2	5.6	7	11.6	5.3	7	11.5
C3450	1131	50	22	3.5	6	9.5	4	5.9	9.8	2.9	4.1	9.2
C432	206	36	7	1.5	3.1	6.9	1.7	3.5	7.2	3.8	5.4	10.1
C499	532	41	32	2.2	4.2	8.8	2.9	4.5	9	5	6.5	12

Unimodal Anomaly detection. In this Section, we evaluate the performance of the unimodal anomaly detection over the three modalities under study. We study three scenarios: (i) a Trojan-free circuit, (ii) one extra NAND2 gate inserted as a Trojan, and (iii) a 3-gate comparator circuit is inserted. The Trojans are inserted in the empty spaces within the automatic layout generated by the Dragon tool. An important property for unimodal anomaly detection is how the diminishing return property changes. As we discussed earlier, this function should be monotonically decreasing assuming no random perturbations. We have tested the validity of our assumption on multiple benchmark circuits. An example result is shown in Figure 3 for the leakage modality for the C432 benchmark and 100 measurements. The results for the other two modalities are similar.

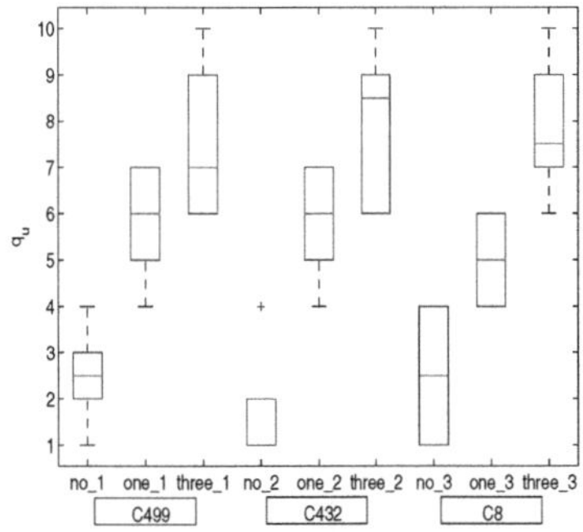

Fig. 2. Boxplots of q_u for Trojan free, 1 Trojan, and 3 Trojan gates

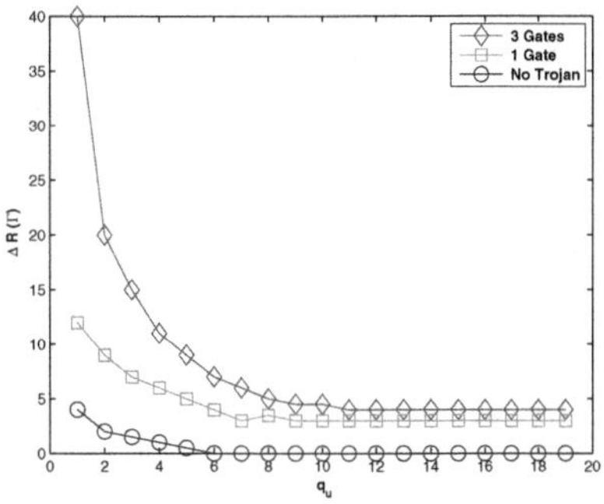

Fig. 3. The stepwise diminishing return improvement for leakage modality

As we can clearly see on the figure, the difference between diminishing returns of two consecutive steps of our algorithm is much higher for the larger Trojan, it becomes lower for the smaller Trojan and it is really low for the Trojan free case. The same phenomena is observed over all the modalities. This result is really important for adjusting the stopping criteria of our algorithm. Basically, when the Trojan circuit reaches the same diminishing return difference as the Trojan-free circuit, no further significant improvement is foreseen. The above results gave us an insight to set the stopping criteria for Algorithm 1. For example, we set it such that the diminishing return is decreased by more than 2 in each step. As can be seen on the Figure 3, this decision would result in an average 1 gate false alarm for this circuit with 206 gates, i.e., $P_{FA} = 1/206 = 0.5\%$ and about 2 gates are not detected in case of the smaller Trojan, about 1%. It should be noted that we detect more anomalous gates than what is inserted by the Trojan because the Trojan gates affect the side channel characteristics of the logically connected gates. This result can be used to help localize the Trojan, however, this is out of the scope of this work. Figure 2 shows the boxplot of the number of iterations (q_u) before our stopping criteria is reached for 100 runs over 3 benchmarks for leakage modality. The number of iterations corresponds

Table 2. The number of gates giving false alarm in a non Trojan circuit

ct	UNA	CON	MAJ	WD
C1355	0/512	4/512	2/512	2/512
c8	0/165	3/165	1/165	2/165
C3450	0/1131	3/1131	3/1131	3/1131
C432	1/206	2/206	1/206	1/206
C499	0/532	3/532	1/532	1/532

to the number of detected anomalous gates. We see that the number of detected anomalous gates increases as the Trojan size increases.

Table 2 shows the P_{FA} (in terms of number of gates giving false alarm in the circuit) for 100 chips with no Trojan. The false negatives ranged from 0 to 4 gates and are largest for the timing modality. The first column shows the name of the benchmark. The rest of the columns show the results for different multimodal methods: unanimous (UNA), conservative (CON), majority (MAJ), and weighed (WD). The unanimous voting performs best in terms of P_{FA}. This is because it can filter out the effect of the modalities that give more false positives.

We also studied the probability of detection P_D using the different voting methods. Our P_D results demonstrate an average of 86%, 99%, 98%, and 98% for unanimous, conservative, majority, and weighted voting respectively. The unanimous voting yields the worst P_D while as we described earlier, it resulted in the best P_{FA}. The conservative voting yields the best P_D at the expense of worsening the false alarm probability P_{FA}. On the other hand, the majority voting and weighed voting result in a good trade-off between the two probabilities. In addition, we observed that the weighted voting gives the best result when we assign the lowest weight to the timing modality. The inefficiency of timing modality is because the small Trojans would only affect a few of the tested timing paths, whereas many more sets of current tests would show the impact of the modified currents. The two power modalities are much more effective in detecting Trojans. Another interesting observation was that even though there is a good amount of independent information in the static and dynamic current tests, the outcomes of the two testing modalities demonstrate an average of 73% correlations on our benchmark circuits.

7 Conclusion

Our work presents a new unified formal framework for IC Trojan detection by noninvasive measurements from multiple test modalities. For each modality, a unimodal anomaly detection is built upon the gate level profiling. Since the problem is extremely complex, we devise an iterative detection and profiling method. Our objective function for detecting the abnormal gate level behavior is shown to be submodular. Because of the objective submodularity, our iterative greedy detection and profiling algorithm achieves a near optimal solution (within a constant fraction of the optimal) in polynomial time. We show a method to calibrate the systematic variations. Our multimodal Trojan detection approach combines the unimodal detection results using a number of different techniques. Experimental evaluations on benchmark circuits using timing, leakage current, and transient currents show the effectiveness of our approach.

References

1. Agrawal, D., Baktir, S., Karakoyunlu, D., Rohatgi, P., Sunar, B.: Trojan detection using ic fingerprinting. In: S&P, pp. 296–310 (2007)
2. Alkabani, Y., Koushanfar, F.: Consistency-based characterization for ic trojan detection. In: ICCAD, pp. 123–127 (2009)
3. Banga, M., Chandrasekar, M., Fang, L., Hsiao, M.: Guided test generation for isolation and detection of embedded trojans in *ICs*. In: GLS-VLSI, pp. 363–366 (2008)
4. Banga, M., Hsiao, M.: A region based approach for the identification of hardware trojans. In: HOST, pp. 43–50 (2008)
5. Cao, Y., Clark, L.T.: Mapping statistical process variations toward circuit performance variability: an analytical modeling approach. In: DAC, pp. 658–663 (2005)
6. Chekuri, C., Pal, M.: A recursive greedy algorithm for walks in directed graphs. In: FOCS, pp. 245–253 (2005)
7. Das, A., Kempe, D.: Algorithms for subset selection in linear regression. In: STOC, pp. 45–54 (2008)
8. Defense Science Board (DSB) study on High Performance Microchip Supply (2005), `http://www.acq.osd.mil/dsb/reports/2005-02-HPMS_Report_Final.pdf`
9. Feige, U.: A threshold of ln n for approximating set cover. Journal of ACM 45(4), 634–652 (1998)
10. Jha, N., Gupta, S.: Testing of Digital Systems. Cambridge University Press, Cambridge (2003)
11. Jin, Y., Makris, Y.: Hardware trojan detection using path delay fingerprint. In: HOST, pp. 51–57 (2008)
12. Koushanfar, F., Boufounos, P., Shamsi, D.: Post-silicon timing characterization by compressed sensing. In: ICCAD, pp. 185–189 (2008)
13. Krause, A., Guestrin, C.: Near-optimal observation selection using submodular functions. In: AAAI, pp. 1650–1654 (2007)
14. Kreinovich, V., Lakeyev, A., Rohn, J., Kahl, P.: Computational Complexity and Feasibility of Data Processing and Interval Computations. Kluwer Academic Publishers, Dordrecht (1997)
15. Leskovec, J., Krause, A., Guestrin, C., Faloutsos, C., VanBriesen, J., Glance, N.: Cost-effective outbreak detection in networks. In: SIGKDD, pp. 420–429 (2007)
16. Li, J., Lach, J.: At-speed delay characterization for IC authentication and trojan horse detection. In: HOST, pp. 8–14 (2008)
17. Liu, F.: A general framework for spatial correlation modeling in VLSI design. In: DAC, pp. 817–822 (2007)
18. Mossel, E., Roch, S.: On the submodularity of influence in social networks. In: STOC, pp. 128–134 (2007)
19. Murakami, A., Kajihara, S., Sasao, T., Pomeranz, I., Reddy, S.M.: A test structure for characterizing local device mismatches. In: ITC, p. 376 (2000)
20. Nelson, M., Nahapetian, A., Koushanfar, F., Potkonjak, M.: Svd-based ghost circuitry detection. In: Katzenbeisser, S., Sadeghi, A.-R. (eds.) IH 2009. LNCS, vol. 5806, pp. 221–234. Springer, Heidelberg (2009)

21. Nemhauser, G., Wolsey, L., Fisher, M.: An analysis of the approximations for maximizing submodular set functions. Math. Programming 14, 265–294 (1978)
22. Potkonjak, M., Nahapetian, A., Nelson, M., Massey, T.: Hardware trojan horse detection using gate-level characterization. In: DAC, pp. 688–693 (2009)
23. Rad, R., Plusquellic, J., Tehranipoor, M.: A sensitivity analysis of power signal methods for detecting hardware trojans under real process and environmental conditions. IEEE Trans. on Very Large Scale Integration (VLSI) Systems 99 (2009)
24. Rad, R., Wang, X., Plusquellic, J., Tehranipoor, M.: Power supply signal calibration techniques for improving detection resolution to hardware trojans. In: ICCAD, pp. 632–639 (2008)
25. Chakravarty, S., Thadikaran, P.: Simulation and generation of iddq tests for bridging faults in combinational circuits. IEEE Trans. on Computers 45(10), 1131–1140 (1996)
26. Sabade, S., Walker, D.: IDDX-based test methods: A survey. ACM Trans. Design Automation of Electronic Systems 9(2), 159–198 (2004)
27. Srivastava, A., Sylvester, D., Blaauw, D.: Statistical Analysis and Optimization for VLSI: Timing and Power. Springer, Heidelberg (2005)
28. Wei, S., Meguerdichian, S., Potkonjak, M.: Gate-level characterization: Foundations and hardware security applications. In: DAC (2010)
29. Yang, K., Cheng, K.T., Wang, L.: TranGen: A SAT-based ATPG for path-oriented transition faults. In: ASPDAC, pp. 92–97 (2004)

A Secure and Robust Approach to Software Tamper Resistance

Sudeep Ghosh, Jason D. Hiser, and Jack W. Davidson

Department of Computer Science, University of Virginia
151 Engineer's Way, Charlottesville, VA-22904
{sudeep,hiser,jwd}@virginia.edu

Abstract. Software tamper-resistance mechanisms have increasingly assumed significance as a technique to prevent unintended uses of software. Closely related to anti-tampering techniques are obfuscation techniques, which make code difficult to understand or analyze and therefore, challenging to modify meaningfully. This paper describes a secure and robust approach to software tamper resistance and obfuscation using process-level virtualization. The proposed techniques involve novel uses of software checksumming guards and encryption to protect an application. In particular, a virtual machine (VM) is assembled with the application at software build time such that the application cannot run without the VM. The VM provides just-in-time decryption of the program and dynamism for the application's code. The application's code is used to protect the VM to ensure a level of circular protection. Finally, to prevent the attacker from obtaining an analyzable snapshot of the code, the VM periodically discards all decrypted code. We describe a prototype implementation of these techniques and evaluate the run-time performance of applications using our system. We also discuss how our system provides stronger protection against tampering attacks than previously described tamper-resistance approaches.

1 Introduction and Overview

Today, software applications have become essential for the correct functionality of many critical systems, e.g., transportation control systems, banking and medical devices. Such critical software systems present potential targets for attacks from adversaries equipped with advanced reverse engineering tools. Any unauthorized modification could lead to extensive disruption of services and loss of life and property.

Software developers have used a variety of schemes to protect software from unauthorized modification [12,7,1,2]. However, current tamper-resistance techniques suffer from a variety of major drawbacks.

- Much of the previous research has targeted making the application hard to analyze statically [14]. For example, an opaque predicate is a predicate that is difficult to analyze statically. However, several runs in a simulator can determine which branches are highly biased and thereby provide the information required to defeat these protections.
- The use of additional hardware is required by some solutions [18]. This extra hardware adds an additional cost that will have to be borne by the end user, and might

R. Böhme, P.W.L. Fong, and R. Safavi-Naini (Eds.): IH 2010, LNCS 6387, pp. 33–47, 2010.

restrict the software to a particular set of platforms. Consequently, adoption of these techniques has been slow.
- Encryption techniques have been used to prevent static disassembly of applications. However, decryption occurs at a coarse level of granularity and as such, there is little protection from dynamic analysis [4].
- A number of schemes have an impractical overhead constraint or provide only a partial solution. The Proteus system involves overheads between 50X-3500X, which is too high for most applications [1]. Remote tamper-proofing techniques, although widely used among embedded devices [19], require strict Quality-of-Service guarantees [6]. A realistic solution must have tolerable overheads otherwise developers will be unwilling to deploy such measures on a large scale.

To address these shortcomings, this work presents a novel scheme for software tamper resistance, using process-level virtualization. The idea involves new and improved techniques for code checksumming and encryption, which are allied with process-level virtualization to impart dynamic tamper resistance to applications. The process-level virtual machine (VM) continually shifts the attack surface of the application, making it hard for the adversary to identify vulnerable locations and mount an attack. The proposed solution provides numerous contributions over previous work.

1. The continuous shifting of the application code provides increased ability to protect against run-time attacks, e.g., checking code in the application is not executed *in-situ* but from randomized locations in memory as the program executes. This dynamism is missing in current techniques.
2. The decryption of program blocks occurs just-in-time on a per-instruction basis. As such, the granularity of decryption is significantly finer than previous software approaches [5,4].
3. Novel uses of checksumming guards provide a strong level of circular protection involving both the application code and the virtual machine's code. Together, these guards are used to achieve dynamism in the virtual machine's protection mechanisms, providing stronger protection than the application could achieve alone.
4. Process-level virtualization, coupled with frequent code cache flushing, provides increased protection against malicious OS attacks with less exposure of the defense mechanism. As described in Section 5.3, the use of virtualization inherently provides security against such attacks without requiring specialized protection techniques.
5. Stealthy schemes for periodic code cache flushing ensure higher levels of entropy in the system by regularly discarding decrypted instructions and relocating application code while maintaining tolerable overhead.

The paper also provides an evaluation of the security techniques and the protection being applied to the application code.

The rest of the paper is structured as follows. Section 2 describes the malicious environment in which software has to run. Section 3 describes the protection mechanisms. The evaluation of these techniques is described in Section 4. Section 5 discusses the relative merits of our approach over previous work done in this area. Section 6 gives a brief overview of previous work, and finally conclusions are presented in Section 7.

2 Threat Model

To understand our approach, it is first necessary to understand the hostile environment in which software must be protected. An adversary can use various tools like debuggers, simulators, and emulators to run, modify, and observe the program in a number of ways. For example, the popular VoIP tool, Skype, was cracked using a debugger to obtain the entire unencrypted code base in memory and then applying standard static analysis techniques on the unencrypted code. Even the operating system can be modified to return inaccurate information [20]. Consequently, we consider all software on the machine as potentially malicious, and the entire application (including any virtual machines distributed with said software) as potentially vulnerable to attack. However, the application is created at a trusted development site (including tools and developers), as such, the threat model does not include attacks during software production.

Furthermore, even hardware can be emulated to return forged results to the application. In essence, this is a white-box attack on the application. The adversary can inspect, modify, or forge any information in the system. Given enough time and resources, the adversary can always succeed in manually inspecting and making modifications to the program. However, human adversaries have difficulty directly solving large problems. As such, they rely on algorithmic solutions to perform various analyses on application packages, including determining instruction locations, disassembling the program, capturing the control-flow or call graphs, to name a few.

3 Protection Model

While being cognizant of the threat model described in Section 2, our solution seeks to address the problems with current technology for tamper resistance and obfuscation. At software creation time, protection features (i.e., guards as described in Section 3.1 and encryption as described in Section 3.2) are applied to a virtual machine and the application, generating a package which is challenging to analyze statically. The protected application (and the VM) is then ready to be distributed into a potentially hostile environment.

The VM provides just-in-time, on-demand decryption of the application instructions. Guards and application code are cached in software, so that the system avoids decrypting frequently executing instructions multiple times. To provide a shifting attack surface, the VM periodically discards any cached (and hence decrypted) instructions (described in Section 3.3).

These approaches make the dynamic analysis techniques challenging to conduct. The following sections provide more details about how the mechanisms work, while Section 5 provides a more in-depth security discussion about the tractability of common attack vectors.

3.1 Guards

We propose novel uses of self-introspective code called *guards* (as described by Chang, *et al.* [7]) to provide tamper resistance. Each guard checks a range of memory addresses,

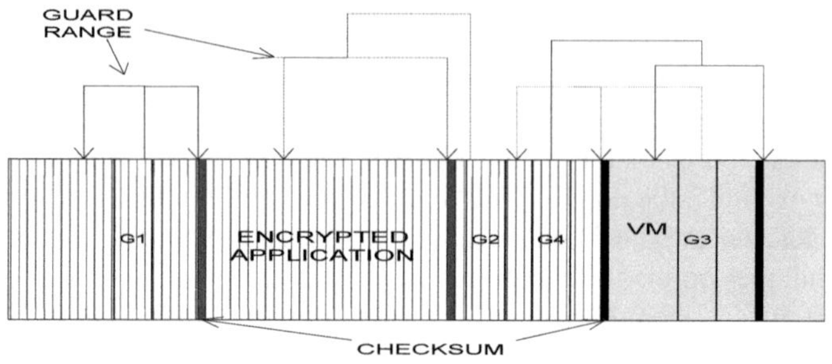

Fig. 1. Layout of an encrypted binary protected by guards

using a lightweight hash function which consists of a few logical and memory operations. Guards are placed such that they mutually protect the application as well as the VM. Unlike previous approaches, this technique binds the protection mechanism of the application closely with that of the VM, hampering the adversary from dissecting them.

In our approach, guards protect the encrypted application instructions. Figure 1 illustrates the layout of a protected binary, with guards ($G1$, $G2$, $G3$, and $G4$) and associated checksums inserted in both the encrypted application and the VM. For example, the guard $G3$, located in the VM, protects encrypted application code (shown in hatching in the figure), whereas the encrypted guard $G4$, located in the application, protects the VM. Initially, guard code templates are inserted probabilistically throughout the application and the VM. The checksums are only calculated after the entire application has been encrypted. This technique offers stronger protection than guarding plaintext code in the case where the adversary is able to locate and adjust the checksums to reflect any malicious modifications. The adversary would now have to encrypt the modifications as well and update the checksums accordingly. Although the guards located in the VM are unencrypted, the cyclic nature of guard protection ensures that such guards are safeguarded from tampering. Implementation details of checksumming guards is described in Section 4.1

An important consequence of using process-level virtualization is that application guards never execute from their original location in the binary. Instead, as illustrated in Figure 2, the code is copied to a separate location and then executed. Therefore, even if the adversary is able to observe the run-time behavior of the guards, their original locations in the application remain secure.

3.2 Encryption

Previous approaches have handled software decryption at a coarse level of granularity (e.g., at the function level [5] or bulk decryption, in which the entire code base is decrypted in memory [4]). Our scheme performs just-in-time, on-demand decryption, as shown in Figure 2. The entire text segment of the application along with its guards, is encrypted using a strong encryption algorithm, as shown by the shading on the right of the figure. At run time, the VM loads the application and starts decrypting and decoding application instructions one basic block at a time (this process is known as *translation*). The block is cached in software and scheduled for execution. After the block executes,

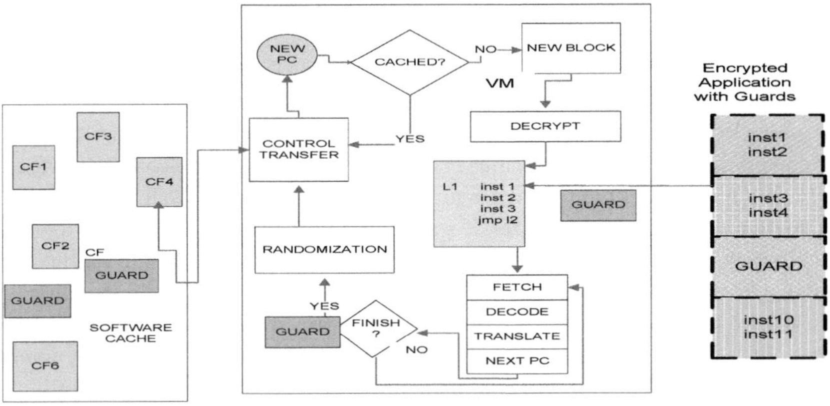

Fig. 2. Encrypted application running on a virtual machine

control again enters the VM. If next instruction to be executed already exists in the code cache, control directly transfers to that instruction. Otherwise, the VM begins decrypting and decoding instructions at the next application address. The translated block is then placed at the next available location in the code cache. To increase the amount of entropy, a random number of instructions are appended to each translated block.

Ensuring the obscurity of the decryption key is important for the success of this scheme. If this key can be determined, the adversary can decrypt the text segment of the application and analyze the contents of the original application. Currently the decryption process uses a single key for encryption and decryption. Chow, *et al.* proposed a solution in which the key of the underlying block cipher is expanded from several bytes to a collection of look up tables with a total size in the order of hundreds of kilobytes [8]. Although an attack on this scheme has been published, it incurs a significant overhead in terms of computation time [3]. The system can be made more secure by using different keys for different parts of the code segment, and encrypting different parts multiple times to make it harder for the adversary to decrypt the application.

3.3 Flushing

Portions of the plaintext application code are cached in software to avoid the overhead of frequent decryption. Over a period of time, the application code will build up in the cache, and the adversary will be able to analyze the application from the cache. A stealthy technique is required to continuously reduce the plaintext code available to the adversary. Flushing the code cache at periodic intervals achieves this goal and prevents the adversary from obtaining a sizable chunk of the plaintext code. Flushing splits up the run-time information into multiple pieces, forcing the adversary to craft them together to fully analyze the application.

Periodic flushing also aids in providing a constantly changing execution profile of the application. It regularly shifts the attack surface of the application, as the protection mechanisms are relocated after each flush. This entropy is furthered increased by adding a random number of instructions to each basic block in the software cache, as described in Section 3.2. Randomized blocks combined with flushing ensure that guards regularly execute from different addresses in the code cache.

Flushing can be triggered by a periodic signal generated by the operating system. However, the threat model includes the case where the application is run on top of a malicious OS. The adversary could modify the OS kernel such that the signal is not delivered to the application at all, thus disabling the flushing completely. Therefore, another technique is required for triggering the code flushing. Ideally, this scheme should depend on an inherent property of the application itself, making it difficult to alter the system via an external agency. The flushing should be triggered stealthily, without alerting the adversary as to when the code cache is cleared of the stored instructions. One such scheme involves counting a particular type of instruction at run time and triggering the flush when a threshold has been crossed. The instruction should be numerous enough and well spread out throughout the application code such that the flushing behavior is approximately periodic. As an example, we chose indirect jumps (including function returns) as our candidate instruction. The application is run in training mode and the average number of indirect jumps executed per second is calculated. Flushing occurs based on some function of this average.

4 Evaluation

We have designed a prototype that implements our techniques, using a combination of binary rewriting and a process-level virtual machine, called Strata [17]. Strata runs as a co-routine with the application, making modifications as it is running. It mediates program execution by dynamically examining and translating program instructions before they are run on the host CPU. Frequently executing instructions are cached in a software code cache. The original application is first linked with the Strata libraries [17]. We then use a link-time binary rewriting tool to introduce guards into the program code (both Strata and original application) and encrypt the application text segment (but not Strata code) using AES.

Our proof-of-concept implementation targets the Intel x86 platform, but the concepts described in this work are platform-independent. This system was evaluated using the C language benchmarks of the SPEC CPU2000 suite. All the performance related plots are normalized to native execution (i.e., no protection techniques) and averaged over five runs. The number of guards inserted into the application and VM is determined by a heuristic. The heuristic takes into account the expected coverage (number of guards covering each byte of the program on average), and the size of the address range, given by $N = \frac{K*P}{S}$ where K is the approximate coverage desired by the user, P is the program text size, S is the average guard range size set by the user and N is the number of guards.

We performed evaluative studies with various guard sizes and coverage rates. For our analytical study, we chose an average guard range of 8KB and set the coverage rate to 4, as this configuration provided good trade off in terms of performance and security. Due to space constraints, we can only present some of the results in the following section.

4.1 Run-Time Protection

The creation and placement of guards are driven by probabilistic models. Each basic block is assigned a probability of obtaining a guard, which is inversely proportional to its profiled execution count. The guard range sizes are determined based on the Gaussian distribution and are independent of the actual location of the guard itself. Experiments showed that guard coverage is highly unbalanced if the starting addresses of the guard ranges are selected randomly. To improve the coverage, addresses are determined sequentially starting from the initial address of the program. The range size is still generated probabilistically, making it hard for the adversary to locate the ranges.

Analyzing the run-time protection afforded by the system is an important aspect of this research. Guards that execute rarely would leave the application vulnerable to modification for a majority of its execution time. Investigation into the run-time behavior of the guard system revealed that the initial phase of the program shows a large number of guards being executed, and this behavior is consistent across all the benchmarks. Our placement scheme is more favorable towards basic blocks which have a low execution count, and application startup code tends to have a high number of such blocks. Once the startup phase comes to a close, the number of guards executing per second decreases but many guards still execute in the steady state. Phase changes will tend to show increased guard execution rate, while steady state behavior (e.g., a tightly run loop) will have a decreased guard rate.

Figure 3 shows the run-time characteristics for a guarded application. Figure 3a shows the frequency of guard execution per second for `176.gcc`. The figure shows that a significant number of guards execute in the first second itself and consequently a large proportion of the code is checked. In an ideal case, each byte of the program should be protected by multiple guards, so that even if an adversary manages to disable some of them, there are other guards offering protection. Therefore, an important metric in security systems is *connectivity* of a program byte, which is defined as the number of unique guards covering that byte. Figure 3b shows that on average about 80% of the application text is covered by three to four guards. This value indicates that on average, to modify a single byte of application code, the adversary will have to disable 3-4 guards, which in turn are protected by 3-4 guards each. Such a distributed protection scheme makes it difficult for the attacker to target any single point of vulnerability.

The strength of a tamper resistance system can be evaluated by measuring the average time delay between successive checks on a program byte. This metric indicates how long modifications can exist in the system before detection. Figure 4 illustrates this metric for all the benchmarks. On average, checks are performed every 3.5 seconds for each byte in the program, indicating that any modification in the code will be detected within 3.5 seconds on average. We believe that this value is sufficient protection for many classes of applications, such as word processors, web browsers and media players. For example, using these techniques to protect media players will prevent the adversary from viewing digital media for any practical amount of time.

`176.gcc` and `254.perlbmk` are the best-protected benchmarks. These two benchmarks are the largest in the benchmark suite, and closely resemble real world applications like word processors and web browsers in having multiple phases of execution. As such, this data indicates our techniques can provide strong tamper resistance

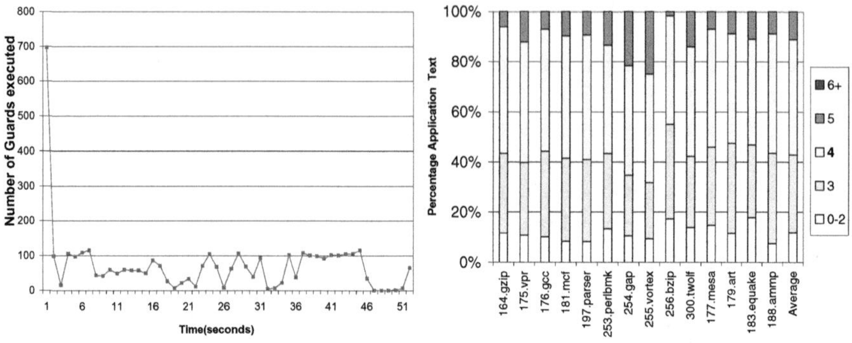

(a) Number of guards executing per second for `176.gcc`

(b) Guard connectivity

Fig. 3. Run-time characteristics for the guarded application

to actual applications while adding tolerable overhead. Applications in which a tight loop runs for a majority of program lifetime need further investigation and will be addressed in future work.

Figure 5 displays the distribution of guards based on the guards' location (VM or application) and the guards' protection target (VM or application), along with the total number of guards executed. There is also a category for guards which simultaneously protect both the VM and the application regardless of their location, called *Combined.* The figure shows that 78% of the guards are located in the application, on average. These guards are the ones encrypted and consequently get exposed to continuous dynamism due to the system's dynamic protection mechanisms. These guards never execute from their original locations in the application. From this data, we conclude that the majority of the guards execute in a stealthy manner, given that the dynamic protection schemes provide sufficient dynamism.

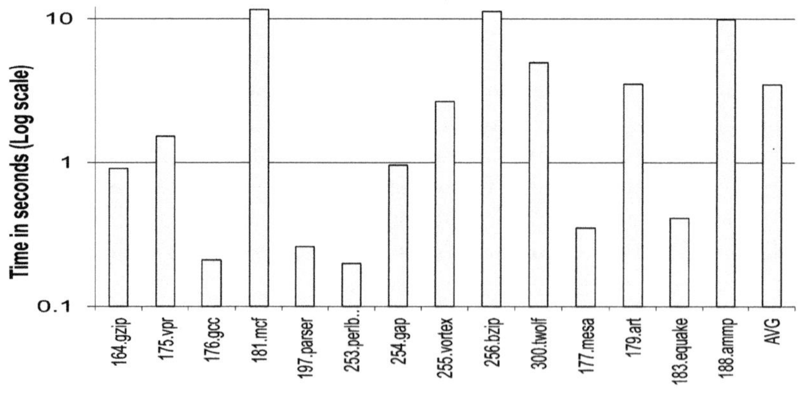

Fig. 4. Average time delay between checks for each byte of the program. This metric gives an indication as to how long a modification can exist undetected.

4.2 Flushing

Guards provide resistance to tampering but flushing is required to reduce leakage of information that an adversary can use to understand the application. This lack of understanding makes it hard for the adversary to disable protection techniques and modify the application in a meaningful manner (i.e., achieving the adversary's goal). Flushing must occur stealthily at a regular interval, forcing the adversary to analyze the code at a finer level of granularity. For example, instead of obtaining the entire plaintext code once the application is loaded in memory, as in Skype, the adversary will have to analyze the code frequently to obtain all the code snapshots. In this section, we analyze the effects of flushing based on the frequency of indirect branches executed on performance and program security. Initially, the application is profiled, using SPEC's training input, to obtain the number of indirect branches executed per second (designated as **R** in the following discussion). The performance overhead for flushing after every 10**R**, **R**, and **R**/10 branches was 25%, 30% and 55% respectively, compared to native execution. The overhead for flushing every **R**/10 branches is quite high but the other two options show promising results. We found the rate of flushing becomes somewhat inconsistent, specially if the threshold is set too low. In particular, flushing performed after every **R/10** branches differs when compared with flushing performed every 0.1 sec. The reason for this behavior is that indirect branches are not temporally uniform but are clumped together in time.

However, the flushing rate by itself does not give any indication of the amount of protection afforded, so we measured the rate at which application text appears in the code cache. Figure 6, which plots the behavior of `176.gcc`, indicates that, even without flushing, only some of the program text is present in the cache at a time. Even flushing every 10 **R** branches per second shows some benefit, as much of the code is used in startup or tear down, after which it can be flushed out of the cache. Flushing every **R** branches does a much better job at keeping a significant portion of the application out of the cache. Flushing **R**/10 branches per second does the best job, as no more than

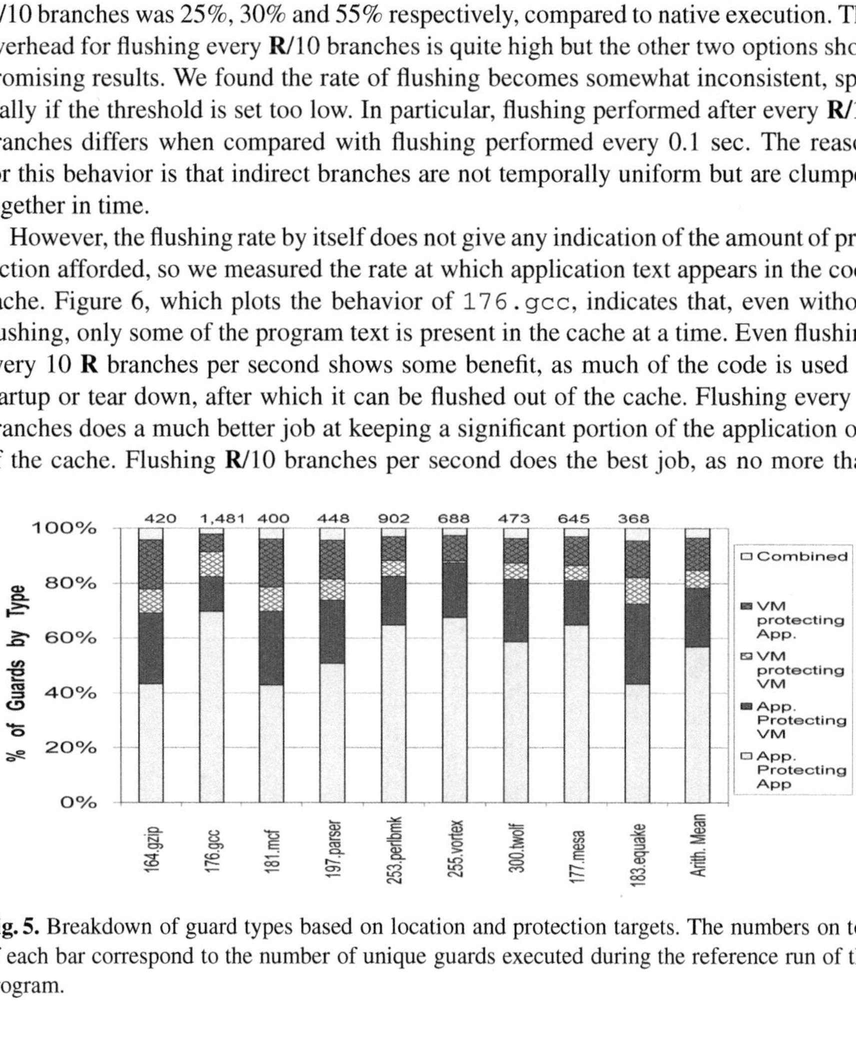

Fig. 5. Breakdown of guard types based on location and protection targets. The numbers on top of each bar correspond to the number of unique guards executed during the reference run of the program.

30% of the code resides in the cache at any point in time, while only adding an overhead of around 20%. This value is much better than bulk decryption, which has 100% application code decryption at startup. Decryption on a per-function basis typically has high overhead (up to 8X), unless functions are encrypted selectively based on profiled information [5].

Flushing aids in relocation during a program run but code needs to execute from different locations across program runs as well. This code shifting hampers the adversary from launching iterative attacks on the application. As such, a small number of random instructions (between 2-8) are appended at the end of each basic block in the software cache. This randomization ensures that there is a high probability that relative distance between any two basic blocks is different across runs. As a consequence, guards will execute from different locations across runs. Figure 7 shows the distribution of cache addresses of guards across 10 runs of the same executable. The software cache was flushed once every second so that even within a single run, guards execute from multiple locations. Each guard had at least 10 distinct locations within the cache.

4.3 Performance Overhead

Figure 8 displays the performance overhead of each individual technique as well as the combined performance compares it with the benchmarks running under Strata without any protection. On average, the overhead of baseline Strata is 16% over native execution while the overhead of running an encrypted application is 17% over native. In spite of using a heavy-weight algorithm (AES), the low overhead can be explained by the fact that most of the time is spent on executing the application from the software cache rather than decryption. The guards add, on average, an overhead of 7% over native execution while flushing adds approximately 10% overhead. Overall, an overhead of 35% is added to the benchmarks compared to native execution. Approximately half of the overhead is due to the VM itself, and research is ongoing into reducing this overhead. The average coverage for each byte of the program has been set to 4. In totality, this data indicates

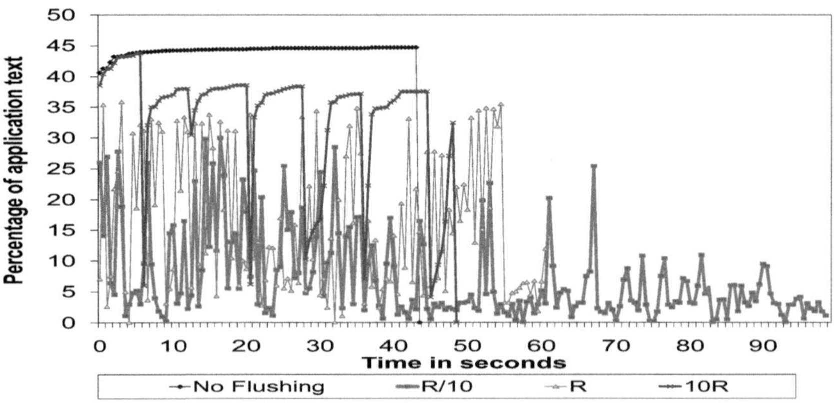

Fig. 6. Rate of plaintext application code due to flushing on indirect branch count

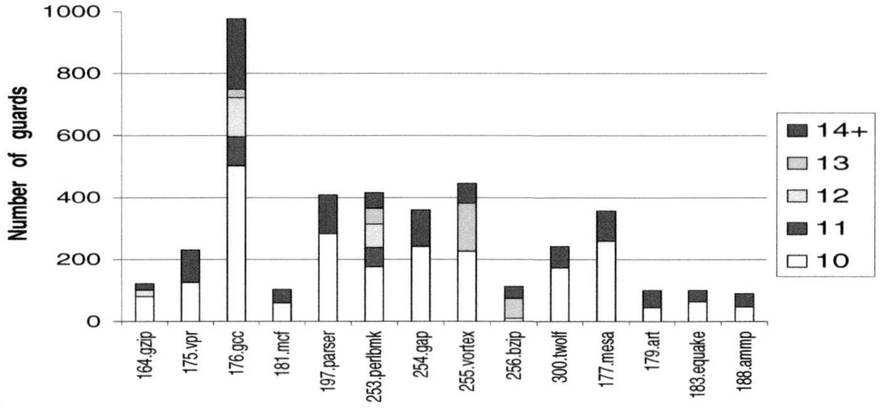

Fig. 7. Distribution of code cache addresses for guards across 10 independent runs. The cache is flushed every second.

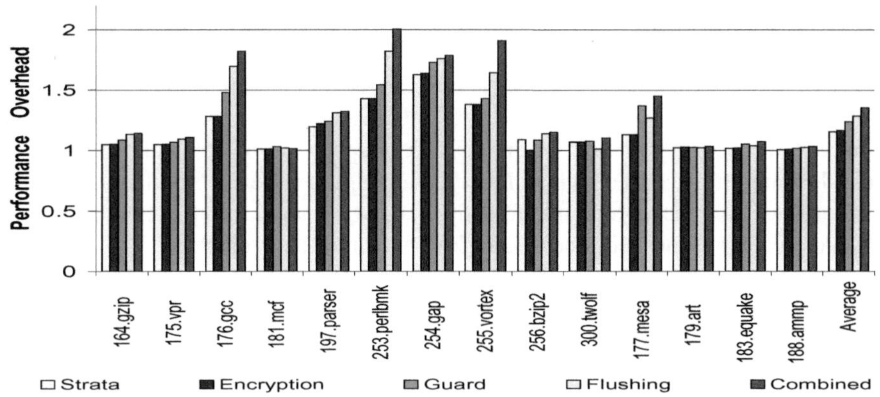

Fig. 8. Run-time overhead for the benchmarks running on the VM (without any protection) and with each individual technique (encryption, guards and flushing). The final bar displays the overhead for the entire framework. The flushing occurs based on instruction counting (approximately once every second). The run time is normalized to native.

that our approach can provide strong protection to real-world applications at a lower overhead than comparable techniques.

5 Security Discussion

5.1 Circular Protection

All the techniques mutually reinforce each other to provide a strong tamper-resistance run-time environment. For example, the guards are inserted into the program and encrypted. They check the encrypted binary as well as the VM code. For an attack to be successful, the adversary will have to find and update all the guards (since each guard is protected by a network of multiple guards). The encryption protects the static location of the guards, while the flushing provides variance for the dynamic location. Likewise, the guards provide protection for the decryption and flushing codes. Similarly, the flushing protects against dynamic tracing techniques, and prevents persistent changes to the code cache. Taken together, these techniques impart missing dynamism to current protection mechanisms. The emphasis of these techniques is the run time, as such, they can be easily combined with existing static protection mechanisms for comprehensive security.

5.2 Effectiveness against Static and Dynamic Analysis

Static analyzers will not function on the application text encrypted with a randomly selected key. The virtual machine is less protected from static analysis than the application code. However, there are encrypted guards in the application protecting the VM. As such, the adversary will not be able to modify the VM using only static analysis. The VM also maintains a static internal data structure that maps original application addresses of basic blocks to their code cache counterparts. To prevent the adversary from exploiting this data structure, its location and organization can be made dynamic on a per-flush basis. Padding the cached basic blocks with random number of bytes makes it highly probable that a mapping is valid only for one cache flush cycle. Disabling the VM will render the application unusable because the VM is responsible for decryption.

Dynamic tools and analyzers use traditional static analysis techniques on the runtime trace of the program and obtain a sizable snapshot of the application code. Periodic flushing of the cached code reduces the amount of code kept in plaintext form in memory by a significant amount. To fully analyze the application, the adversary will have to obtain multiple code cache snapshots and craft them together.

5.3 Effectiveness against OS and VM attacks

Modified OSes have already been used to mount successful attacks against software guard systems. Guard systems work on the assumption that the underlying hardware follows the von Neumann architecture (data reads and instruction fetches go to the same memory structure). Wurster, *et al.* showed a quick workaround to guards: separate data and instruction memory [20]. Each page of the application was duplicated and modifications were applied to it. The kernel was modified such that data reads would go to the unmodified application, whereas instruction fetches would bring in instructions from the tampered copy. Our solution defeats this attack in two ways. First, the application code is encrypted on disk and decrypted on an on-demand basis, as such creating a tampered copy is not trivial. Secondly, VMs continuously flush and execute instructions from the code cache, which defeats this split-memory attack [11]. Any attempt to tamper with the underlying memory system would render the VM and consequently, the application, unusable.

An attacker might consider replacing the VM with a copy where flushing has been disabled. However, VM replacement requires extracting the decryption key, which is believed to be expensive [3]. Such extraction can be made arbitrarily difficult through the use of multiple decryption keys, multiple binary encodings and decodings, and multiple VMs.

5.4 Effectiveness against Skype Attack

Skype is a popular Voice-over-IP tool, whose protection mechanism included a combination of encryption, guards and run-time code generation [4]. However, there were flaws in these techniques, which led to the binary being cracked. The code decryption and run-time code generation operated at coarse levels of granularity, enabling the adversary to overcome these protections at run time. The location of the checking guards was determined statically, making them vulnerable to collusion attacks.

Our techniques would have thwarted this attack in multiple ways. The VM decrypts the application in an on-demand manner and flushes periodically, as such the application code is never exposed completely. Therefore, analysis can only proceed in a piecemeal manner, making it harder for the adversary. The VM constantly rearranges and randomizes the code locations in the cache, making it hard for the adversary to use any systematic means to locate the guards. So, even if two or more instances of the application are compared, there will be no correlation between guards locations in each instance. Finally, even if the adversary is able to make any meaningful changes to the code cache, periodic flushing discards any modification within a short amount of time.

6 Related Work

Considering the importance of tamper resistance and obfuscation, a significant amount of work has been done in these areas [13,7,2,1,5]. Code obfuscation protects applications against reverse engineering, insofar as to make the algorithms harder to understand [9]. Tamper resistance strives to make software immune to malicious modifications. Realistic, usable solutions must provide practical overhead or software developers will be unwilling to adopt the security measures.

Tamper resistance protects the authenticity of application code. Aucsmith introduced a system called Integrity Verification Kernels (IVK) [2]. Significant work in tamper resistance is based on one or more of Aucsmith's ideas. Perhaps the most well known work in this area is software guards. Chang, *et al.* first introduced this idea [7], and this was further extended by Horne *et al.* [12]. However, neither work was evaluated extensively over a standard benchmark suite. Jacob, *et al.* developed another technique called *oblivious hashing* [13]. None of these techniques prevent attackers from using dynamic techniques to identify and remove the protections. Adapting hardware-based solutions, to production environments is non-trivial, since the threat model requires linking software with specialized hardware, increasing the overall cost of the system [18]. Tamper-resistance solutions created via a Virtual Machine Monitor(VMM) often leave the VMM vulnerable [10,16]. Our system provides provides protection for the entire package. An important prerequisite for modifying applications is to understand the underlying algorithms of the program. Techniques which make it difficult to understand a program are known as *obfuscations*, both statically [9], and by using dynamic techniques [15]. Code encryption hampers software tampering and static analysis to a great extent, however, previous techniques have either incurred high performance overhead [5], or used extra hardware for decryption.

7 Conclusions

We have described and evaluated a unique technique which provides resistance to tampering and code obfuscation, using process-level virtualization. We make use of software guards that checksum portions of the application text. For added protection, these guards are encrypted using a strong encryption algorithm using a random key. The encrypted application is then executed under control of a VM, whose code cache is periodically flushed to reduce information leakage. Therefore, with an additional overhead of approximately 14% over the VM (35% over native execution), these techniques can be applied to applications, resulting in a more secure execution environment, even on malicious hosts.

This research was supported in part by the National Science Foundation under grants CNS-0716446 and CCF-0811689 and in part by DOD AFOSR MURI Grant FA9550-07-1-0532. We also gratefully acknowledge the support of Christopher Milner and SAIC for support of this research. Any opinions, findings, and conclusions or recommendations expressed in this material are those of the authors and do not necessarily reflect the views of the National Science Foundation or the Department of Defense.

References

1. Anckaert, B., Jakubowski, M., Venkatesan, R.: Proteus: virtualization for diversified tamper-resistance. In: DRM 2006: ACM Workshop on Digital Rights Management, pp. 47–58. ACM, New York (2006)
2. Aucsmith, D.: Tamper resistant software:An implementation. In: 1st International Workshop on Information Hiding, pp. 317–333. Springer, London (1996)
3. Billet, O., Gilbert, H., Ech-Chatbi, C.: Cryptanalysis of a white box AES implementation. In: Handschuh, H., Hasan, M.A. (eds.) SAC 2004. LNCS, vol. 3357, pp. 227–240. Springer, Heidelberg (2004)
4. Biondi, P., Fabrice, D.: Silver needle in the skype. In: Black Hat Europe, Amsterdam, the Netherlands (2006)
5. Cappaert, J., Preneel, B., Anckaert, B., Madou, M., Bosschere, K.D.: Towards tamper resistant code encryption: Practice and experience. In: Chen, L., Mu, Y., Susilo, W. (eds.) ISPEC 2008. LNCS, vol. 4991, pp. 86–100. Springer, Heidelberg (2008)
6. Castelluccia, C., Francillon, A., Perito, D., Soriente, C.: On the difficulty of software-based attestation of embedded devices. In: CCS 2009: 16th ACM Conference on Computer and Communications Security, pp. 400–409. ACM, New York (2009)
7. Chang, H., Atallah, M.: Protecting software code by guards. In: ACM Workshop on Security and Privacy in Digital Rights Management, pp. 160–175 (2000)
8. Chow, S., Eisen, P.A., Johnson, H., Oorschot, P.C.v.: White-box cryptography and an AES implementation. In: Nyberg, K., Heys, H.M. (eds.) SAC 2002. LNCS, vol. 2595, pp. 250–270. Springer, Heidelberg (2003)
9. Collberg, C., Thomborson, C., Low, D.: Manufacturing cheap, resilient and stealthy opaque constructs. In: POPL 1998: 25th ACM SIGPLAN-SIGACT Symposium on Principles of Programming Languages, pp. 184–196. ACM Press, New York (1998)
10. Garfinkel, T., Pfaff, B., Chow, J., Rosenblum, M., Boneh, D.: Terra: a virtual machine-based platform for trusted computing. In: SOSP 2003: 19th ACM Symposium on Operating Systems Principles, pp. 193–206. ACM Press, New York (2003)

11. Giffin, J.T., Christodorescu, M., Kruger, L.: Strengthening software self-checksumming via self-modifying code. In: ACSAC 2005: 21st Annual Computer Security Applications Conference, pp. 23–32. IEEE Computer Society, Washington (2005)
12. Horne, B., Matheson, L.R., Sheehan, C., Tarjan, R.E.: Dynamic self-checking techniques for improved tamper resistance. In: Sander, T. (ed.) DRM 2001. LNCS, vol. 2320, pp. 141–159. Springer, Heidelberg (2002)
13. Jacob, M., Jakubowski, M.H., Venkatesan, R.: Towards integral binary execution: implementing oblivious hashing using overlapped instruction encodings. In: MM&Sec 2007: 9th Workshop on Multimedia & Security, pp. 129–140. ACM, New York (2007)
14. Linn, C., Debray, S.: Obfuscation of executable code to improve resistance to static disassembly. In: CCS 2003: 10th ACM Conference on Computer and Communications Security (CCS), pp. 290–299. ACM Press, Washington (2003)
15. Madou, M., Anckaert, B., Moseley, P., Debray, S., De Sutter, B., De Bosschere, K.: Software protection through dynamic code mutation. In: Song, J.-S., Kwon, T., Yung, M. (eds.) WISA 2005. LNCS, vol. 3786, pp. 194–206. Springer, Heidelberg (2005)
16. Quynh, N.A.: Hijacking (Xen) virtual machine for fun and profit. In: Bellua Security Conference, Jakarta, Indonesia (2007)
17. Scott, K., Kumar, N., Velusamy, S., Childers, B., Davidson, J.W., Soffa, M.L.: Retargetable and reconfigurable software dynamic translation. In: CGO 2003: International Symposium on Code Generation and Optimization, pp. 36–47. IEEE Computer Society, Washington (2003)
18. Lie, D., Thekkath, C., Mitchell, M., Lincoln, P., Boneh, D., Mitchell, J., Horowitz, M.: Architectural support for copy and tamper resistant software. In: ASPLOS 2000: 9th International Conference on Architectural Support for Programming Languages and Operating Systems, vol. 35, pp. 168–177. ACM, New York (2000)
19. Seshadri, A., Luk, M., Shi, E., Perrig, A., van Doorn, L., Khosla, P.: Pioneer: Verifying code integrity and enforcing untampered code execution on legacy systems. In: SOSP 2005: 20th ACM Symposium on Operating Systems Principles, vol. 39, pp. 1–16. ACM Press, New York (December 2005)
20. Wurster, G., Oorschot, P.C.v., Somayaji, A.: A generic attack on checksumming-based software tamper resistance. In: SP 2005: 2005 IEEE Symposium on Security and Privacy, pp. 127–138. IEEE Computer Society, Washington (2005)

Security with Noisy Data

(Extended Abstract of Invited Talk)

Boris Škorić

Eindhoven University of Technology

An overview was given of security applications where noisy data plays a substantial role. Secure Sketches and Fuzzy Extractors were discussed at tutorial level, and two simple Fuzzy Extractor constructions were shown. One of the latest developments was presented: quantum-readout PUFs.

Biometrics and Physical Unclonable Functions. Noise plays an essential role in a number of security applications based on biometrics and Physical Unclonable Functions (PUFs). In biometric authentication, the best way to preserve user privacy would be to store a one-way *hash* of the biometric, just like the password hashes in a unix password file. However, the noise inherent in biometric measurements prevents one from directly applying a hash function. A PUF [6,7] is a complex piece of material. Depending on the properties (physical and/or mathematical unclonability/opaqueness/many challenge-response pairs) PUFs can be used for a variety of purposes such as authentication, anti-counterfeiting, tamper evidence, software-to-hardware binding and read-proof key storage. In the latter case the term Physically Obfuscated Key (POK) is used [3]. An overview can be found in [8]. Just as with biometrics, measurements are noisy, which poses a problem when the data has to serve as a cryptographic key or has to be hashed.

Secure Sketches and Fuzzy Extractors. A special form of error correction is needed: It is prudently assumed that the adversary is an insider who has access to any database keys and therefore to all enrollment data. The error correction redundancy must not leak confidential information. Security primitives achieving these goals were first developed in [4,5,2]. A Secure Sketch (SS) allows for exact reconstruction of a discrete variable X, with the use of helper data W that minimally leaks about X. A Fuzzy Extractor (FE) allows for exact reproduction of a derived secret S (from discrete or continuous X), where W does not leak about S.

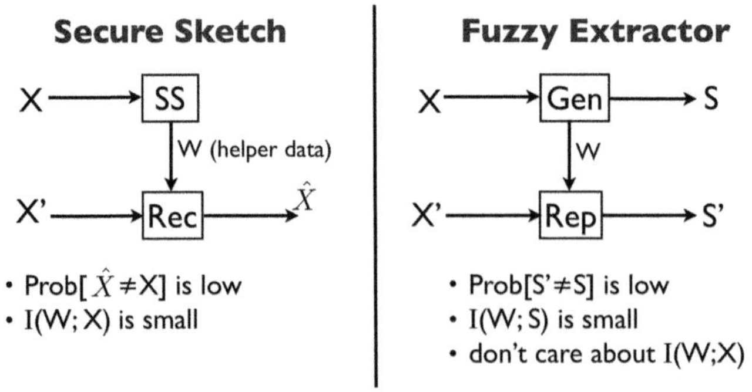

R. Böhme, P.W.L. Fong, and R. Safavi-Naini (Eds.): IH 2010, LNCS 6387, pp. 48–50, 2010.

A SS can always be turned into a FE by using Universal Hash Functions (UHFs) [1]. This involves a large entropy loss, which in the case of continuous variables can be avoided [9] by the use of equiprobable discretization (right Fig. below).

Application	*privacy of X ?*	*uniform secret?*	*Technique*
password authent.	■		One-Way Function
biometric authent.	■		Secure Sketch + OWF
anticounter-feiting PUF	■		Secure Sketch + OWF
anticounter-feiting PUF			---
PUF authent. w/o MACs			---
PUF authent. with MACs		■	Fuzzy Extractor
POK		■	Fuzzy Extractor

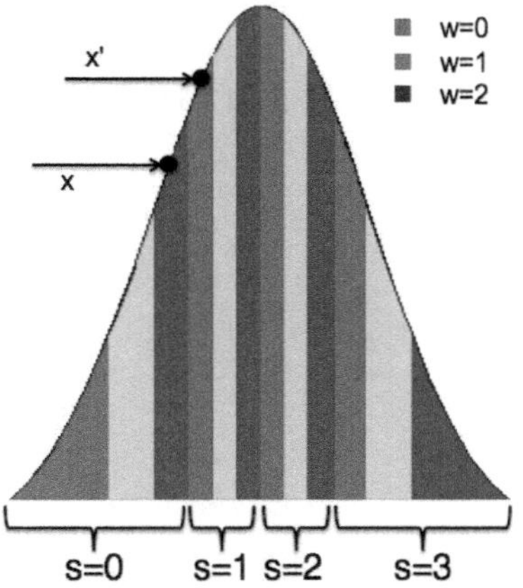

Quantum Readout of PUFs. Recently it was realized [10] that a PUF's physical unclonability can be combined with quantum unclonability: if the PUF challenges and responses are single quanta, tamper evident readout is automatically achieved, even if the PUF is in hostile territory. This readout can serve as an authenticated quantum channel for Quantum Key Distribution. (Usually in QKD the classical channel is authenticated.)

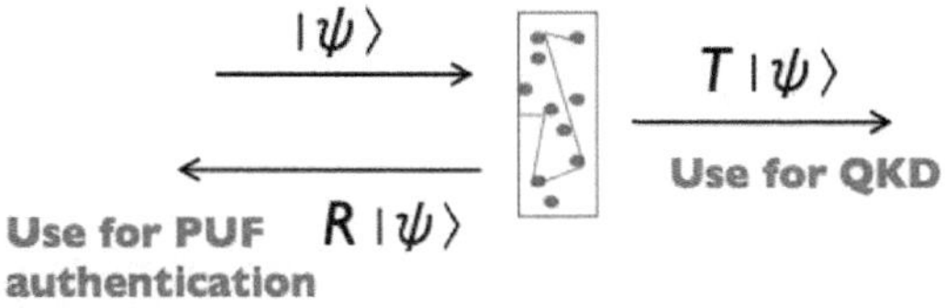

References

1. Carter, J.L., Wegman, M.N.: Universal classes of hash functions. J. of Computer and System Sciences 18(2), 143–154 (1979)
2. Dodis, Y., Reyzin, M., Smith, A.: Fuzzy Extractors: How to generate strong keys from biometrics and other noisy data. In: Cachin, C., Camenisch, J.L. (eds.) EUROCRYPT 2004. LNCS, vol. 3027, pp. 523–540. Springer, Heidelberg (2004)
3. Gassend, B., Clarke, D.E., van Dijk, M., Devadas, S.: Silicon Physical Unknown Functions. In: ACM CCS 2002, pp. 148–160 (2002)
4. Juels, A., Wattenberg, M.: A fuzzy commitment scheme. In: ACM CCS 1999, pp. 28–36 (1999)
5. Linnartz, J.P.M.G., Tuyls, P.: New shielding functions to enhance privacy and prevent misuse of biometric templates. In: Kittler, J., Nixon, M.S. (eds.) AVBPA 2003. LNCS, vol. 2688, pp. 238–250. Springer, Heidelberg (2003)
6. Pappu, R.: Physical One-Way Functions. PhD thesis, MIT (2001)

7. Pappu, R., Recht, B., Taylor, J., Gershenfeld, N.: Physical One-Way Functions. Science 297, 2026–2030 (2002)
8. Tuyls, P., Škorić, B., Kevenaar, T.: Security with Noisy Data. Springer, Heidelberg (2007)
9. Verbitskiy, E.A., Tuyls, P., Obi, C., Schoenmakers, L.A.M., Škorić, B.: Key extraction from general non-discrete signals. IEEE Trans. Inf. Forensics and Security 5(2), 269–279 (2010)
10. Škorić, B.: Quantum Readout of Physical Unclonable Functions. In: Bernstein, D.J., Lange, T. (eds.) AFRICACRYPT 2010. LNCS, vol. 6055, pp. 369–386. Springer, Heidelberg (2010)

Detection of Copy-Rotate-Move Forgery Using Zernike Moments

Seung-Jin Ryu, Min-Jeong Lee, and Heung-Kyu Lee

Department of Computer Science,
Korea Advanced Institute of Science and Technology,
Daejeon, Republic of Korea
{sjryu,mjlee,hklee}@mmc.kaist.ac.kr

Abstract. As forgeries have become popular, the importance of forgery detection is much increased. Copy-move forgery, one of the most commonly used methods, copies a part of the image and pastes it into another part of the the image. In this paper, we propose a detection method of copy-move forgery that localizes duplicated regions using Zernike moments. Since the magnitude of Zernike moments is algebraically invariant against rotation, the proposed method can detect a forged region even though it is rotated. Our scheme is also resilient to the intentional distortions such as additive white Gaussian noise, JPEG compression, and blurring. Experimental results demonstrate that the proposed scheme is appropriate to identify the forged region by copy-rotate-move forgery.

Keywords: Digital Forensics, Copy-Move Forgery, Copy-Rotate-Move Forgery, Zernike Moments.

1 Introduction

As the image processing softwares have been developed, even people who are not experts in image processing can easily alter digital images. It brings about great benefits, but also side effects: a number of tampered images have recently been distributed or have even been published by major newspapers. Therefore, it is important to verify the authenticity of digital images. Among forgery techniques using typical image processing tools, copy-move forgery is one of the most commonly used methods. The copy-move forgery copies a part of the image and pastes it into another part of the image to conceal an evidence or deceive people. Figure 1 shows an example of the altered photograph released by Iran and published by western media including The New York Times, The Los Angeles Times, BBC News, and *etc.* on July 9, 2008 [1]. In Fig. 1(a), two major sections (encircled in black) appear to be replicated from other sections (encircled in white). Actually Fig. 1(a) was released on the front pages of those of newspapers and lately corrected to the original image as Fig. 1(b).

The first method for detecting copy-move forgery was suggested by Fridrich *et al.* [2]. They lexicographically sorted quantized discrete cosine transform (DCT) coefficients of small blocks and then checked whether the adjusted blocks are

R. Böhme, P.W.L. Fong, and R. Safavi-Naini (Eds.): IH 2010, LNCS 6387, pp. 51–65, 2010.

(a) (b)

Fig. 1. An example of copy-move forgery [1]: (a) the forged image with four missiles and (b) the original image with three missiles

similar or not. On the other hand, Popescu *et al.* employed principal component analysis (PCA) to extract important feature vectors and checked the similarity of blocks [3]. Similarly, Li *et al.* calculated the similarity of blocks based on discrete wavelet transform and singular vector decomposition (DWT-SVD) and Luo *et al.* measured block characteristics vector from each block [4,5]. Mahdian *et al.* exploited blur invariant moments to detect duplicated regions [6]. Since they used the property invariant to blur, their scheme has robustness against post-processing such as blur degradation, additional noise, and arbitrary contrast changes.

Copy-move forgery as depicted in Fig. 1 usually means that the copied part of the image is pasted into another part of the image without any geometric change. However, people easily modify the geometry of the copied part so that the forged image seems to be original. Among the geometric modifications, rotation is commonly used to provide spatial synchronization between the copied region and its neighbors. In this paper, therefore, the forgery technique which copies a region and rotates it before pasting is named as copy-rotate-move (CRM) forgery. Figure 2 shows an example of CRM forgery. Fig. 2(a) is an original image and Fig. 2(b) and Fig. 2(c) are the forged images. In Fig. 2(b), the left aircraft (encircled in white) is copied and pasted into the image with no change. In Fig. 2(c), by contrast, the copied aircraft (encircled in black) is slightly rotated before pasting into the middle region. As seen with the naked eye, the rotated aircraft in Fig. 2(c) looks more natural than the duplicated aircraft in Fig. 2(b).

There are several papers for figuring out CRM forgery. Bayram *et al.* applied Fourier-Mellin transform to the block [7]. However, according to their experimental results, the scheme performed well when the degree of rotation is small. Bravo-Solorio *et al.* suggested to represent each block in log-polar coordinates [8]. Then they defined 1-D descriptor as summation of angle values to achieve rotational invarance. Since the method depends on the pixel values, it is sensitive to the change of the pixel values. There are some approaches that extracted interest points on the whole image by scale-invariant feature transform (SIFT) [9,10,11]. Due to the fact that SIFT keypoints guarantee geometric invariance,

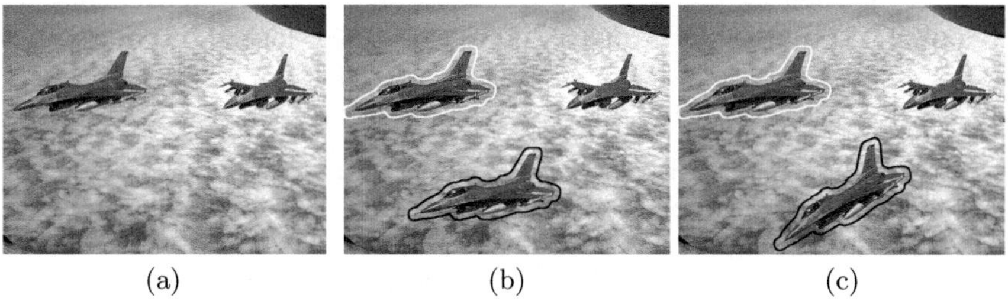

Fig. 2. An example of copy-rotate-move forgery: (a) the original image with two aircrafts, (b) the forged image with three aircrafts by copy-move forgery, and (c) the forged image with three aircrafts by copy-rotate-move (CRM) forgery

their method enables to detect rotated duplication. However, these schemes still have a limitation on detection performance since it is only possible to extract the keypoints from peculiar points of the image.

In this paper, we propose detection scheme for copy-rotate-move (CRM) forgery using Zernike moments. Since the magnitude of Zernike moments are algebraically invariant against rotation, the proposed method can detect the forged region even though it is rotated before pasting. The proposed scheme also performs well when white Gaussian noise is added to the image, the image is compressed in JPEG format, and even blurred.

The rest of the paper is structured as follows. We first overview the Zernike moments in Sec. 2. The details of proposed method are explained in Sec. 3. Experimental results are then exhibited in Sec. 4 and Sec. 5 concludes.

2 Zernike Moments

Moments and invariant functions of moments have been extensively used for invariant feature extraction in a wide range of pattern recognition, digital watermark applications and *etc.* [12,13]. Of various types of moments defined in the literature, Zernike moments have been shown to be superior to the others in terms of their insensitivity to image noise, information content, and ability to provide faithful image representation [13,14,15]. In this section, we describe Zernike moments mathematically. Some of the materials in the following are based on [13,15].

2.1 Definition

The Zernike moments [16] of order n with repetition m for a continuous image function $f(x, y)$ that vanishes outside the unit circle are

$$A_{nm} = \frac{n+1}{\pi} \int\int_{x^2+y^2 \leq 1} f(x,y) V_{nm}^*(\rho, \theta) dxdy \ , \tag{1}$$

where n a nonnegative integer and m an integer such that $n-|m|$ is nonnegative and even. The complex-valued functions $V_{nm}(x,y)$ are defined by

$$V_{nm}(x,y) = V_{nm}(\rho,\theta) = R_{nm}(\rho)\exp(jm\theta)\ , \tag{2}$$

where ρ and θ represent polar coordinates over the unit disk and R_{nm} are polynomials of ρ (Zernike polynomials) given by

$$R_{nm}(\rho) = \sum_{s=0}^{(n-|m|)/2} \frac{(-1)^s[(n-s)!]\rho^{n-2s}}{s!(\frac{n+|m|}{2}-s)!(\frac{n-|m|}{2}-s)!}\ . \tag{3}$$

Note that $R_{n,-m}(\rho) = R_{nm}(\rho)$. These polynomials are orthogonal and satisfy

$$\begin{aligned} &\int\int_{x^2+y^2\leq 1} [V_{nm}^*(x,y)] \times V_{pq}(x,y)dxdy = \tfrac{\pi}{n+1}\delta_{np}\delta_{mq}\ , \\ &\qquad\qquad \text{where } \delta_{ab} = \begin{cases} 1, & a=b \\ 0, & \text{otherwise}\ . \end{cases} \end{aligned} \tag{4}$$

For a digital image, the integrals are replaced by summations. To compute the Zernike moments of a given block, the center of the block is taken as the origin and pixel coordinates are mapped to the range of the unit circle. Those pixels falling outside the unit circle are not used in the computation. Note that $A_{nm}^* = A_{n,-m}$.

2.2 Rotational Invariance of Zernike Moments

This section proves algebraic invariance of Zernike moments against rotation. Consider a rotation of the image through angle α. If the rotated image is denoted by f', the relationship between the original and rotated image in the same polar coordinate is

$$f'(\rho,\theta) = f(\rho,\theta-\alpha)\ . \tag{5}$$

From Eq. (1) and (2), we can construct

$$\begin{aligned} A_{nm} &= \tfrac{n+1}{\pi}\int_0^{2\pi}\int_0^1 f(\rho,\theta)V_{nm}^*(\rho,\theta)\rho\ d\rho\, d\theta \\ &= \tfrac{n+1}{\pi}\int_0^{2\pi}\int_0^1 f(\rho,\theta)R_{nm}(\rho)\exp(-jm\theta)\rho\ d\rho\, d\theta\ . \end{aligned} \tag{6}$$

Therefore, the Zernike moment of the rotated image in the same coordinate is

$$A'_{nm} = \tfrac{n+1}{\pi}\int_0^{2\pi}\int_0^1 f(\rho,\theta-\alpha)R_{nm}(\rho)\exp(-jm\theta)\rho\ d\rho\, d\theta\ . \tag{7}$$

By a change of variable $\theta_1 = \theta - \alpha$,

$$\begin{aligned} A'_{nm} &= \tfrac{n+1}{\pi} \int_0^{2\pi} \int_0^1 f(\rho,\theta_1) R_{nm}(\rho) \exp(-jm(\theta_1+\alpha))\rho \; d\rho \, d\theta_1 \\ &= \left[\tfrac{n+1}{\pi} \int_0^{2\pi} \int_0^1 f(\rho,\theta_1) R_{nm}(\rho) \exp(-jm\,\theta_1)\rho \; d\rho \, d\theta_1\right] \exp(-jm\alpha) \\ &= A_{nm} \exp(-jm\alpha) \; . \end{aligned} \tag{8}$$

Equation (8) shows that each Zernike moment acquires a phase shift on rotation. Thus $|A_{nm}|$, the magnitude of the Zernike moment, can be used as a rotation invariant feature of the image. Therefore we calculate the magnitude of the Zernike moments to uniquely describe each block regardless of the rotation.

3 Copy-Rotate-Move (CRM) Forgery Detection

In order to detect CRM forgery, it is reminded that the proposed scheme should satisfy the property of Eq. (5) from the algebraic point of view. Moreover, it should be insensitive to additive noise or blurring since a forger might slightly manipulate the tampered region to conceal clues of forgery. In this perspective, we adopt Zernike moments which have desirable properties such as rotation invariance, robustness to noise, and multi-level representation [14].

We first divide the suspicious image f of $M \times N$ into overlapped sub-blocks of $L \times L$ to calculate Zernike moments. Each block is denoted as B_{ij}, where i and j indicates the starting point of the block's row and column, respectively.

$$B_{ij}(x,y) = f(x+j, y+i) \; , \tag{9}$$
$$\text{where } x, y \in \{0, ..., L-1\},\ i \in \{0, ..., M-L\}, \text{and } j \in \{0, ..., N-L\}$$

Hence, we are able to obtain N_{blocks} of overlapped sub-blocks from the suspicious image.

$$N_{blocks} = (M-L+1) \times (N-L+1) \tag{10}$$

We assume that the pre-defined size of block is smaller than the tampered region. After that, the Zernike moments $\mathbf{A}_{ij}$ of particular degree n are calculated from each block and vectorized. The entire number of moments in the vector is

$$N_{moments} = \sum_{i=0}^{n} \left(\left\lfloor \frac{i}{2} \right\rfloor + 1 \right) . \tag{11}$$

After that, we can construct $\mathbf{Z}$, a set of vectorized magnitude values of the moments $\mathbf{A}_{ij}$.

$$\mathbf{Z} = \begin{bmatrix} |\mathbf{A}_{00}| \\ \cdots \\ |\mathbf{A}_{(M-L)(N-L)}| \end{bmatrix} \tag{12}$$

The set $\mathbf{Z}$ is then lexicographically sorted since each element of $\mathbf{Z}$ is a vector. The sorted set is denoted as $\hat{\mathbf{Z}}$. From the set $\hat{\mathbf{Z}}$, the Euclidean distance between adjacent pairs of $\hat{\mathbf{Z}}$ is calculated. If the distance is smaller than the pre-defined threshold D_1, we consider the inquired blocks as a pair of candidates for the forgery.

$$\begin{gathered}\hat{\mathbf{Z}}_p = \left(\hat{z}_1^p, \hat{z}_2^p, ..., \hat{z}_{N_{moments}-1}^p, \hat{z}_{N_{moments}}^p\right) , \\ \hat{\mathbf{Z}}_{p+q} = \left(\hat{z}_1^{p+q}, \hat{z}_2^{p+q}, ..., \hat{z}_{N_{moments}-1}^{p+q}, \hat{z}_{N_{moments}}^{p+q}\right) , \\ \sqrt{\sum_{r=1}^{N_{moments}} \left(z_r^p - z_r^{p+q}\right)^2} < D_1\end{gathered} \tag{13}$$

Due to the fact that the neighboring blocks might result in relatively similar Zernike moments, we calculate the distance between the actual blocks of the image as follows:

$$\begin{gathered}\sqrt{(i-k)^2 + (j-l)^2} > D_2 , \\ \text{where } \hat{\mathbf{Z}}_p = \ |\mathbf{A}_{ij}| \text{ and } \hat{\mathbf{Z}}_{p+q} = \ |\mathbf{A}_{kl}| .\end{gathered} \tag{14}$$

We determine whether the investigated blocks are duplicated or not according to the Eq. (13) and Eq. (14).

3.1 Complexity Analysis

This section analyzes time complexity of the proposed method. We first calculate $N_{moments}$ of Zernike polynomials from Eq. (2). It roughly takes

$$O(N_{moments}) .$$

After that, we should compute Zernike moments from each overlapped block using the polynomials. Since a moment is calculated by the pointwise multiplication of the polynomial and the overlapped block, we need $O(L^2)$ time to attain the moment value. Therefore, we entirely need about

$$O(N_{blocks} \times N_{moments} \times L^2)$$

time to quantify all the moments. The following component to consider is time complexity of the lexicographical sorting of N_{blocks} data with the length of $N_{moments}$. It approximately takes

$$O(N_{moments} \times N_{blocks} \times \log N_{blocks}) .$$

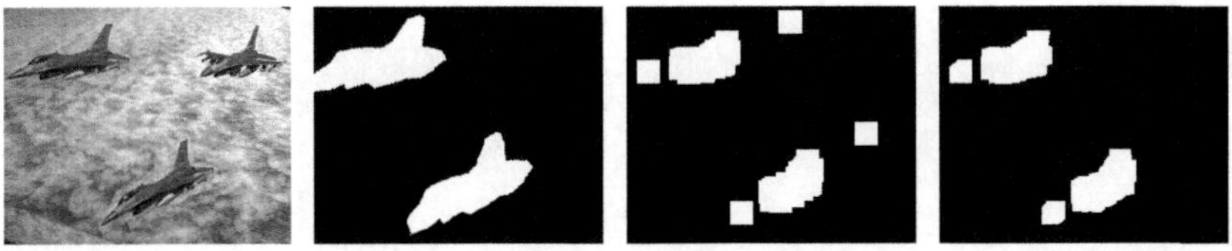

Fig. 3. Examples of CRM forgery and its detection result: (a) the forged image by CRM forgery of 10°, (b) *Forged Region*, (c) *Detected Region*, and (d) (*Forged Region* ∩ *Detected Region*)

Since L is relatively small, $O(N_{blocks} \times N_{moments} \times L^2)$ takes similar time to $O(N_{moments} \times N_{blocks} \times \log N_{blocks})$. To sum up, total time complexity is around

$$O(N_{moments}) + O(N_{blocks} \times N_{moments} \times L^2) + O(N_{moments} \times N_{blocks} \times \log N_{blocks}) .$$

In the actual experiment with the machine of 2.4 GHz quadcore processor, 4 GB RAM, coded by C++, and the condition of Sec. 4, it takes about 5 seconds to process one image.

4 Experimental Results

4.1 Measuring the Forgery

For a detection of copy-rotate-move or copy-move forgery, we need appropriate measures to evaluate the performance of the method. In this paper, we adopt *Precision*, *Recall*, and F_1-*measure* which are often-used measures in the field of information retrieval [17].

Precision and *Recall*, corresponding to exactness and completeness of the method, respectively, are defined as

$$Precision = \frac{True\,Positive}{True\,Positive + False\,Positive} , \tag{15}$$

$$Recall = \frac{True\,Positive}{True\,Positive + False\,Negative} . \tag{16}$$

From Equations (15) and (16), we see that high *Precision* values indicate low a $FalsePositive$ rate, whereas high *Recall* values correspond to a low $FalseNegative$ rate. More specifically, the *Precision* denotes the ratio of *True Positive* components to elements categorized into the positive class after investigating. In summary, the *Precision* is a measure for the probability that a detected region is correct. In our perspective, the *Precision* in percentage terms is represented as below.

$$Precision = \frac{(Forged\,Region\ \cap\ Detected\,Region)}{Detected\,Region} \times 100\,[\%] \tag{17}$$

Fig. 4. Images used in the experiments

On the other hand, the *Recall* is the ratio of *True Positive* components to elements inherently ranked as the positive class. It means that the *Recall* is a measure for the probability that a correct region is detected. In this context, the *Recall* in percentage terms is

$$Recall = \frac{(Forged\ Region\ \cap\ Detected\ Region)}{Forged\ Region} \times 100\,[\%]\ . \tag{18}$$

However, there is a trade-off between *Precision* and *Recall*. Greater *Precision* might decrease *Recall* and *vice versa*. To consider both *Precision* and *Recall* together, we compute the F_1-*measure*, the harmonic-mean of $Precision(P)$ and $Recall(R)$.

$$F_1\text{-}measure = \frac{2}{\frac{1}{P} + \frac{1}{R}} = \frac{2PR}{P+R} \tag{19}$$

Figure 3 shows examples of CRM forgery and its detection result. We can calculate *Precision*, *Recall*, and F_1-*measure* from the forged and detected region.

4.2 Experimental Setup

We firstly conducted our experiments with 12 TIFF images from a personal collection and [3,6]. Using these images, copy-move forgery with various manipulations such as rotation, JPEG compression, AWGN, blurring and combined

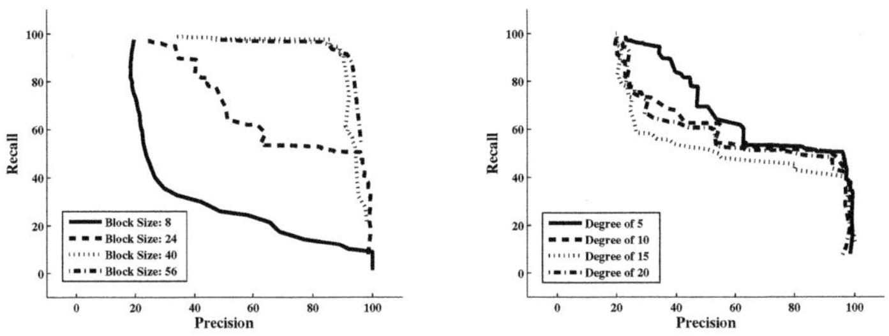

(a) (b)

Fig. 5. *Precision-Recall* Curves for Fig. 4(a) : (a) the curves for varying block size, (b) the curves for varying degree of Zernike moments

attacks was performed. Figure 4 shows the images used in the experiments. We carried out the proposed method for every test image and consequently *Precision*, *Recall*, and F_1-*measure* were evaluated. Moreover, we conducted our experiments with an extended dataset, which consists of 100 images from the National Geographic [18].

Since the targets to be investigated are normally color images, there exist two options for operating the method: 1) calculate Zernike moments from each color channel and subsequently concatenate the moment values, 2) simply convert the RGB image into a gray image. Since each individual color channel undergoes the same copy-move forgery, we choose the latter method.

All duplications were performed with regions of size 100×70 and a translation of (100, 50). To decide on the block size we drew *Precision-Recall* curve for various block sizes by changing the threshold D_1. Figure 5(a) shows *Precision-Recall* curves for different block sizes for the image depicted in Fig. 4(a). We notice that larger block sizes result in the higher detectability. However the high detectability with large blocks is dominated by the size of duplicated region. We also notice that a small block size almost does not detect the copy-move forgery. Therefore we set the block size L to 24 in all our following experiments. As mentioned in Eq. (10), the number of blocks in a suspicious image is $N_{blocks} = (M - L + 1) \times (N - L + 1)$. Since L is relatively smaller than M or N, the complexity of the method is dominated by the image size. We define M, and N as 400, and 320, respectively. Therefore, total number of blocks to be dealt with is $(400 - 24 + 1) \times (320 - 24 + 1) = 111969$. We furthermore analyzed the influence of the degree of Zernike moments. Figure 5(b) depicts *Precision-Recall* curves for different degrees for the image depicted in Fig. 4(a). We can observe that the degree of Zernike moments almost does not affect to the detectability. Therefore, each block is represented by the Zernike moments of 5 indicating $N_{moments} = 12$ by Eq. (11). Finally, we need to define decision thresholds D_1 and D_2, which represent the similarity between two blocks and the distance of them, respectively. From the *Precision-Recall* curves with the block size of 24 and the degree of 5 depicted in Fig. 5, we calculated F_1-*measures* for varying

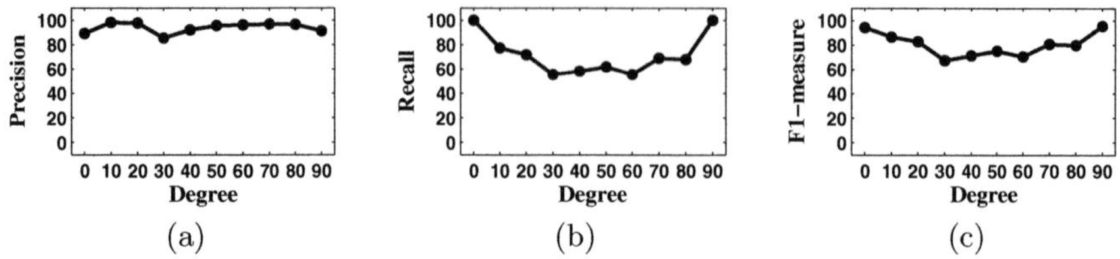

Fig. 6. CRM forgery detection results for Fig. 4(a) : (a) *Precision*, (b) *Recall*, and (c) F_1-*measure*

threshold D_1. We set the D_1 to 300 from the result of F_1-*measures*. Since the adjacent blocks might have similar moment values, the distance threshold D_2 is defined as 50.

Under these conditions, the following sub-sections analyze the performance of the proposed scheme in three perspectives. First of all, we take account of CRM forgery. After that, we present the robustness against intended distortions such as JPEG compression, AWGN, and blurring. Finally, combined attacks are considered.

4.3 Test for CRM Forgery

In this experiment, we conducted CRM with rotations in the range of 0° to 90°, applied in steps of 10°. Figure 6 depicts *Precision*, *Recall*, and F_1-*measure* of various degrees for Fig. 4(a). Even though the proposed scheme is theoretically invariant against rotation, the actual results have lower performance than expected as shown in Fig. 6. There might be two reasons for the degradation. At first, Zernike moments calculated on the discrete domain have inherent quantization error since the moments are originally defined on the continuous domain. Secondly, the interpolation caused by the rotation step can also increase the error rate. In this experiment, we used cubic kernel for the interpolation. Nevertheless, the experiments confirm that the *Precision* is relatively high, which means most part of detected region is correct. Table 1 shows experimental results for

Table 1. Detection rates for CRM of 30° for 12 images

Image	Measures (%)			Image	Measures (%)		
	P	R	F_1		P	R	F_1
(a)	85.41	55.50	67.28	(g)	83.76	85.49	84.62
(b)	92.76	68.67	78.91	(h)	86.60	82.01	84.24
(c)	66.78	91.10	77.06	(i)	73.43	83.54	78.16
(d)	97.84	62.87	76.55	(j)	79.31	83.97	81.57
(e)	67.96	71.66	69.76	(k)	83.57	85.68	84.51
(f)	98.80	65.33	74.71	(l)	86.88	83.76	85.29
Average					83.59	76.63	78.89

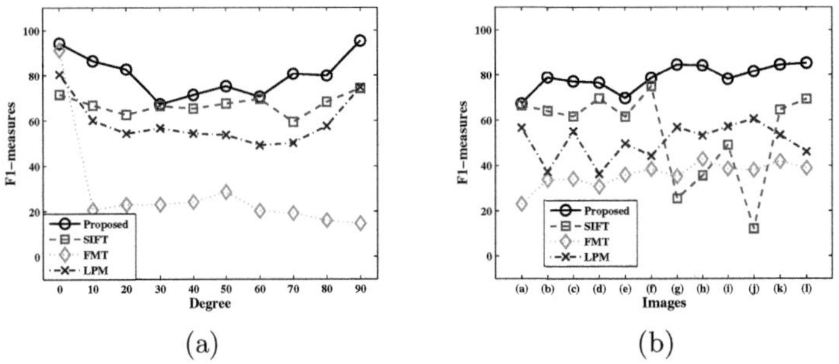

Fig. 7. Detection results for CRM forgery among proposed, SIFT, FMT, and LPM detector: (a) F_1-*measure* of various degrees for Fig. 4(a), (b) F_1-*measure* for 12 images undergoing CRM of 30°

12 images undergoing CRM of 30°. The average rate of *Precision*, *Recall*, and F_1-*measure* were 83.59%, 76.63%, and 78.89%, respectively.

We also compared our method with several CRM detectors: SIFT [10], FMT [7], and LPM [8]. Except for the SIFT based detector, we lexicographically sorted the extracted features from overlapping blocks to find adequate pairs of similar blocks. Since the SIFT is a kind of region descriptor, constructed with a set of matched points, it is hard to define where detected area is. Therefore we constructed a maximum polygonal convex inside the detected cluster. Then we calculated F_1-*measure* of each detector to measure quantitative performance. Figure 7(a) shows F_1-*measure* of various rotational degrees for Fig. 4(a) by 4 detectors. Similarly, Fig. 7(b) represents the experimental results for 12 images undergoing CRM of 30° by the detectors. We observe that the proposed detector provides higher F_1-*measure* than the others regardless of the amount of rotation or the concrete image. It is noticeable that the SIFT based method shows low detectability for the image (g) and (j) in Fig. 7(b) since the number of matched points are reduced for the image with less prominent structures.

To ensure the reliability of the proposed method, we also tested the method with the extended dataset. Figure 8 shows detectability of CRM of 30° for 100 images. Boxes represent F_1-*measures* between lower quartile and upper quartile. The red line inside the box indicates the median value. Whiskers extend from each end of the box to the adjacent values in the data; the most extreme values within 1.5 times the interquartile range from the ends of the box. Outliers are data with values beyond the ends of the whiskers. Outliers are displayed with a red + sign. From the result of Fig. 8, we notice that the proposed method performs better than the other detectors. We also notice that the SIFT based method shows several outliers, which represent lower detectability. As a consequence, the experimental results suggest that the proposed method is indeed capable of detecting CRM forgeries.

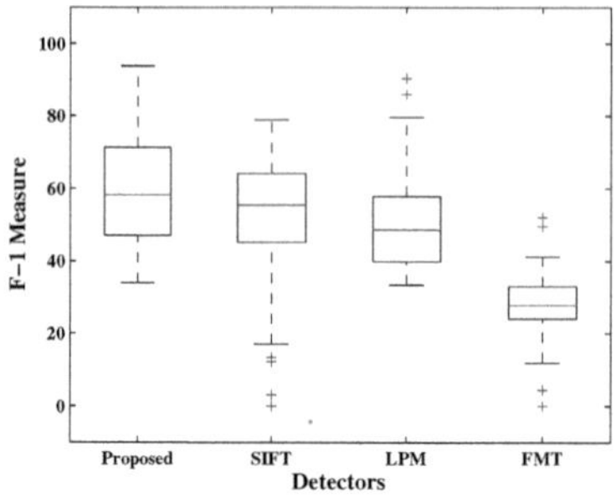

Fig. 8. Detection results for CRM of 30° with extended dataset

4.4 Test for Intended Distortions

Through this section, we present the detectability of copy-move forgery without rotation against intended distortions such as JPEG compression, AWGN, and blurring. We added Gaussian noise to the copied region or performed blurring before pasting into another part of the image. In case of JPEG compression, we compressed the whole image and not only the copied part. Figure 9 shows detection results of forgeries for Fig. 4(a) under several circumstances. We regularly changed the strength of each attack and analyzed the result.

By concentrating on the graphs for *Recall* in Fig. 9, we notice that the *Recall* values decrease considerably as a function of image quality. From these results, we conclude that severe attacks cause low detectability. Therefore we restrict our analysis in the following to attacks where the PSNR of the distorted region is above 30 dB. For example, we concentrate on noisy images with a variance of the Gaussian noise less than or equal to 0.003, since a distortion with $N(0, 0.003)$ amounts to about 31 dB. Similarly, the blurring with the radius larger than 2 or the quality factor for JPEG compression smaller than 60% is not considered in this test. Table 2 shows experimental results for intended distortions without rotation for 12 images. The average rate of F_1-*measure* for each case was 81.20%, 72.50%, and 93.67%, respectively. The experiments demonstrate that the proposed method is reasonably robust against intended distortions.

Table 2. Detection rates for intended distortions without rotation for 12 images

Image	F_1-$Measures$(%) JPEG (QF=70%)	F_1-$Measures$(%) AWGN (var=0.003)	F_1-$Measures$(%) Blurring ($radius$=1)	Image	F_1-$Measures$(%) JPEG (QF=70%)	F_1-$Measures$(%) AWGN (var=0.003)	F_1-$Measures$(%) Blurring ($radius$=1)
(a)	85.50	71.46	97.47	(g)	74.34	60.06	92.01
(b)	88.74	82.07	94.83	(h)	72.55	60.59	89.85
(c)	82.82	69.05	96.19	(i)	72.75	75.55	94.67
(d)	78.58	70.75	92.50	(j)	68.14	61.16	92.89
(e)	91.05	55.50	95.70	(k)	85.53	88.95	94.45
(f)	88.38	93.95	85.62	(l)	86.06	80.94	97.85
Average					81.20	72.50	93.67

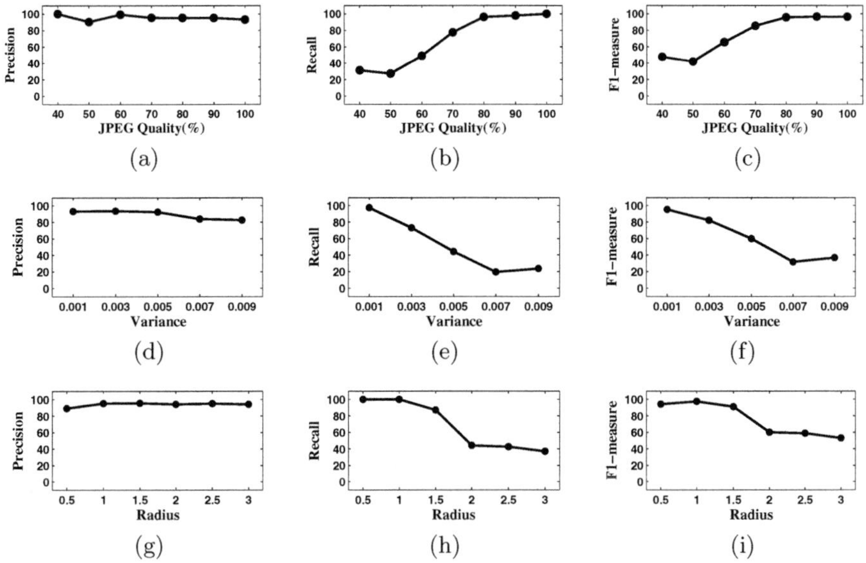

Fig. 9. Detection rates for intended distortions without rotation for Fig. 4(a): (a)∼(c): detection rate against JPEG quality factor, (d)∼(f): detection rate against AWGN with different variances, and (g)∼(i): detection rate against blurring with different radius

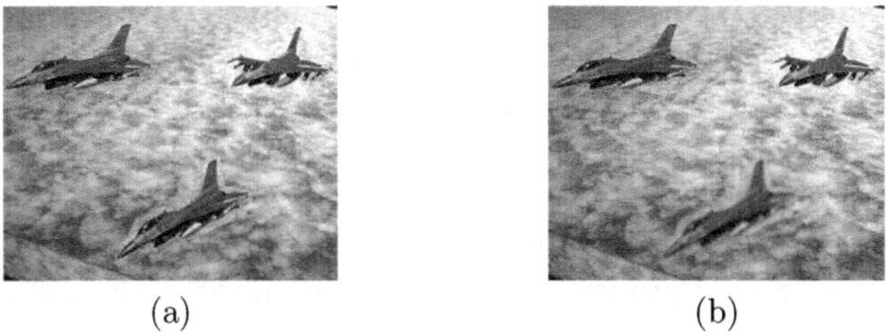

Fig. 10. Two scenarios of combined manipulation: (a) CRM of 10°, AWGN with $var =$ 0.003, and JPEG re-compression (QF=80%), (b) CRM of 10°, blurring with $radius =$ 1, and JPEG re-compression (QF=80%)

4.5 Test for Combined Manipulation

Finally, we present the robustness of proposed CRM detection scheme against combined manipulation. There might be two scenarios of CRM when a forger tampers an image. The forger would spread additional noise to eliminate the clues for manipulation after CRM. And then he or she will recompress the forged image. Similarly, the forger would blur the altered region instead of adding noise in the second scenario. Figure 10 depicts the detailed scenarios. Furthermore, Figure 11 represents detectability of various detectors of the scenario. We observe that the detectability of the SIFT based method in Fig. 11 has become worse compared with Fig. 7(b). This is so because the number of matched points by

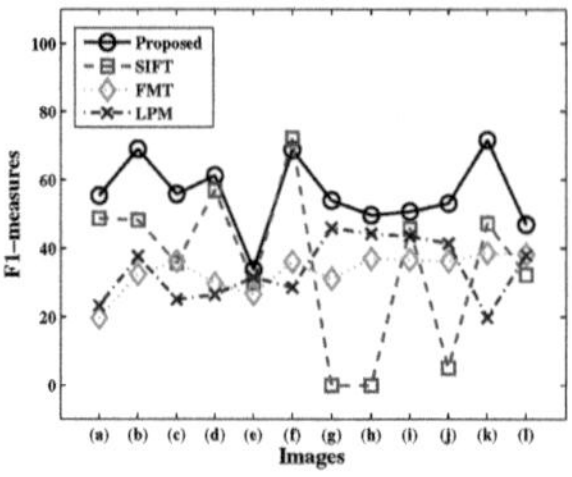

(a)

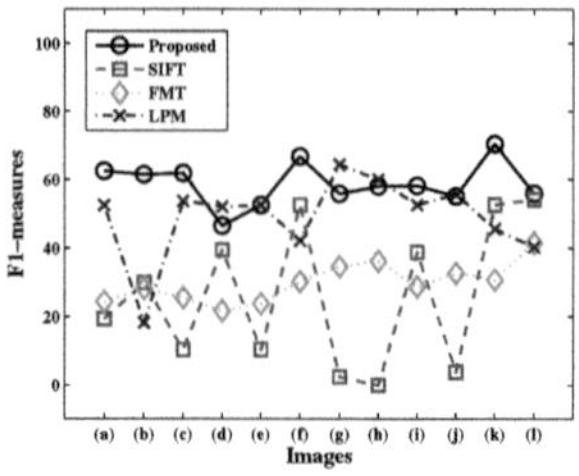

(b)

Fig. 11. Detectability for combined attacks among proposed, SIFT, FMT, and LPM detector: (a) F_1-*measure* for 12 images undergoing the scenario of Fig. 10(a), (b) F_1-*measure* for 12 images undergoing the scenario of Fig. 10(b)

the SIFT method is reduced as we manipulate the image. On the other hand, the results confirm the reliability of the proposed scheme even after combined manipulation. Through the experiments, it proves that the proposed detector performs better than others as well.

5 Conclusion

With the rapid progress of image processing technology, an appropriate forensic application has become more important. In this paper, we proposed copy-rotate-move (CRM) detection scheme for a suspicious image. To extract feature vectors of a given block, we calculated the magnitude of Zernike moments. The vectors were then sorted in lexicographical order. We investigated the similarity of adjacent vectors after that. Finally, the suspected regions were measured by *Precision*, *Recall*, and F_1-*measure*. Experimental results supported that the proposed method was appropriate to identify and localize the CRM region even though the region had been manipulated intentionally. However, in spite of an algebraic invariant of rotation, detection errors occurred due to the quantization and interpolation error. Though we concerned several attacks, our method is still weak against scaling or the other tampering based on Affine transform. Thus, we need to improve the proposed method so that it is robust against those of attacks. Additionaly, there exist many efficient data structures to represent nearest neighbors. Therefore, our future work concentrates on establishing an appropriate data structure as well.

Acknowledgments. We are grateful to Matthias Kirchner for many helpful advices and suggestions. This work was partially supported by Defense Acquisition Program Administration and Agency for Defense Development under the contract. (UD060048AD)

References

1. In an Iranian image, a missile too many, `http://thelede.blogs.nytimes.com/2008/07/10/in-an-iranian-image-a-missile-too-many/`
2. Fridrich, J., Soukal, D., Lukáš, J.: Detection of copy-move forgery in digital images. In: Digital Forensic Research Workshop (2003)
3. Popescu, A.C., Farid, H.: Exposing digital forgeries by detecting duplicated image regions. In: Technical Report, TR2004-515, Dartmouth College, Computer Science (2004)
4. Li, G., Wu, Q., Tu, D., Sun, S.: A sorted neighborhood approach for detecting duplicated regions in image forgeries based on DWT and SVD. In: IEEE International Conference on Multimedia and Expo. IEEE Press, New York (2007)
5. Luo, W., Huang, J., Qiu, G.: Robust detection of region-duplication forgery in digital image. In: The 18th International Conference on Pattern Recognition (2006)
6. Mahdian, B., Saic, S.: Detection of copy-move forgery using a method based on blur moment invariants. Forensic Science International 171, 180–189 (2007)
7. Bayram, S., Sencar, H.T., Memon, N.: An efficient and robust method for detecting copy-move forgery. In: IEEE International Conference on Acoustics, Speech, and Signal Processing. IEEE Press, New York (2009)
8. Bravo-Solorio, S., Nandi, A.K.: Passive method for detecting duplicated regions affected by reflection, rotation and scaling. In: 17th European Signal Processing Conference (2009)
9. Huang, H., Guo, W., Zhang, Y.: Detection of copy-move forgery in digital images using SIFT algorithm. In: Pacific-Asia Conference on Computational Intelligence and Industrial Applications (2008)
10. Amerini, I., Ballan, L., Caldelli, R., Bimbo, A.D., Serra, G.: Geometric tampering estimation by means of a SIFT-based forensic analysis. In: IEEE International Conference on Acoustics, Speech, and Signal Processing. IEEE Press, New York (2010)
11. Pan, X., Lyu, S.: Detecting image region duplication using SIFT features. In: IEEE International Conference on Acoustics, Speech, and Signal Processing. IEEE Press, New York (2010)
12. Hu, M.K.: Visual pattern recognition by moment invariants. IEEE Trans. Information Theory 8, 179–187 (1962)
13. Kim, H.S., Lee, H.K.: Invariant image watermark using Zernike moments. IEEE Trans. Circuits and Systems for Video Technology 13(8), 766–775 (2003)
14. Teh, C.H., Chin, R.T.: On image analysis by the methods of moments. IEEE Trans. Pattern Analysis and Machine Intelligence 10(4), 496–513 (1988)
15. Khotanzad, A., Hong, Y.H.: Invariant image recognition by Zernike moments. IEEE Trans. Pattern Analysis and Machine Intelligence 12(5), 489–497 (1990)
16. Zernike, F.: Beugungstheorie des Schneidenverfahrens und seiner verbesserten Form, der Phasenkontrastmethode. Physica 1, 689–704 (1934)
17. Manning, C.D., Raghavan, P., Schütze, H.: An Introduction to Information Retrieval. Cambridge University Press, Cambridge (2009)
18. National Geographic, `http://photography.nationalgeographic.com`

Scene Illumination as an Indicator of Image Manipulation

Christian Riess* and Elli Angelopoulou

Pattern Recognition Lab
University of Erlangen-Nuremberg
{riess,angelopoulou}@i5.cs.fau.de
http://www5.informatik.uni-erlangen.de

Abstract. The goal of blind image forensics is to distinguish original and manipulated images. We propose illumination color as a new indicator for the assessment of image authenticity. Many images exhibit a combination of multiple illuminants (flash photography, mixture of indoor and outdoor lighting, etc.). In the proposed method, the user selects illuminated areas for further investigation. The illuminant colors are locally estimated, effectively decomposing the scene in a map of differently illuminated regions. Inconsistencies in such a map suggest possible image tampering. Our method is physics-based, which implies that the outcome of the estimation can be further constrained if additional knowledge on the scene is available. Experiments show that these illumination maps provide a useful and very general forensics tool for the analysis of color images.

Keywords: Blind image forensics, scene analysis, physics-based illuminant color estimation.

1 Introduction

The goal of image forensics is to assess image authenticity. This can be achieved by actively embedding a security scheme in the image, like a digital watermark. However, current hardware does not typically provide such signatures. Therefore, *blind* image forensics aims at assessing image authenticity and origin without having an embedded security scheme available. In the past few years, different branches of blind image forensics have evolved.

For a more complete overview on the methods in blind image forensics, see e.g. [37,33].Some of the existing approaches are classification based [6,22]. Unfortunately, the outcome of these algorithms is often hard to interpret for non-technical surveyors, e.g. in court. Other approaches search for artifacts of a specific tampering operation, like the algorithms for copy-move forgery detection (see e.g. [20,2,3,32,31]), resampling detection [35], or methods for the analysis of double JPEG compression (see e.g. [29,15,7]). In general, these methods are suitable for an automated analysis of an image. Unfortunately, researchers are also

* Corresponding author.

R. Böhme, P.W.L. Fong, and R. Safavi-Naini (Eds.): IH 2010, LNCS 6387, pp. 66–80, 2010.

working on methods to hide image manipulation artifacts, thus effectively counteracting the aforementioned methods [23]. A third family of algorithms aims at verifying expected image-sensing artifacts. One prominent example of these methods is the recovery of the characteristic noise pattern of camera sensors [30]. Other methods estimate the camera response function [17], demosaicing [12], or lens properties [21].

Lastly, the examination of scene properties is another approach in image forensics. Unlike the aforementioned methods, it often involves (to a limited extend) user interaction, especially if human knowledge about the scene content is required. In this respect, methods for the assessment of scene properties often serve as a computational tool for a human surveyor. The advantage of such techniques is that it is frequently not straightforward to hide traces of tampering from these methods. Disguising scene inconsistencies is typically a tedious manual process, that may additionally require high algorithmic knowledge. Thus, scene consistency assessment can provide powerful tools for forensic analysis. To our knowledge, only a small amount of work has been done in this direction. For instance, Johnson and Farid demonstrated the recovery of the illumination direction of objects [19] and compared the estimated position of light sources in the scene [18]. Yu *et al.* examined specularity distributions for recapturing detection [40]. Lalonde and Efros used color distributions in pictures in order to detect spliced images [25].

We propose a new method for the assessment of illumination-color consistency over the scene by extracting local illumination estimates. To our knowledge, no similar approach has been proposed in image forensics. Our method is based on an extension of an illumination estimation method that is grounded on the physical principles of image formation. In contrast, most state-of-the-art methods for illuminant color estimation are machine-learning based. However, it is our belief that deviations from the expected result can be easier explained using a physics foundation than by machine-learning results, as detailed in Sec. 4.2. We believe this is a highly desirable property in forensics applications. Depending on the number of light sources of the scene, we show that these local estimates can provide further insights on the scene construction. For instance, if a photographer took an image at night using flashlight (which is typically a relatively bluish light source), we can obtain a rough relative depth estimate from the decay of the blue channel in the illuminant estimates. Inconsistencies in the illumination distribution can be used to distinguish original and spliced images.

The contributions of this paper are:

1. The development of a physics-based method for the recovery of the illuminant color for different objects in the scene.
2. The introduction of an **illumination map** based on a **distance measure** on the estimated results.
3. The demonstration of the **feasibility** of employing this illuminant map in **forensic analysis**.

2 Overview of the Method

We present a system for the assessment of the illuminant color consistency over the image. The method involves the following steps.

1. The image is segmented in regions of approximately the same object color. These segments are called *superpixels*. A superpixel is required to a) be directly illuminated by the light sources under examination and b) roughly adhere to the physical model presented in Sect. 3.
2. A user selects such superpixels whose incident illuminant he wants to further investigate. Every group of superpixels represents one illuminant color under investigation.
3. Estimation of the illuminant color is performed twice. First, the estimation is done on every superpixel separately. Second, the estimation is done on the user-selected superpixel groups for greater robustness.
4. The user-selected groups form the reference illuminants. A distance measure from these illuminants to every superpixel estimate is computed. We visualize these per-superpixel distances in what we call a *distance map* to support the analysis of the illumination color consistency.

In special cases, this method can be fully automated. On the other hand, since the estimation of the illuminant color is an underconstrained problem, there will always exist scenes that can not be correctly processed. We believe that a limited degree of human interaction is a valid tradeoff between the accuracy of the method and its usability.

3 Estimation of the Illuminant Color

There is a large body of work on the estimation of illuminant color. Most of these methods process the entire scene globally for the recovery of a single dominant illumination color. To overcome the fact that illuminant color estimation is an underconstrained problem, many (especially physics-based) techniques make restrictive assumptions that limit their applicability (e.g. [26,13,24]). As a result, machine learning methods have been more successful in processing arbitrary images, e.g. [4,9,10,14,28]. However, since illuminant estimation is an underconstrained problem, every method has its individual failure points. We chose a physics-based method, since the failures typically result from broken underlying assumptions. As such, they can (by human observers) more easily be predicted and explained, which we consider highly important for forensics applications.

3.1 Inverse-Intensity Chromaticity Space

We extend a physics-based approach that was originally proposed by Tan *et al* [39] so that:

1. it can be applied on a wider range of real-world images, and
2. it can be applied locally, so that illuminants at selected regions can be independently estimated.

The foundation of [39] is the dichromatic reflectance model [38], which states that the amount of light reflected from a point, $\boldsymbol{x}$, of a dielectric, non-uniform material is a linear combination of diffuse reflection and specular reflection. Further assumptions are that the color of the specularities approximates the color of the illuminant, and that the camera response is linear.

When an image is taken with a trichromatic camera, the sensor response $I_c(\mathbf{x})$ for each color filter c, $c \in \{R, G, B\}$ is:

$$I_c(\mathbf{x}) = w_d(\mathbf{x})B_c(\mathbf{x}) + w_s(\mathbf{x})G_c(\mathbf{x}) \ , \tag{1}$$

where $w_d(\mathbf{x})$ and $w_s(\mathbf{x})$ are geometric parameters of diffuse and specular reflection respectively. $B(\mathbf{x})$ and $G_c(\mathbf{x})$ are the sensor responses for diffuse and specular reflectance. Note that in [39], G_c does not depend on $\mathbf{x}$ due to the assumption that the specular color is globally constant. In this paper, we estimate illumination locally and thus write $G_c(\mathbf{x})$. Let σ_c the image chromaticity, i.e.

$$\sigma_c(\mathbf{x}) = \frac{I_c(\mathbf{x})}{\sum_i I_i(\mathbf{x})} \text{ where } i \in \{R, G, B\} \ . \tag{2}$$

For the remainder of the paper, we define $i \in \{R, G, B\}$ and use this index for summing over the color channels. In a similar manner, we can define the diffuse chromaticity $\Lambda_c(\mathbf{x})$ and the specular chromaticity $\Gamma_c(\mathbf{x})$ as

$$\Lambda_c(\mathbf{x}) = \frac{B_c(\mathbf{x})}{\sum_i B_i(\mathbf{x})} \ , \tag{3}$$

$$\Gamma_c(\mathbf{x}) = \frac{G_c(\mathbf{x})}{\sum_i G_i(\mathbf{x})} \ . \tag{4}$$

Equation (1) can be rewritten as

$$I_c(\mathbf{x}) = m_d(\mathbf{x})\Lambda_c(\mathbf{x}) + m_s(\mathbf{x})\Gamma_c(\mathbf{x}) \ , \tag{5}$$

where

$$m_d(\mathbf{x}) = w_d(\mathbf{x}) \sum_i B_i(\mathbf{x}) \ , \tag{6}$$

$$m_s(\mathbf{x}) = w_s(\mathbf{x}) \sum_i G_i(\mathbf{x}) \ . \tag{7}$$

Tan *et al.* [39] showed, that there exists a linear relationship between diffuse, specular and image chromaticities,

$$\sigma_c(\mathbf{x}) = p_c(\mathbf{x}) \frac{1}{\sum_i I_i(\mathbf{x})} + \Gamma_c(\mathbf{x}) \ , \tag{8}$$

where

$$p_c(\mathbf{x}) = m_d(\mathbf{x})(\Lambda_c(\mathbf{x}) - \Gamma_c(\mathbf{x})) \ . \tag{9}$$

$p_c(\mathbf{x})$ is the slope of a line with intercept $\Gamma_c(\mathbf{x})$, i.e. the specular chromaticity, which is also the illuminant chromaticity. The domain of the line is determined

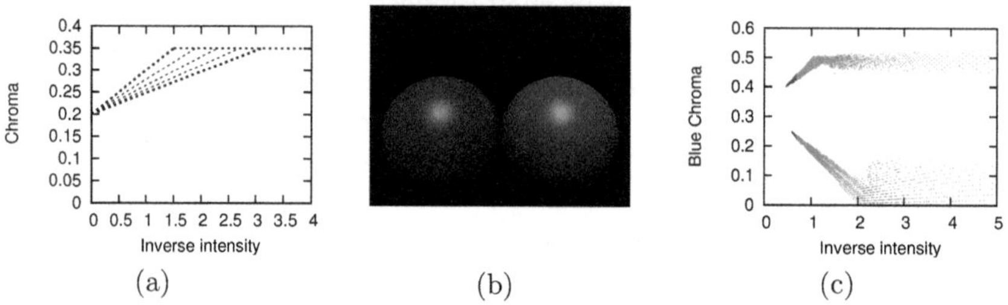

Fig. 1. Sample pixel distributions in IIC space (blue chromaticity). Left: ideal image, middle: synthetic image (violet and green bowls). Right: specular pixels converge towards to the blue portion of the illuminant color (recovered at at the y-axis intercept). Highly specular pixels are shown in red.

by $1/\sum_i I_i(\mathbf{x})$ and the range is given by $0 \leq \sigma_c \leq 1$. Domain and range together form the *inverse-intensity chromaticity (IIC) space* [39].

This space can be used to estimate the illumination color in an image. Every color band $c \in \{R, G, B\}$ is projected in a separate IIC diagram. One can then obtain estimates for each of these three color channels. All three channels together form the illuminant estimate. Fig. 1 shows a synthetic example for the blue channel. The x-axis corresponds to the inverse-intensity $1/\sum_i I_i(\mathbf{x})$, and the y-axis to σ_c. In Fig. 1 the y-axis shows the blue chromaticities. An idealized distribution is shown in Fig. 1(a). On a uniformly colored surface, pixels with a mixture of specular and diffuse chromaticities form roughly a triangle that points towards the illuminant chromaticity (the y-axis intercept). Purely diffuse pixels form a straight line. Fig. 1(b) shows a rendered image of two balls of distinct colors, and Fig. 1(c) shows the distributions of the balls-image in the blue chromaticity IIC space. In this synthetic setup, the triangles can be clearly captured, as well as the diffuse horizontal lines. Any method for finding the intersection of the triangle with the y-axis gives then the illuminant chromaticity estimate in the respective channel. The final estimate of the illuminant color is obtained by considering all three color bands red, green and blue. In [39], this is done by first segmenting highly specular pixels in the image (marked red in Fig. 1(c)) and then performing a Hough transform on these specular pixels.

3.2 Local Analysis of Pixel Distributions

On real-world images, the automatic extraction of highly specular pixels is a very challenging task with unreliable performance (see e.g. [36]). Furthermore, the basic method [39] does not handle cases with multiple light sources. We analyzed and extended the exploitation of pixel distributions in IIC space, so that it can overcome these two weaknesses.

In order to avoid specularity segmentation, we chose to perform simple shape checks on the pixel distributions in IIC space in order to identify specular regions. Instead of examining the entire pixel distribution, we perform the analysis over small connected image regions of roughly uniform object color (albedo). Depending on the outcome of our shape analysis, we can either use this local region to obtain an illuminant estimate, or reject it if it does not seem to fulfill the underlying assumptions of the proposed model. Using local regions allows us to incorporate multiple sampling and voting in the estimation of local illuminants. Ultimately, this improved exploitation of the IIC space makes the method more robust to real-world analysis and also enables us to examine multiple illuminants. More specifically, the proposed algorithm works as follows.

1. For every dominant illuminant in the scene, select regions that a) follow the dichromatic reflectance model and b) are mostly lit by that light source.
2. Segment these regions in superpixels with roughly uniform chromaticity.
3. Further subdivide these superpixels in a rectangular grid. We call each such rectangular subregion a *patch*.
4. Transform every patch to inverse intensity space.
5. Apply tests on the shape of the patch's pixel distribution. If the distribution passes, obtain a local illuminant color estimate for this patch.
6. Obtain a color estimate for each dominant illuminant, based on a majority vote on local estimates of the user-selected regions.

For the superpixel segmentation, we used the publicly available code by Felzenszwalb and Huttenlocher [8] on the image chromaticities, though any segmentation method could be used. We choose the segmentation parameters $0.1 \leq \sigma \leq 0.3$ and $100 \leq k \leq 300$. Typically $\sigma = 0.3$ and $k = 300$ gave satisfying results, dividing the image in not too small regions of similar object color. The grid size is adaptive to the image size, typically between 16 and 32 pixels in the horizontal and vertical directions.

Pixel distributions that are assumed to exhibit a combination of specular and diffuse reflectance should have: a) a minimum x-axis elongation in IIC space, and b) a non-horizontal slope (as long as the object albedo is different from the illuminant color, which is typically the case). Thus, we use the following criteria to examine whether a patch satisfies these two properties and hence has an increased probability of providing a reasonable estimate.

- The superpixel segmentation is performed for increasing the probability of uniform underlying albedo.
- The elongation of a patch in IIC space is tested by computing the eccentricity of the distribution of the pixels in the patch,

$$\mathrm{ecc}(P_{\mathrm{IIC}}) = \sqrt{1 - \frac{\sqrt{\lambda_1}}{\sqrt{\lambda_2}}} \;, \tag{10}$$

  where λ_1 and λ_2 are the largest and second largest eigenvalues, respectively. A value close to 1 is matches the model best, while lower values lead to more estimates. In our experiments, we chose as a limit 0.6.

- The slope of a patch in IIC space is approximated by the direction of the eigenvector λ_1, and should be slightly larger than 0. In our experiments, we chose a minimum slope of 0.003.

The vote for a patch is computed as the intercept of the eigenvector of λ_1 with the y-axis. Note that duplicate entries in the IIC diagram are discarded from these computations, as well as pixels that are very close to the limits of the camera sensor response.

4 Illuminant Color for Image Forensics

Once the illuminant color estimates for the user-annotated regions are computed, the whole image can be examined for illumination color inconsistencies, as described in Sect. 4.1. Since the estimation of the illuminant color is a severely underconstrained problem, we briefly discuss failure cases and possible workarounds in Sect. 4.2. Please note that, as will be shown in Sect. 5, our illumination estimation method performs comparably to other state-of-the-art single illuminant estimation methods.

4.1 Detecting Inconsistencies in Illumination

The same process (see Sect. 3.2) that was used in computing the illuminant color estimates at the user-specified regions is now extended to the entire image. The voting, however, is now performed for every superpixel. Thus, every superpixel contains an individual illuminant estimate. We store these illuminant estimates in a new image, where each superpixel is colored according to its estimated illuminant color $\Gamma_I(\mathbf{x})$. We call this new image *illumination map*, see Fig. 2. This map gives already quite meaningful results for the analysis.

For forensic analysis, we aim to quantify the relationship between the illuminant estimates. In a scene with truly one dominant illuminant, this can be done by comparing the angular errors of the individual illuminant estimates. However, most real-world scenes contain a mixture of illuminants. Their influence on the scene is closely connected to the positions of the objects relative to the positions of the light sources. Since the geometric composition of the scene is typically unknown, we resort to developing a tool for supporting the visual assessment of the scene, which we call *distance map*.

The distance map captures how well the illuminant estimation at each superpixel fits to the estimated dominant illuminants. For improved clarity, we assume two dominant illuminants I_1 and I_2 that were obtained from two user-selected regions. The methodology can however easily generalize to more illumination sources. We aim to create a grayscale-image that depicts the relative influence of both light sources. The distance map is created by assigning the value 0 (black) to the user-defined region corresponding to illuminant I_1. Similarly, the second user-defined region, which gave rise to dominant illuminant I_2, is assigned the value 1 (white). Then, for all the remaining pixels, the distance value $I_d(\mathbf{x})$ is computed as

$$I_d(\mathbf{x}) = (\Gamma_I(\mathbf{x}) - I_1) \circ (I_2 - I_1) \ , \tag{11}$$

where $\circ$ denotes scalar multiplication. The distance map is then a grayscale image with values in the range $[0, 1]$. Such a map captures the relative influence of both light sources in this pixel.

The illumination map and the distance map are used together for the analysis of the image. In order to be consistent, a local illuminant estimate in an image must a) either exhibit a relative illuminant contribution that fits in the spatial layout of the scene or b) fail to fulfill the underlying physical model. In the latter case, it must be ignored for the analysis.

Fig. 2. Original Image, illumination map and distance map for the image under examination. Foreground persons are estimated with a bluish color, probably due to flashlight, while persons in the background are increasingly red illuminated. The distance map between foreground and background illumination spots captures this relationship as a black-to-white transition.

By adjusting the values of the criteria on the pixel distributions, it is possible to obtain fewer estimates that fit the physical model better (at the expense of larger regions with sparse or no estimates). Alternatively, less strict parameters lead to a more complete map, where also more outliers are expected. In general, we preferred in our experiments lenient settings. For the slope we set a lower bound of 0.003, and for the eccentricity 0.5. A stricter set of values, i.e. 0.01 for the slope and 0.95 for the eccentricity, typically results in fewer outliers.

4.2 Caveats and Workarounds

In some cases, the estimation of the illuminant color can not be successfully applied. Fortunately, for a physics-based method like the proposed one, the reasoning about failure cases is often easier than for machine-learning methods. While failures in the latter case often arise due to limitations of the training data or algorithm-dependent assumptions on the color distributions, physics-based methods mainly fail due to violations of the assumed reflectance model. This makes it possible to argue about possible problems and look for workarounds.

(a) (b) (c) (d)

Fig. 3. Failure cases for the proposed illuminant color estimation method. Figures 3(a) and 3(c) are the original images, Figures 3(b) and 3(d) the respective illumination maps. In Fig. 3(b), the illuminant estimate in the shadowed area under the head of the left actor is biased towards the object color. In Fig. 3(d), the fluorescent suit of the actor overproportionally pushes the illuminant estimate towards extreme values.

We present some cases where our method is problematic. First, the camera response is assumed to be linear. This is leveraged by the fact that we exploit only the relationship between illuminant estimates, and do not consider absolute estimates. Nevertheless, a gamma estimation method, e.g. [27], can be used to normalize the image. Some non-dielectric surfaces are especially difficult to handle, e.g. fluorescent materials (see Fig. 3) and metals. Other failure cases involve areas that are mostly diffuse, or highly textured, or in shadow (see Fig. 3). Finally, the method is inherently limited by the assumption that the color of the specularity closely approximates the color of the illuminant.

We found, that by visual inspection it is often possible to distinguish failure cases from real inconsistencies. However, it is possible to follow specific rules to minimize the risk of misjudging the scene under observation. The most robust approach is to use only identical or very similar materials for the analysis, e.g. faces in a crowded scene. We reflect this by demanding the user to select regions that a) are of interest for the examination and b) roughly adhere to the model.

5 Experiments

Section 5.1 demonstrates the effectiveness of our core algorithm in accurately recovering the color of the illumination. In Sect. 5.2, we demonstrate its usefulness in forensics applications. The code is publicly available on our webpage[1].

5.1 Evaluation on Benchmark Databases

Until today, the color research community focused mostly on single-illuminant scenes. Therefore, to the best of our knowledge, no ground-truth illuminant

[1] `http://www5.informatik.uni-erlangen.de/code`

Fig. 4. Examples of benchmark laboratory images by [1]

Table 1. Algorithm performance on benchmark laboratory images (left) and on real-world images (right). The results are taken from [14,28].

Method	Median e
Gamut mapping	**3.1**°
Gray-World	8.8°
White-Patch	5.0°
Color-by-Correlation	8.6°
Proposed method	4.4°

Method	Median e
Regular gamut with offset-model	5.7°
Gray-World	7.0°
White-Patch	6.7°
Color-by-Correlation	6.5°
1st-order Gray-Edge	5.2° (∗)
2nd-order Gray-Edge	5.4° (∗)
Tan *et al.* [39]	5.6
Proposed method	**4.4**°

color dataset exists for scenes containing multiple illuminants. We evaluated the proposed illuminant estimation method on two widely used publicly available ground-truth databases. The error between ground truth Γ_l and estimated illuminant color Γ_e is typically measured as angular error in RGB-space, defined as

$$e = \cos^{-1}\left(\frac{\Gamma_l \cdot \Gamma_e}{\|\Gamma_l\|\|\Gamma_e\|}\right) \tag{12}$$

Somewhat unusual compared to other fields of computer vision, the success of illuminant-estimation methods is typically measured using the median over multiple images [16].

The first dataset, introduced by Barnard *et al.* [1], contains high-quality laboratory images. We used the "dielectric specularities" part of the dataset. It

Fig. 5. Examples of benchmark real-world images

Fig. 6. Tampered image. Illumination map as well as distance map show a clear difference between the first two and the third person. Since the three stand close together in the image, it can be assumed that this difference is due to tampering.

Fig. 7. Original image (top left) and tampered image (top right). A comparison of the skin regions of the people exposes the inserted man in the distance map.

Fig. 8. Original image (top left) and tampered image (top right). At first glance, illumination map and distance map show plausible results on the tampered image. However, the illuminant estimates are obtained from the front of the woman, which is turned away from the restaurant lights. Therefore, the expected illumination should be more bluish, like e.g. at the back of the person in the middle.

contains 9 scenes, each under 11 different illuminants, since this part contains a mixture of specular and diffuse pixels. Example images are shown in Fig. 4. The proposed method performs comparable to other state-of-the art illuminant estimation methods, as shown in Table 5.1 (left). The results from the other methods were taken from [34].

The second dataset, presented by Ciurea and Funt [5], contains a wide variety of real-world scenes. The ground truth is estimated from a fixed matte gray ball that is mounted in front of the camera, as shown in Fig. 5. For the evaluation of the methods, the ball has to be masked out. Table 5.1 (right) shows that the proposed method is highly competitive on this dataset. The results marked with (*) are taken from [28], the remaining results are from [14].

5.2 Exposing Digital Forgeries

For qualitative results on multiple illuminants, we collected from various sources, mostly flickr [11], approximately 430 images containing scenes with multiple illuminants or unusual single-illuminant setups. Besides these images, which were assumed (or known, respectively) to be original, 10 forgeries have also been examined using the proposed method. In the following, we present three cases where image geometry and illumination create discontinuities. Fig. 6 shows a case where the change in the illumination color is barely explicable with the scene setup. Both the illumination map as well as the distance map exhibit a sharp transition between the two persons in the foreground and the third in the back, which could only be feasible if there was a greater distance between them.

The example in Fig. 7 shows outdoor illumination with one dominant illuminant. Again, we compare the skin regions of the people, in order to have roughly comparable object materials. The selected regions are the directly lit skin of the inserted person versus the directly lit skin of other guests. The illumination map shows blueish estimates for the inserted man. The distance map makes this difference even more visible. Note that the estimates of the coast line in the background should be ignored (although they fit well, in this particular case). The underlying pixels must be assumed to be purely diffuse, and thus do not satisfy our underlying assumptions.

Fig. 8 contains a more complex case. The woman in the right is inserted in the image. Illumination map and distance map are plausible, compared to the people that stand similarly close to the restaurant. However, adding again the scene geometry gives a strong clue that this scene is not original. Since the woman is turned away from the restaurant, the illuminant color on the woman's chest should share greater similarity with the body parts of the other people that are turned away from the restaurant lights.

6 Conclusion

We presented a method that estimates the color of the illuminant locally and applied it to the detection of tampered images. A user interactively marks regions whose illuminants should be further investigated. On these regions, the illuminant color is estimated. Then, the local illuminant estimation process is extended to the whole image. In scenes with multiple illuminants, one can typically observe a transition between these illuminants that is consistent with the scene geometry. In order to verify this, we introduced the *illuminant map*, consisting of all local illuminant estimates, and a *distance map*, that captures the influence of every illuminant. If an image has been manipulated, the transition between these illuminants should accordingly be disturbed.

This is preliminary work. In the future, we will extend this approach in two directions. First, the user interaction can further be reduced by postprocessing of the illuminant map and the distance map. Secondly, we aim to develop more rigorous methods for a more detailed analysis of inconsistencies in the illumination map and the distance map.

References

1. Barnard, K., Martin, L., Funt, B., Coath, A.: A Data Set for Color Research. Color Research and Application 27(3), 147–151 (2002)
2. Bayram, S., Sencar, H., Memon, N.: An efficient and robust method for detecting copy-move forgery. In: Acoustics, Speech, and Signal Processing, pp. 1053–1056 (2009)

3. Bravo-Solorio, S., Nandi, A.K.: Passive Forensic Method for Detecting Duplicated Regions Affected by Reflection, Rotation and Scaling. In: European Signal Processing Conference (2009)
4. Cardei, V.C., Funt, B., Barnard, K.: Estimating the Scene Illumination Chromaticity Using a Neural network. Journal of the Optical Society of America A 19(12), 2374–2386 (2002)
5. Ciurea, F., Funt, B.: A Large Image Database for Color Constancy Research. In: Color Imaging Conference, pp. 160–164 (2003)
6. Dirik, A.E., Bayram, S., Sencar, H.T., Memon, N.: New features to identify computer generated images. In: IEEE International Conference on Image Processing, pp. 433–436 (2007)
7. Farid, H.: Exposing Digital Forgeries from JPEG Ghosts. IEEE Transactions on Information Forensics and Security 1(4), 154–160 (2009)
8. Felzenszwalb, P.F., Huttenlocher, D.P.: Efficient Graph-based Image Segmentation. International Journal of Computer Vision 59(2), 167–181 (2004)
9. Finlayson, G.D., Hordley, S.D., Hubel, P.M.: Color by Correlation: A Simple, Unifying Framework for Color Constancy. IEEE Transactions on Pattern Analysis and Machine Intelligence 23(11), 1209–1221 (2001)
10. Finlayson, G.D., Hordley, S.D., Tastl, I.: Gamut Constrained Illuminant Estimation. International Journal of Computer Vision 67(1), 93–109 (2006)
11. Flickr, `http://www.flickr.com`
12. Gallagher, A., Chen, T.: Image Authentication by Detecting Traces of Demosaicing. In: Computer Vision and Pattern Recognition Workshops, pp. 1–8 (2008)
13. Geusebroek, J.M., Boomgaard, R., Smeulders, A., Gevers, T.: Color Constancy from Physical Principles. Pattern Recognition Letters 24(11), 1653–1662 (2003)
14. Gijsenij, A., Gevers, T., van de Weijer, J.: Generalized Gamut Mapping using Image Derivative Structures for Color Constancy. International Journal of Computer Vision 86(2-3), 127–139 (2010)
15. He, J., Lin, Z., Wang, L., Tang, X.: Detecting Doctored JPEG Images Via DCT Coefficient Analysis. In: Leonardis, A., Bischof, H., Pinz, A. (eds.) ECCV 2006. LNCS, vol. 3953, pp. 423–435. Springer, Heidelberg (2006)
16. Hordley, S.D., Finlayson, G.D.: Re-evaluating Color Constancy Algorithm Performance. Journal of the Optical Society of America A 23(5), 1008–1020 (2006)
17. Hsu, Y., Chang, S.: Image Splicing Detection using Camera Response Function Consistency and Automatic Segmentation. In: International Conference on Multimedia and Expo., pp. 28–31 (2007)
18. Johnson, M., Farid, H.: Exposing Digital Forgeries through Specular Highlights on the Eye. In: Furon, T., Cayre, F., Doërr, G., Bas, P. (eds.) IH 2007. LNCS, vol. 4567, pp. 311–325. Springer, Heidelberg (2007)
19. Johnson, M.K., Farid, H.: Exposing Digital Forgeries by Detecting Inconsistencies in Lighting. In: Workshop on Multimedia and Security, pp. 1–10 (2005)
20. Johnson, M.K., Farid, H.: Exposing Digital Forgeries through Chromatic Aberration. In: Multimedia and Security, pp. 48–55 (2006)
21. Johnson, M.K., Farid, H.: Exposing Digital Forgeries through Chromatic Aberration. In: ACM Workshop on Multimedia and Security, pp. 48–55 (2006)
22. Kharrazi, M., Sencar, H.T., Memon, N.: Blind Source Camera Identification. In: IEEE International Conference on Image Processing, pp. 709–712 (2004)
23. Kirchner, T., Böhme, R.: Hiding Traces of Resampling in Digital Images. Information Forensics and Security 3(4), 582–592 (2008)
24. Klinker, G.J., Shafer, S.A., Kanade, T.: The Measurement of Highlights in Color Images. International Journal of Computer Vision 2(1), 7–26 (1992)

25. Lalonde, J.F., Efros, A.A.: Using Color Compatibility for Assessing Image Realism. In: IEEE International Conference on Computer Vision (2007)
26. Lee, H.C.: Method for Computing the Scene-Illuminant Chromaticity from Specular Highlights. Journal of the Optical Society of America A 3(10), 1694–1699 (1986)
27. Lin, S., Gu, J., Yamazaki, S., Shum, H.Y.: Radiometric Calibration from a Single Image. In: IEEE Conference on Computer Vision and Pattern Recognition, pp. 938–945 (2004)
28. Lu, R., Gijsenij, A., Gevers, T., Nedovic, V., Xu, D., Geusebroek, J.M.: Color Constancy using 3D Scene Geometry. In: IEEE International Conference on Computer Vision (2009)
29. Lukáš, J., Fridrich, J.: Estimation of Primary Quantization Matrix in Double Compressed JPEG Images. In: Digital Forensics Research Workshop (2003)
30. Lukáš, J., Fridrich, J., Goljan, M.: Digital Camera Identification From Sensor Pattern Noise. Information Forensics and Security 1(2), 205–214 (2006)
31. Luo, W., Huang, J., Qiu, G.: Robust Detection of Region-Duplication Forgery in Digital Images. Pattern Recognition 4, 746–749 (2006)
32. Mahdian, B., Saic, S.: Detection of Copy-Move Forgery using a Method Based on Blur Moment Invariants. Forensic Science International 171(2), 180–189 (2007)
33. Ng, T., Chang, S., Lin, C., Sun, Q.: Passive-Blind Image Forensics. In: Multimedia Security Technologies for Digital Rights, ch. 15, pp. 383–412. Academic Press, London (2006)
34. Personal Communication: Arjan Gijsenij, University of Amsterdam
35. Popescu, A., Farid, H.: Exposing Digital Forgeries by Detecting Traces of Resampling. Signal Processing 53(2), 758–767 (2005)
36. Riess, C., Angelopoulou, E.: Physics-Based Illuminant Color Estimation as an Image Semantics Clue. In: International Conference on Image Processing (2009)
37. Sencar, H., Memon, N.: Overview of State-of-the-art in Digital Image Forensics. In: Algorithms, Architectures and Information Systems Security, pp. 325–344 (2008)
38. Shafer, S.A.: Using Color to Separate Reflection Components. Journal Color Research and Application 10(4), 210–218 (1985)
39. Tan, R., Nishino, K., Ikeuchi, K.: Color Constancy through Inverse-Intensity Chromaticity Space. Journal of the Optical Society of America A 21(3), 321–334 (2004)
40. Yu, H., Ng, T.T., Sun, Q.: Recaptured Photo Detection Using Specularity Distribution. In: IEEE International Conference on Image Processing, pp. 3140–3143 (2008)

Capacity of Collusion Secure Fingerprinting — A Tradeoff between Rate and Efficiency
(Extended Abstract of Invited Talk)

Gábor Tardos

School of Computing Science
Simon Fraser University, Canada
and
Rényi Institute, Budapest, Hungary
tardos@cs.sfu.ca

Abstract. The talk presented a short history of collusion resistant fingerprinting concentrating on recent results that represent joint work with Ehsan Amiri.

1 The Model

To ensure protection of their copyright, content producers often make each copy of their productions unique by embedding a distinct code in each. For this to work they have to be able to hide the positions where the code is embedded. A *collision attack* is performed by a group of malicious users (the *pirates*), who compare their copies and identify the positions where they differ as a position of the embedded code. They can then arbitrarily change the code in these positions. We assume however that they do not notice the positions of the hidden code where all their codes agreed and therefore they cannot alter these positions. This is the *marking assumption.*

A (collusion resistant) *fingerprinting code* consists of a randomized procedure to choose codewords (the *code generation*) and a *tracing algorithm* that finds one of the pirates based on all these codewords and and the *forged codeword* read from the unauthorized copy made by the pirates. We say that the code (or the tracing algorithm) *errs* if it falsely accuses an innocent user or outputs no accused user at all. This should happen with small probability. The mathematical definition was first given by Boneh and Shaw [4] and can also be found in most of the papers cited below. Here we do not go beyond the slightly informal explanation given above, but in order to be able to speak about the result we have to list the numerous parameters a fingerprinting code has. These are

- *alphabet size.* The codewords are are sequences over a fixed alphabet Σ. Most of the research on fingerprinting codes concentrates on the binary alphabet $\Sigma = \{0, 1\}$, but fingerprinting is worth studying over larger alphabets too and the size $|\Sigma|$ of the alphabet is an important parameter.
- *codelength.* This is the length of the codewords, usually denoted by n.

R. Böhme, P.W.L. Fong, and R. Safavi-Naini (Eds.): IH 2010, LNCS 6387, pp. 81–85, 2010.

- *number of users.* Usually denoted by N, this is also the number of codewords.
- *number of pirates.* In most codes one has to assume a bound t on the number of pirates responsible for the collusion attack. If the actual number of pirates is t or less, the tracing algorithm should perform with small error probability. In many of the fingerprinting codes the probability of accusing an innocent user is small even if the number of pirates exceeds t, only the probability of tracing algorithm failing by not producing any accusations increases in this case.
- *error probability.* We say that a a code is ϵ-secure against t pirates or it is an ϵ-secure t-fingerprinting code, if the probability of the error of the tracing algorithm is at most ϵ for any set of at most t pirates performing an arbitrary pirate strategy to produce the forged codeword provided that they obey the marking assumption.
- *rate.* The rate R of a fingerprinting code is computed from the number N of codewords and the length n as $R = \log N/n$, where the logarithm is binary. In the trivial $t = 1$ case (no collusion) it is enough to ensure that the codewords are pairwise distinct and thus a rate of $R = \log|\Sigma|$ is achievable ($R = 1$ for binary codes), the reciprocal of the rate gives how much longer the codewords are compared to these trivial binary codewords.

To simplify the large number of parameters we concentrate on maximizing the rate of a sequence of fingerprinting codes (a *fingerprinting scheme*) subject to the conditions that the number of users and the length should go to infinity, the error probability should go to zero while the number of pirates and and the alphabet is fixed. The *t-fingerprinting capacity* for alphabet size $|\Sigma|$ is the maximum achievable limit rate of such fingerprinting schemes.

The goal of fingerprinting research is to find efficient and secure fingerprinting codes. The paramount problem in the application of fingerprinting codes is the high cost of embedding every single digit of the code. This makes it important to design secure fingerprinting codes that are short, or equivalently, have high rate. In particular, recent research focused on finding or estimating the t-fingerprinting capacity for various values of t (mostly considering the binary alphabet).

2 History of Results

Boneh and Shaw [4] were first to define fingerprinting secure against collusion attacks. They proposed such binary codes of length $O(t^4 \log(N/\epsilon)\log(1/\epsilon))$ for N users that are ϵ-secure against t pirates. This translates to a t-secure fingerprinting scheme with rate of $\Omega(1/t^4)$. The same paper gave a lower bound of $\Omega(t\log(1/(t\epsilon)))$ for the length of the fingerprinting code with the same parameters. This does not quite translate to an upper bound on the binary t-fingerprinting capacity but still roughly correspond to an $O(1/t)$ bound.

The paper [11] introduced a new and more efficient codes. The rate of these binary t-secure codes are $1/(100t^2)$ and the same paper also gave a lower bound on the length of t-secure fingerprinting codes that roughly translates to an $O(1/t^2)$

upper bound on the rate for any alphabet size. This settled the order of magnitude for the t-fingerprinting capacity: it is $\Theta(1/t^2)$. The large constant factor between the lower and upper bound motivated further research and many subsequent papers (e.g., [3,7,8,9,10]) managed to improve the constant 100 in the construction with optimizing various parameters in the construction and/or better analysis in the estimate of the error probability.

Amiri and Tardos [1] and independently Huang and Moulin [5,6] (for a broader class of models including the marking assumption model surveyed here) designed fundamentally different fingerprinting codes in an attempt to find the optimal rate, the fingerprinting capacity. While these new codes in the two papers are slightly different they are very similar and achieve identical rates. I conjecture that the rates achieved are optimal (i.e., they achieve the t-fingerprinting capacity) to the best of my knowledge this has not been fully proved yet.

3 Comparison of the Techniques

Bias based code generation was introduced in [11]. This is a two phase process for generating the code words starting with picking iid. *biases* for each position from 1 to n and continuing with picking each digit of each codeword independently with the bias determined by the position of the digit. In the binary case a bias for position i is a real number $0 \leq p_i \leq 1$ and the digit i of a codeword x is picked with $P[x_i = 1] = p_i$. For larger alphabets the bias is an arbitrary distribution on Σ. The same code generation was used later in [1], but with a different distribution to choose the biases from.

The tracing algorithm in [11] is simple and efficient, whether a user is accused or not is determined by simple linear constraint, most notably it is determined by the codeword of the user, the forged codeword and the biases without regard to all the other codewords. This makes the tracing algorithm linear time in the size of the code matrix. In contrast, in the schemes of [1,5] and also in the scheme of the earlier paper [2] of Anthapadmanabhan, Barg and Dumer the tracing algorithm has to consider each t-tuple of users to decide which user to accuse, thus the accusation of a user depends on the codewords of all other users too. This makes tracing rather inefficient.

4 Further Research Directions

An obvious research direction is to combine the efficiency of the tracing algorithm in [11] with the higher rates of [2] (for $t = 2$ pirates) and of [1,6] (for any number of pirates). With Ehsan Amiri we achieved partial results in this direction. These results are yet to be published.

For the first non-trivial case when the code should be secure against only $t = 2$ pirates we can design a more efficient tracing algorithm for the fingerprinting code of [2,1] (these have the same very simple code generation procedure for $t = 2$: each user receives an independent uniform random binary codeword). Although accusation of a user still depends on codewords of all other users too (this is unavoidable), the new tracing algorithm is linear time in the size of the codematrix. So in the case of two pirates the shorter codes and the faster tracing algorithms can be achieved simultaneously without having to compromise in either.

For more than two pirates our results are much more modest. First, we can achieve a small speedup of the tracing algorithm. Instead of considering all t-tuples of users for a tracing algorithm with N^t as the leading term in its running time we can slightly improve the exponent. This represent a slightly more efficient, but still optimal length fingerprinting codes. Another possible approach is to insist on a much faster tracing algorithm for the price of a slightly longer code (i.e., lower rates). We devised a sequence of intermediate fingerprinting codes representing a gradual trade-off between efficiency (speed of tracing) and length (rate). It is not clear however, whether this trade-off is inherent or just the result of our imperfect techniques.

Another important research direction is to study how the alphabet size influences the fingerprinting capacity. By the results of [11,1] the $O(1/t^2)$ upper bound for the t-fingerprinting capacity holds with an absolute constant independent of the alphabet size, but our preliminary calculations show that the actual t-fingerprinting capacity grows substantially with moving from binary to larger alphabets. It would be important to determine the limit rates achievable as the alphabet size grows.

References

1. Amiri, E., Tardos, G.: High rate fingerprinting codes and the fingerprinting capacity. In: Proceedings of the 20th Annual ACM-SIAM Symposium on Discrete Algorithms (SODA 2009), pp. 336–345 (2009)
2. Anthapadmanabhan, N.P., Barg, A., Dumer, I.: Fingerprinting capacity under the marking assumption. IEEE Transactions on Information Theory 54(6), 2678–2689 (2008); Preliminary version appeared in the Proceedings of the 2007 IEEE International Symposium on Information Theory, ISIT 2007 (2007)
3. Blayer, O., Tassa, T.: Improved versions of Tardos' fingerprinting scheme. Designs, Codes and Cryptography 48, 79–103 (2008)
4. Boneh, D., Shaw, J.: Collusion-secure fingerprinting for digital data. IEEE Transactions of Information Theory 44, 480–491 (1988)
5. Huang, Y., Moulin, P.: Saddle-point solution of the fingerprinting capacity game under the marking assumption. In: Proceedings of the IEEE International Symposium on Information Theory, ISIT 2009 (2009)
6. Moulin, P.: Universal fingerprinting: capacity and random-coding exponents. In: Proceedings of the IEEE International Symposium on Information Theory (ISIT 2008), pp. 220–224 (2008)
7. Nuida, K., Fujitsu, S., Hagiwara, M., Kitagawa, T., Watanabe, H., Ogawa, K., Imai, H.: An improvement of the discrete Tardos fingerprinting codes. Cryptology ePrint Archive, Report 2008/338
8. Nuida, K., Hagiwara, M., Watanabe, H., Imai, H.: Optimization of Tardos's fingerprinting codes in a viewpoint of memory amount. In: Furon, T., Cayre, F., Doërr, G., Bas, P. (eds.) IH 2007. LNCS, vol. 4567, pp. 279–293. Springer, Heidelberg (2008)

9. Škorić, B., Katzenbeisser, S., Celik, M.U.: Symmetric Tardos fingerprinting codes for arbitrary alphabet sizes. Designs, Codes and Cryptography 46(2), 137–166 (2008)
10. Škorić, B., Vladimirova, T.U., Celik, M., Talstra, J.C.: Tardos fingerprinting is better than we thought. IEEE Transactions on Information Theory 54(8), 3663–3676 (2008)
11. Tardos, G.: Optimal probabilistic fingerprint codes. Journal of the ACM (to appear); Preliminary version appeared in Proceedings of the 35th Annual ACM Symposium on Theory of Computing, (STOC 2003), pp. 116–125

Short Collusion-Secure Fingerprint Codes against Three Pirates

Koji Nuida

Research Center for Information Security (RCIS), National Institute of Advanced Industrial Science and Technology (AIST), Akihabara-Daibiru Room 1003, 1-18-13 Sotokanda, Chiyoda-ku, Tokyo, 101-0021, Japan
k.nuida@aist.go.jp

Abstract. In this article, we propose a new construction of probabilistic collusion-secure fingerprint codes against up to three pirates and give a theoretical security evaluation. Our pirate tracing algorithm combines a scoring method analogous to Tardos codes with an extension of parent search techniques of some preceding 2-secure codes. Numerical examples show that our code lengths are significantly shorter than (about 30% to 40% of) the shortest known c-secure codes by Nuida et al. (Des. Codes Cryptogr., 2009) with $c = 3$. Some preliminary proposal for improving efficiency of our tracing algorithm is also given.

1 Introduction

1.1 Background and Related Works

Recently, digital content distribution services have been widespread by virtue of progress of information technology. Digitization of content distribution has improved convenience for ordinary people. However, the digitization also enables malicious persons to perform more powerful attacks, and the amount of illegal content redistribution is increasing very rapidly. Hence technical countermeasures for such illegal activities are strongly desired.

Fingerprint code is a possible solution for such problems. It supposes that each copy of a content is divided into several segments (common to all copies), in each of which a bit of an encoded user ID is embedded by the content provider by using watermarking technique. The embedded encoded ID (fingerprint) provides traceability of an adversarial user (pirate) when an unauthorized copy of the content is distributed. Such a scheme aims at tracing some pirates, without falsely tracing any innocent user, from the fingerprint embedded in the pirated content with an overwhelming probability. It has been noticed that a coalition of pirates can perform certain strong attacks (so called collusion attacks) against the fingerprint, therefore any effective fingerprint code should be secure against collusion attacks, called collusion-secure codes. In particular, if the code is secure against collusion attacks by up to c pirates, then the code is called c-secure [2].

Among several constructions of collusion-secure codes proposed so far, the one proposed by Tardos [13] is "asymptotically optimal", in the sense that the

R. Böhme, P.W.L. Fong, and R. Safavi-Naini (Eds.): IH 2010, LNCS 6387, pp. 86–102, 2010.

order of his code length with respect to the allowable number c of pirates is theoretically the lowest (which is quadratic in c). For improvements of Tardos codes, the constant factor of the asymptotic code length has been reduced by c-secure codes given by Nuida et al. [9] to approximately 5.35% of Tardos codes, which is the smallest value so far provable without any additional assumption. On the other hand, after the article of Tardos, several collusion secure codes have been proposed, e.g., [1,3,6,8,10], which restrict the number of pirates to $c = 2$ but achieve further short code lengths. Such constructions of short c-secure codes for a small c would have not only theoretical but also practical importance; for example, when the users are less anonymous for the content provider (e.g., the case of secret documents distributed in a company), it seems infeasible to make a large coalition confidentially. The aim of this article is to extend such a "compact" construction to the next case $c = 3$.

For related works, we notice that there is an earlier work by Sebé and Domingo-Ferrer [12] for 3-secure codes. On the other hand, there is another work by Kitagawa et al. [5] on construction of 3-secure codes, in which very short code lengths are proposed but its security is evaluated only by computer experiments for some special attack strategies.

1.2 Our Contribution

In this article, we propose a new construction of 3-secure codes and give a theoretical security evaluation. The codeword generation is just a bit-wise random sampling, which has been used by many preceding works as well. The novel point of our construction is in the pirate tracing algorithm, which combines the use of score computation analogous to Tardos codes [13] with an extension of "parent search" technique of preceding works against two pirates [1,6,10]. Intuitively, the scoring method works well when the parts of fingerprint in the pirated content are chosen not evenly from the codewords of pirates, while the extended "parent search" technique works well when the fingerprint is evenly chosen from codewords of pirates, hence their combination is effective in any case.

Under some parameter choices, our code lengths are approximately 3% to 4% shorter than 3-secure codes by Sebé and Domingo-Ferrer [12], and approximately 30% to 40% shorter than c-secure codes by Nuida et al. [9] for $c = 3$. This shows that our code length is even significantly shorter than the shortest known c-secure codes [9].

In fact, Kitagawa et al. [5] claimed that their 3-secure code provides almost the same security level as our code for the case of 100 users and 128-bit length. However, they evaluated the security by only computer experiments for the case of some special attack algorithms, while in this article we give a theoretical security evaluation against arbitrary attack algorithms under the standard Marking Assumption (cf., [2]). (One may think that the perfect protection of so-called undetectable positions required by Marking Assumption is not practical. However, this is in fact not a serious problem, as a general conversion technique recently proposed by Nuida [7] can supply robustness against erasure of a bounded number of undetectable bits.)

Moreover, for efficiency of our tracing algorithm, we also discuss an implementation method for the algorithm. By an intuitive observation, it seems more efficient for an average case than the naive implementation. A detailed evaluation of the proposed implementation method will be a future research topic.

1.3 Notations

In this article, log denotes the natural logarithm. We put $[n] = \{1, 2, \ldots, n\}$ for an integer n. Unless some ambiguity emerges, we often abbreviate a set $\{i_1, i_2, \ldots, i_k\}$ to $i_1 i_2 \cdots i_k$. Let $\delta_{a,b}$ denote Kronecker delta, i.e., we have $\delta_{a,b} = 1$ if $a = b$ and $\delta_{a,b} = 0$ if $a \neq b$. For a family $\mathcal{F}$ of sets, let $\bigcup \mathcal{F}$ and $\bigcap \mathcal{F}$ denote the union and the intersection, respectively, of all members of $\mathcal{F}$.

1.4 Organization of the Article

Section 2 gives a formal definition of collusion-secure fingerprint codes. In Sect. 3, we describe our codeword generation and pirate tracing algorithms, state main results on the security of our 3-secure codes, and give numerical examples for comparison to preceding works. Section 4 summarizes the outline of the security proof; details are given in the appendix, where some parts are omitted due to the page limitation and will appear in the forthcoming full version of this article. Finally, in Sect. 5, we discuss an implementation issue of our tracing algorithm.

2 Collusion-Secure Fingerprint Codes

In this section, we introduce a formal definition for fingerprint codes (Definition 1). First we explain notations in the definition. Let N be the total number of users, m the code length, and $1 \leq c \leq N$ an integer parameter. The set $U = [N]$ signifies the set of all users, the subset U_{P} in Step 2 is the coalition of pirates, and $U_{\mathrm{I}} = U \setminus U_{\mathrm{P}}$ is the set of innocent users (the innocent users are not concerned in Definition 1, as they play passive roles only). Fix a symbol '?' different from '0' and '1', called *erasure symbol.* The codeword w_i in Step 3 is the fingerprint for user $i \in U$, and the word y in Step 4, called *attack word,* is the fingerprint embedded in the pirated content. The set Acc in Step 5 consists of the users traced from the pirated content. The events $\mathsf{Acc} \cap U_{\mathrm{P}} = \emptyset$ and $\mathsf{Acc} \not\subseteq U_{\mathrm{P}}$ are referred to as *false-negative* and *false-positive* (or *false-alarm*), respectively, and both of them are called *tracing error.* Now the definition is as follows:

Definition 1. *Given the parameters N, m and c, we define the following game, which we refer to as* pirate tracing game. *The players of the game is a* provider *and* pirates, *and the game is proceeded as follows:*

1. Provider *generates an $N \times m$ binary matrix $W = (w_{i,j})_{i \in [N], j \in [m]}$ and an element* st *called* state information.
2. Pirates *generate $U_{\mathrm{P}} \subseteq U = [N]$, $1 \leq |U_{\mathrm{P}}| \leq c$, without knowing W and* st.
3. Pirates *receive the codeword $w_i = (w_{i,1}, \ldots, w_{i,m})$ for every $i \in U_{\mathrm{P}}$.*

4. Pirates *generate a word* $y = (y_1, \dots, y_m)$ *on* $\{0, 1, ?\}$ *under a certain restriction (e.g., Definition 2 below), and send* y *to* provider.
5. Provider *generates* $\mathsf{Acc} \subseteq U$ *from* y, W, *and* st, *without knowing* U_{P}.
6. *Then* pirates *win if* $\mathsf{Acc} \cap U_{\mathrm{P}} = \emptyset$ *or* $\mathsf{Acc} \not\subseteq U_{\mathrm{P}}$, *and otherwise* provider *wins.*

Let Gen, Reg, ρ, and Tr denote the algorithms used in Steps 1, 2, 4, and 5, respectively. We call Gen, Reg, ρ, and Tr *codeword generation algorithm*, *registration algorithm*, *pirate strategy*, and *tracing algorithm*, respectively. We refer to the pair $\mathcal{C} = (\mathsf{Gen}, \mathsf{Tr})$ as a *fingerprint code*, and the following quantity

$$\begin{aligned} Pr[(W, \mathsf{st}) \leftarrow \mathsf{Gen}();\ U_{\mathrm{P}} \leftarrow \mathsf{Reg}();\ y \leftarrow \rho(U_{\mathrm{P}}, (w_i)_{i \in U_{\mathrm{P}}});\\ \mathsf{Acc} \leftarrow \mathsf{Tr}(y, W, \mathsf{st}) : \mathsf{Acc} \cap U_{\mathrm{P}} = \emptyset \text{ or } \mathsf{Acc} \not\subseteq U_{\mathrm{P}}] \end{aligned} \tag{1}$$

(i.e., the overall probability that pirates win) is called an *error probability* of $\mathcal{C}$.

For $j \in [m]$, j-th column in codewords is called *undetectable* if j-th bits $w_{i,j}$ of the codewords w_i of pirates $i \in U_{\mathrm{P}}$ coincide with each other; otherwise the column is called *detectable*. In this article, we put the following standard assumption on y in Step 4, called *Marking Assumption* [2]:

Definition 2. *The* Marking Assumption *states the following: For every undetectable column* j, *we have* $y_j = w_{i,j}$ *for some (or equivalently, all)* $i \in U_{\mathrm{P}}$.

We say that a fingerprint code $\mathcal{C}$ is *collusion-secure* if the error probability of $\mathcal{C}$ is sufficiently small for any Reg and ρ under Marking Assumption. More precisely, we say that $\mathcal{C}$ is *c-secure* (*with* ε*-error*) [2] if the error probability is not higher than a sufficiently small value ε under Marking Assumption.

3 Our 3-Secure Codes

Here we propose a codeword generation algorithm Gen and a tracing algorithm Tr for 3-secure codes ($c = 3$). The security property will be discussed below.

The algorithm Gen, with parameter $1/2 \leq p < 1$, is the codeword generation algorithm of Tardos codes [13] but the probability distribution of biases is different: For each (say, j-th) column, each user's bit $w_{i,j}$ is independently chosen by $Pr[w_{i,j} = 1] = p_j$, where $p_j = p$ or $1 - p$ with probability 1/2 each. Then Gen outputs $W = (w_{i,j})_{i \in [N], j \in [m]}$ and $\mathsf{st} = (p_j)_{j \in [m]}$.

To describe the algorithm Tr, we introduce some notations. For binary words $w_1, \dots, w_k$ of length m, we define the *envelope* of $w_1, \dots, w_k$ by

$$\mathcal{E}(w_1, \dots, w_k) = \{y \in \{0,1\}^m \mid y_j \in \{w_{1,j}, \dots, w_{k,j}\} \text{ for every } j \in [m]\} \ . \tag{2}$$

Then for a binary word y of length m and a collection $W = (w_{i,j})$ of codewords of users, we define

$$\mathcal{T}(y) = \{i_1 i_2 i_3 \subseteq U \mid i_1 \neq i_2 \neq i_3 \neq i_1,\ y \in \mathcal{E}(w_{i_1}, w_{i_2}, w_{i_3})\} \tag{3}$$

(see Sect. 1.3 for the notation $i_1 i_2 i_3$). A key property implied by Marking Assumption is that if the attack word y contains no erasure symbols, then y belongs to the envelope of the codewords of pirates and, if furthermore $|U_{\mathrm{P}}| = 3$, the family $\mathcal{T}(y)$ contains the set of three pirates. By using these notations, we define the algorithm Tr as follows, where the words $y, w_1, \dots, w_N$ and the state information $\mathsf{st} = (p_j)_{j \in [m]}$ are given:

1. Replace each erasure symbol '?' in y with '0' or '1' independently in the following manner. If $y_j = ?$, then it is replaced with '1' with probability p_j, and with '0' with probability $1 - p_j$. Let y' denote the resulting word.
2. Calculate a threshold parameter $Z = Z_{y'}$ as specified below.
3. For each $i \in U$, calculate the score $S(i)$ of i by

$$S(i) = \sum_{j \in [m]\,;\, y'_j = 1} \delta_{w_{i,j}, y'_j} \log \frac{1}{p_j} + \sum_{j \in [m]\,;\, y'_j = 0} \delta_{w_{i,j}, y'_j} \log \frac{1}{1 - p_j} \ . \tag{4}$$

4. If $S(i) \geq Z$ for some $i \in U$, then output every $i \in U$ such that $S(i) \geq Z$, and halt.
5. Calculate $\mathcal{T}' = \{T \in \mathcal{T}(y') \mid T \cap T' \neq \emptyset \text{ for every } T' \in \mathcal{T}(y')\}$. If $\mathcal{T}' = \emptyset$, then output nobody, and halt.
6. If $\bigcap \mathcal{T}' \neq \emptyset$, then output every member of $\bigcap \mathcal{T}'$, and halt.
7. Calculate $\mathcal{P} = \{P = i_1 i_2 \subseteq U \mid i_1 \neq i_2, P \cap T \neq \emptyset \text{ for every } T \in \mathcal{T}'\}$. Let $\mathcal{P}_k$ be the set of all $i \in U$ such that $|\{P \in \mathcal{P} \mid i \in P\}| = k$.
8. If $\mathcal{P}_1 \neq \emptyset$, then output every $i \in U$ such that $ii' \in \mathcal{P}$ for some $i' \in \mathcal{P}_1$, and halt.
9. If $|\mathcal{P}| = 7$, then output every $i \in U$ such that $ii' \in \mathcal{P}$ for some $i' \in \mathcal{P}_2$, and halt.
10. If $|\mathcal{P}| = 6$, then output every $i \in \mathcal{P}_3$, and halt.
11. If $|\mathcal{P}| = 5$ and $\mathcal{T}'' = \{i_1 i_2 i_3 \in \mathcal{T}' \mid i_1 i_2, i_2 i_3, i_1 i_3 \in \mathcal{P}\} \neq \emptyset$, then output every member of $\mathcal{P}_2 \cap (\bigcup \mathcal{T}'')$, and halt.
12. If $|\mathcal{P}| = 5$ and $\mathcal{T}'' = \emptyset$, then output every $i \in \bigcup \mathcal{P}$ such that $ii' \notin \mathcal{P}$ for some $i' \in \bigcup \mathcal{P}$, and halt.
13. If $|\mathcal{P}| = 4$, then output every $i \in \bigcup \mathcal{P}$ such that $T \in \mathcal{T}'$ and $T \subseteq \bigcup \mathcal{P}$ imply $i \in T$, and halt.
14. If $|\mathcal{P}| = 3$, then output every $i \in \bigcup \mathcal{P}$, and halt.
15. Output nobody, and halt.

This algorithm is divided into two parts; Steps 1–4 and the remaining steps. The former part aims at performing coarse tracing to defy "unbalanced" pirate strategies; namely, if some pirates' codewords contribute to generate y at too many columns than the other pirates, then it is very likely that scores of such pirates exceed the threshold and they are correctly accused by Step 4.

On the other hand, the latter complicated part aims at performing more refined tracing. First, the algorithm enumerates the collections of three users such that y' can be made (under Marking Assumption) from their codewords, in other words, the collection is a candidate of the actual triple of pirates. Steps 5 and 6 are designed according to an intuition that a pirate would be very likely to be contained in much more candidate triples than an innocent user. When the tracing algorithm did not halt until Step 6, the possibilities of "structures" of the set

$\mathcal{T}'$ are mostly limited, even allowing us to enumerate all the possibilities. However, it is space-consuming to enumerate them and determine suitable outputs in a case-by-case manner. Instead, we give an explicit algorithm (Steps 7–15) to determine a suitable output, which is artificial but not too space-consuming. Some examples of the possibilities of $\mathcal{T}'$ are given in Fig. 1, where 1, 2, 3 are the pirates, i_j are innocent users and the members of $\mathcal{T}'$ are denoted by triangles.

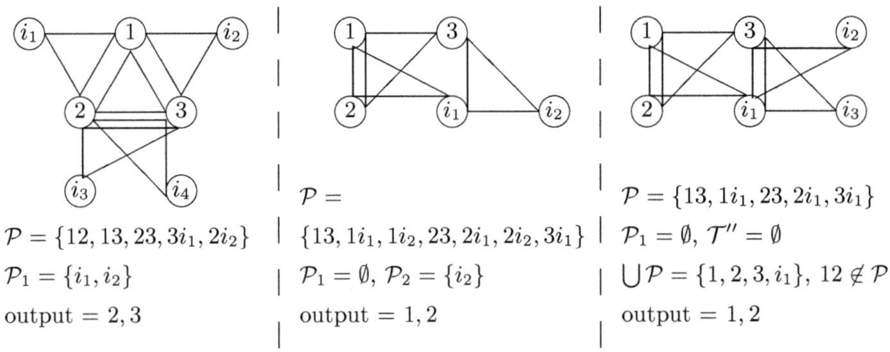

Fig. 1. Examples of the sets $\mathcal{T}'$ and $\mathcal{P}$

For the latter part, the tracing tends to fail in the case that the set $\mathcal{T}(y')$ contains much more members other than the triple of the pirates, which tends to occur when the contributions of the pirates' codewords to y was too unbalanced. However, such an unbalanced attack is defied by the former part, therefore the latter part also works well. More precisely, an upper bound of the error probability at the latter part will be derived by using the property that scores of pirates are lower than the threshold (as otherwise the tracing halts at the former part); cf., Appendix E. Our scoring function (4), which is different from the ones for Tardos codes [13] and its symmetrized version [11], is adopted to simplify the derivation process. Although it is possible that the true error probability is reduced by applying the preceding scoring functions, a proof of a bound of error probability with those scoring functions requires another evaluation technique and would be much more involved, which is a future research topic.

Note that, for the case $p = 1/2$, it is known that the "minority vote" by three pirates for generating y cancels the mutual information between y and a single codeword, therefore the pirates are likely to escape from the former part of Tr. However, even by such a strategy the pirates are unlikely to escape from the latter part of Tr, as *collections* of users rather than individual users are considered there.

The threshold parameter $Z = Z_{y'}$ in Step 2 is determined as follows. Let A_{H} be the set of column indices j such that $(p_j, y'_j) = (p, 1)$ or $(1-p, 0)$, i.e., the occurrence probability of the bit $y'_j \in \{0,1\}$ at j-th column is $p \geq 1/2$, and let $A_{\mathrm{L}} = [m] \setminus A_{\mathrm{H}}$. Put $a_{\mathrm{H}} = |A_{\mathrm{H}}|$ and $a_{\mathrm{L}} = |A_{\mathrm{L}}|$. Choose a parameter $\varepsilon_0 > 0$ which is smaller than the desired bound ε of error probability. Then choose $Z = Z_{y'}$ satisfying the following condition:

$$\sum_{\substack{k_{\mathrm{H}},k_{\mathrm{L}} \geq 0 \\ k_{\mathrm{H}} \log \frac{1}{p} + k_{\mathrm{L}} \log \frac{1}{1-p} \geq Z}} \binom{a_{\mathrm{L}}}{k_{\mathrm{L}}} p^{a_{\mathrm{L}}-k_{\mathrm{L}}}(1-p)^{k_{\mathrm{L}}} \binom{a_{\mathrm{H}}}{k_{\mathrm{H}}} p^{k_{\mathrm{H}}}(1-p)^{a_{\mathrm{H}}-k_{\mathrm{H}}} \leq \frac{\varepsilon_0}{N} \ . \quad (5)$$

An example of a concrete choice of Z satisfying the condition (5) is as follows:

$$\begin{aligned} Z_0 &= a_{\mathrm{H}} p \log \frac{1}{p} + a_{\mathrm{L}}(1-p) \log \frac{1}{1-p} \\ &+ \sqrt{\frac{1}{2}\left(\left(\log \frac{1}{p}\right)^2 a_{\mathrm{H}} + \left(\log \frac{1}{1-p}\right)^2 a_{\mathrm{L}}\right) \log \frac{N}{\varepsilon_0}} \end{aligned} \quad (6)$$

(see Appendix A for the proof). From now, we suppose that the threshold Z satisfies the condition (5) and $Z \leq Z_0$.

For the security of the proposed fingerprint code, first we present the following result, which will be proven in Sect. 4:

Theorem 1. *By the above choice of ε_0 and Z, if the number of pirates is three, then the error probability of the proposed fingerprint code is lower than*

$$\begin{aligned} &\varepsilon_0 + \binom{N-3}{3}(1 - 3p^2 + 10p^3 - 15p^4 + 12p^5 - 4p^6)^m \\ &+ 3(N-3)(N-4)\left(p^2(1-p)^2(\sqrt{p} + \sqrt{1-p}) + 1 - p - p^2 + 4p^3 - 2p^4\right)^m \\ &+ (N-3)(1-p)^{-3\sqrt{(m/2)\log(N/\varepsilon_0)}}\left(p^{4-3p}(p^2 - 3p + 3) + (1-p)^{3p+1}(p^2 + p + 1)\right)^m \ . \end{aligned} \quad (7)$$

Some numerical analysis suggests that the choice $p = 1/2$ would be optimal (or at least pretty good) to decrease the bound of error probabilities specified in Theorem 1. In fact, an elementary analysis shows that the second term in the sum, which seems dominant (cf., Theorem 2 below), takes the minimum over $p \in [1/2, 1)$ at $p = 1/2$. Hence we use $p = 1/2$ in the following argument. Now it is shown that the error probability against less than three pirates also has the same bound under a condition (9) below (which seems trivial in practical situations), therefore we have the following (which will be proven in Sect. 4):

Theorem 2. *By using the value $p = 1/2$, the proposed fingerprint code is 3-secure with error probability lower than*

$$\begin{aligned} &\varepsilon_0 + \binom{N-3}{3}\left(\frac{7}{8}\right)^m + 3(N-3)(N-4)\left(\frac{10+\sqrt{2}}{16}\right)^m \\ &+ (N-3)8^{\sqrt{(m/2)\log(N/\varepsilon_0)}}\left(\frac{7\sqrt{2}}{16}\right)^m \end{aligned} \quad (8)$$

provided

$$m \geq 8 \log(N/\varepsilon_0) \left(1 + (16 \log(N/\varepsilon_0))^{-1}\right)^2 . \tag{9}$$

Note that when $p = 1/2$, the score $S(i)$ of a user i is equal to $\log 2$ times the number of columns in which the words w_i and y' coincide. Hence the calculation of scores can be made easier by using the "normalized" score $S(i)/\log 2$ instead, which is equal to m minus the Hamming distance of w_i from y', together with the "normalized" threshold $Z_0/\log 2 = m/2 + \sqrt{(m/2)\log(N/\varepsilon_0)}$.

Table 1 shows comparison of our code lengths (numerically calculated by using Theorem 2) with 3-secure codes by Sebé and Domingo-Ferrer [12]. Table 2 shows the comparison with c-secure codes by Nuida et al. [9] for $c = 3$. The values of N and ε and the corresponding code lengths are chosen from those articles. The tables show that our code lengths are much shorter than the codes in [12], and even significantly shorter than the codes in [9] which are in fact the shortest c-secure codes known so far (improving the celebrated Tardos codes [13]). On the other hand, recently Kitagawa et al. [5] proposed another construction of 3-secure codes, and evaluated the security against some typical pirate strategies in the case $N = 100$ and $m = 128$ by computer experiment. The resulting error probability was $\varepsilon = 0.009$. For the same error probability, our code length (with parameter $\varepsilon_0 = \varepsilon/2$) is $m = 135$. Therefore our code, which is *provably secure* in contrast to their code, has almost the same length as their code.

Table 1. Comparison of code lengths with the codes in [12]

	$N = 128$ $\varepsilon = 0.14 \times 10^{-6}$	$N = 256$ $\varepsilon = 0.15 \times 10^{-13}$	$N = 512$ $\varepsilon = 0.19 \times 10^{-27}$
Sebé & Domingo-Ferrer [12]	6985	14025	28105
Our code ($\varepsilon_0 =$)	282 $(1/2)\varepsilon$	502 $(7/10)\varepsilon$	934 $(7/10)\varepsilon$
ratio	4.04%	3.58%	3.32%

Table 2. Comparison of code lengths with the codes in [9] ($c = 3$)

	$N = 300$, $\varepsilon = 10^{-11}$	$N = 10^9$, $\varepsilon = 10^{-6}$	$N = 10^6$, $\varepsilon = 10^{-3}$
Nuida et al. [9]	1309	1423	877
Our code ($\varepsilon_0 =$)	420 $(9/10)\varepsilon$	556 $(1/100)\varepsilon$	349 $(1/100)\varepsilon$
ratio	32.1%	39.1%	39.8%

4 Security Proof

In this section, we present an outline of the proof of Theorems 1 and 2. Omitted details of the proof will be supplied in the appendix. First, we present some properties of the threshold $Z = Z_{y'}$, which will be proven in Appendix A:

Proposition 1. *1. If Z satisfies the condition (5), then the conditional probability that $S(\mathsf{I}) \geq Z$ for some $\mathsf{I} \in U_{\mathrm{I}}$, conditioned on the choice of y', is not higher than $(N-1)\varepsilon_0/N$.*
2. The value $Z = Z_0$ in (6) satisfies the condition (5).

To prove Theorem 1, we consider the case that $|U_{\mathrm{P}}| = 3$. By symmetry, we may assume that $U_{\mathrm{P}} = \{1, 2, 3\}$. Put $T_{\mathrm{P}} = 123$, therefore we have $T_{\mathrm{P}} \in \mathcal{T}(y')$ by Marking Assumption. Now we consider the following four kinds of events:

Type I error: $S(\mathsf{I}) \geq Z$ for some innocent user $\mathsf{I} \in U_{\mathrm{I}}$.
Type II error: $T \cap T_{\mathrm{P}} = \emptyset$ for some $T \in \mathcal{T}(y')$.
Type III error: There are $T_1, T_2 \in \mathcal{T}(y')$ such that $\emptyset \neq T_1 \cap T_2 \subseteq U_{\mathrm{I}}$, $|T_1 \cap T_{\mathrm{P}}| = 1$ and $|T_2 \cap T_{\mathrm{P}}| = 1$.
Type IV error: $S(i) < Z$ for every $i \in \{1, 2, 3\}$, and there is an innocent user I such that $12\mathsf{I} \in \mathcal{T}(y')$, $13\mathsf{I} \in \mathcal{T}(y')$ and $23\mathsf{I} \in \mathcal{T}(y')$.

Then we have the following property, which will be proven in Appendix B:

Proposition 2. *If the number of pirates is three, then tracing error occurs only when one of the Type I, II, III and IV errors occurs.*

By this proposition, the error probability is bounded by the sum of the probabilities of Type I–IV errors. By Proposition 1, the probability of Type I error is bounded by ε_0. Now Theorem 1 is proven by combining this with the following three propositions, which will be proven in Appendices C–E, respectively:

Proposition 3. *If the number of pirates is three, then the probability of Type II error is not higher than $\binom{N-3}{3}(1 - 3p^2 + 10p^3 - 15p^4 + 12p^5 - 4p^6)^m$.*

Proposition 4. *If the number of pirates is three, then the probability of Type III error is not higher than*

$$3(N-3)(N-4)\left(p^2(1-p)^2(\sqrt{p} + \sqrt{1-p}) + 1 - p - p^2 + 4p^3 - 2p^4\right)^m \ . \quad (10)$$

Proposition 5. *If the number of pirates is three and the threshold Z is chosen so that the condition (5) holds and $Z \leq Z_0$, then the probability of Type IV error is lower than*

$$(N-3)(1-p)^{-3\sqrt{(m/2)\log(N/\varepsilon_0)}}\left(p^{4-3p}(p^2 - 3p + 3) + (1-p)^{3p+1}(p^2 + p + 1)\right)^m \ . \quad (11)$$

To prove Theorem 2, we set $p = 1/2$. Then the bound of error probability given by Theorem 1 is specialized to the value specified in Theorem 2. Hence our remaining task is to evaluate the error probabilities for the case that the number of pirates is one or two.

First we consider the case that there are exactly two pirates, say, $1, 2 \in U$. The key property is the following, which will be proven in Appendix F:

Proposition 6. *In this situation, if the condition (9) is satisfied, then the probability that $S(1) < Z$ and $S(2) < Z$ is lower than ε_0/N.*

By this proposition, when the condition (9) is satisfied, at least one of the two pirates is output in Step 4 of the tracing algorithm with probability not lower than $1 - \varepsilon_0/N$. On the other hand, by Proposition 1, some innocent user is output in Step 4 with probability not higher than $(N-1)\varepsilon_0/N$. Hence in Step 4, at least one pirate and no innocent users are output with probability not lower than $1-\varepsilon_0$. This implies that the error probability is bounded by ε_0 in this case.

Secondly, we consider the case that there is exactly one pirate, say, $1 \in U$. In this case, Marking Assumption implies that $y' = w_1$. Now an easy calculation shows that $S(1) \geq Z_0 \geq Z$ provided $m \geq 2\log(N/\varepsilon_0)$ (note that $p = 1/2$), which follows from the condition (9). Therefore the pirate is always output in Step 4 of the tracing algorithm, and by the same argument as the previous paragraph, the error probability is bounded by ε_0 in this case as well. Hence the proof of Theorem 2 is concluded.

5 On Implementation of the Tracing Algorithm

In this section, we discuss some implementation issue of the tracing algorithm Tr of the proposed 3-secure code. More precisely, we consider the calculation of the set $\mathcal{T}(y')$ appeared in Step 5 of Tr. By a naive calculation method based on the definition (3) of $\mathcal{T}(y')$, we need to check the condition $y' \in \mathcal{E}(w_{i_1}, w_{i_2}, w_{i_3})$ for every triple $i_1i_2i_3$ of users, therefore the time complexity with respect to the user number N is inevitably $\Omega(N^3)$. As this complexity is larger than tracing algorithms of many other c-secure codes such as Tardos codes [13], it is important to reduce the complexity of calculation of $\mathcal{T}(y')$.

To calculate the collection $\mathcal{T}(y')$, we consider the following algorithm, with codewords $w_1, \dots, w_N$ and the m-bit word y' as input:

1. Set $\mathcal{L}_1^{(1)} = \{i \in [N] \mid w_{i,1} = y'_1\}$ and $\mathcal{L}_2^{(1)} = \mathcal{L}_3^{(1)} = \emptyset$.
2. For each $2 \leq j \leq m$, construct $\mathcal{L}_1^{(j)}$, $\mathcal{L}_2^{(j)}$ and $\mathcal{L}_3^{(j)}$ inductively, in the following manner. (At the beginning, set $\mathcal{L}_1^{(j)} = \mathcal{L}_2^{(j)} = \mathcal{L}_3^{(j)} = \emptyset$.)
   (a) Put $C_j = \{i \in [N] \mid w_{i,j} = y'_j\}$.
   (b) Set $\mathcal{L}_1^{(j)} = \mathcal{L}_1^{(j-1)} \cap C_j$.
   (c) Add the pair $\{\mathcal{L}_1^{(j-1)} \setminus C_j, C_j \setminus \mathcal{L}_1^{(j-1)}\}$ of subsets of $[N]$ to $\mathcal{L}_2^{(j)}$.
   (d) For each pair $\{K_1, K_2\}$ of subsets of $[N]$ in $\mathcal{L}_2^{(j-1)}$,
      - add two pairs $\{K_1 \cap C_j, K_2\}$ and $\{K_1 \setminus C_j, K_2 \cap C_j\}$ to $\mathcal{L}_2^{(j)}$;
      - add the triple $\{K_1 \setminus C_j, K_2 \setminus C_j, C_j \setminus (K_1 \cup K_2)\}$ of subsets of $[N]$ to $\mathcal{L}_3^{(j)}$.
   (e) For each triple $\{K_1, K_2, K_3\}$ of subsets of $[N]$ in $\mathcal{L}_3^{(j-1)}$, add three triples $\{K_1 \cap C_j, K_2, K_3\}$, $\{K_1 \setminus C_j, K_2 \cap C_j, K_3\}$, $\{K_1 \setminus C_j, K_2 \setminus C_j, K_3 \cap C_j\}$ to $\mathcal{L}_3^{(j)}$.
   (f) Remove from $\mathcal{L}_2^{(j)}$ every pair $\{K_1, K_2\}$ with K_1 or K_2 being empty, and from $\mathcal{L}_3^{(j)}$ every triple $\{K_1, K_2, K_3\}$ with K_1, K_2 or K_3 being empty.
3. Output the collection of the triples $T = i_1i_2i_3$ of distinct numbers i_1, i_2, i_3 satisfying one of the following conditions:

- we have $i_1 \in \mathcal{L}_1^{(m)}$ and i_2, i_3 are arbitrary;
- for some $\{K_1, K_2\} \in \mathcal{L}_2^{(m)}$, we have $i_1 \in K_1$, $i_2 \in K_2$ and i_3 is arbitrary;
- for some $\{K_1, K_2, K_3\} \in \mathcal{L}_3^{(m)}$, we have $i_1 \in K_1$, $i_2 \in K_2$ and $i_3 \in K_3$.

An inductive argument shows that, for each $j \in [m]$ and each triple of distinct i_1, i_2, i_3, the j-bit initial subword $(y'_1, \dots, y'_j)$ of y' is in the envelope of the j-bit initial subwords of $w_{i_1}, w_{i_2}, w_{i_3}$ if and only if one of the following conditions is satisfied (note that the order of members of a pair or triple is ignored):

- we have $i_1 \in \mathcal{L}_1^{(j)}$ and i_2, i_3 are arbitrary;
- for some $\{K_1, K_2\} \in \mathcal{L}_2^{(j)}$, we have $i_1 \in K_1$, $i_2 \in K_2$ and i_3 is arbitrary;
- for some $\{K_1, K_2, K_3\} \in \mathcal{L}_3^{(j)}$, we have $i_1 \in K_1$, $i_2 \in K_2$ and $i_3 \in K_3$.

By setting $j = m$, it follows that the above algorithm outputs $\mathcal{T}(y')$ correctly.

Now for each $2 \le j \le m$, complexity of computing $\mathcal{L}_1^{(j)}$, $\mathcal{L}_2^{(j)}$, and $\mathcal{L}_3^{(j)}$ from $\mathcal{L}_1^{(j-1)}$, $\mathcal{L}_2^{(j-1)}$, and $\mathcal{L}_3^{(j-1)}$ is approximately proportional to N times the total number of members of $\mathcal{L}_2^{(j-1)}$ and $\mathcal{L}_3^{(j-1)}$. Hence the total complexity of the algorithm is approximately proportional to Nm times the average of total number of members in $\mathcal{L}_2^{(j)}$ and $\mathcal{L}_3^{(j)}$ over all $1 \le j \le m-1$. This implies that the order (with respect to N) of complexity of calculating $\mathcal{T}(y')$ can be reduced from $\Theta(N^3)$ if the average number of pairs and triples in $\mathcal{L}_2^{(j)}$ and $\mathcal{L}_3^{(j)}$ is sufficiently small. The author guesses that the latter average number is indeed sufficiently small in most of the practical cases, as the size of $\mathcal{T}(y')$ would be not large in average case (provided the code length m is long enough to make the error probability of the fingerprint code sufficiently small). A detailed analysis of this calculation method will be a future research topic. Instead, here we show some experimental data for running time of the above algorithm, which was implemented on a usual PC with 1.83GHz Intel Core 2 CPU and 2Gbytes memory. We chose parameters $N = 1000$, $m = 180$, $\varepsilon_0 = 0.001$, and adopted minority vote attack as pirate strategy. Then the average running time of the algorithm over 10 trials was 4331.5 seconds, i.e., about 1 hour and 13 minutes, where the calculation of running times was restricted to the case that scores of all users are less than the threshold, as otherwise the tracing algorithm halts before Step 5.

6 Conclusion

In this article, we proposed a new construction of probabilistic 3-secure codes and presented a theoretical evaluation of their error probabilities. A characteristic of our tracing algorithm is to make use of both score comparison and search of the triples of "parents" for a given pirated fingerprint word. Some numerical examples showed that code lengths of our proposed codes are significantly shorter than the previous provably secure 3-secure codes. Moreover, for the sake of improving efficiency of our tracing algorithm, we also proposed an implementation method for the algorithm, which seems indeed more efficient for an average case than the naive implementation. A detailed evaluation of the proposed implementation method will be a future research topic.

Acknowledgments. The author would like to express his deep gratitude to Dr. Teddy Furon, who gave several invaluable comments and suggestions as the shepherd of pre-proceedings revision of this article. Also, the author would like to thank the anonymous referees for their precious comments.

References

1. Blakley, G.R., Kabatiansky, G.: Random Coding Technique for Digital Fingerprinting Codes. In: Proc. IEEE ISIT 2004, p. 202. IEEE, Los Alamitos (2004)
2. Boneh, D., Shaw, J.: Collusion-Secure Fingerprinting for Digital Data. IEEE Trans. Inform. Th. 44, 1897–1905 (1998)
3. Cotrina-Navau, J., Fernandez, M., Soriano, M.: A Family of Collusion 2-Secure Codes. In: Barni, M., Herrera-Joancomartí, J., Katzenbeisser, S., Pérez-González, F. (eds.) IH 2005. LNCS, vol. 3727, pp. 387–397. Springer, Heidelberg (2005)
4. Hoeffding, W.: Probability Inequalities for Sums of Bounded Random Variables. J. Amer. Statist. Assoc. 58, 13–30 (1963)
5. Kitagawa, T., Hagiwara, M., Nuida, K., Watanabe, H., Imai, H.: A Group Testing Based Deterministic Tracing Algorithm for a Short Random Fingerprint Code. In: Proc. ISITA 2008, pp. 706–710 (2008)
6. Nuida, K.: An Improvement of Short 2-Secure Fingerprint Codes Strongly Avoiding False-Positive. In: Katzenbeisser, S., Sadeghi, A.-R. (eds.) IH 2009. LNCS, vol. 5806, pp. 161–175. Springer, Heidelberg (2009)
7. Nuida, K.: Making Collusion-Secure Codes (More) Robust against Bit Erasure. Cryptology ePrint Archive 2009/549. IACR (2009)
8. Nuida, K., Fujitsu, S., Hagiwara, M., Imai, H., Kitagawa, T., Ogawa, K., Watanabe, H.: An Efficient 2-Secure and Short Random Fingerprint Code and Its Security Evaluation. IEICE Trans. Fundamentals E92-A, 197–206 (2009)
9. Nuida, K., Fujitsu, S., Hagiwara, M., Kitagawa, T., Watanabe, H., Ogawa, K., Imai, H.: An Improvement of Discrete Tardos Fingerprinting Codes. Des. Codes Cryptogr. 52, 339–362 (2009)
10. Nuida, K., Hagiwara, M., Kitagawa, T., Watanabe, H., Ogawa, K., Fujitsu, S., Imai, H.: A Tracing Algorithm for Short 2-Secure Probabilistic Fingerprinting Codes Strongly Protecting Innocent Users. In: Proc. IEEE CCNC 2007, pp. 1068–1072. IEEE, Los Alamitos (2007)
11. Škorić, B., Katzenbeisser, S., Celik, M.U.: Symmetric Tardos fingerprinting codes for arbitrary alphabet sizes. Des. Codes Cryptogr. 46, 137–166 (2008)
12. Sebé, F., Domingo-Ferrer, J.: Short 3-Secure Fingerprinting Codes for Copyright Protection. In: Batten, L.M., Seberry, J. (eds.) ACISP 2002. LNCS, vol. 2384, pp. 316–327. Springer, Heidelberg (2002)
13. Tardos, G.: Optimal Probabilistic Fingerprint Codes. J. ACM 55, 1–24 (2008)

Appendix: Proofs of the Propositions

A Proof of Proposition 1

First, we prove the claim 1 of Proposition 1. For each $\mathsf{I} \in U_{\mathrm{I}}$ and $\sigma \in \{\mathrm{H}, \mathrm{L}\}$, let $K_\sigma = \{j \in A_\sigma \mid w_{\mathsf{I},j} = y'_j\}$. We have $S(\mathsf{I}) = |K_{\mathrm{H}}| \log(1/p) + |K_{\mathrm{L}}| \log(1/(1-p))$.

Now note that the choice of y' is independent of w_{I}. This implies that we have $Pr[w_{\mathsf{I},j} = y'_j \mid y'] = p$ for each $j \in A_{\mathrm{H}}$, and we have $Pr[w_{\mathsf{I},j} = y'_j \mid y'] = 1 - p$ for each $j \in A_{\mathrm{L}}$. Hence the conditional probability that $|K_{\mathrm{H}}| = k_{\mathrm{H}}$ and $|K_{\mathrm{L}}| = k_{\mathrm{L}}$, conditioned on this y', is $\binom{a_{\mathrm{L}}}{k_{\mathrm{L}}}(1-p)^{k_{\mathrm{L}}}p^{a_{\mathrm{L}}-k_{\mathrm{L}}}\binom{a_{\mathrm{H}}}{k_{\mathrm{H}}}p^{k_{\mathrm{H}}}(1-p)^{a_{\mathrm{H}}-k_{\mathrm{H}}}$. This implies that $Pr[S(\mathsf{I}) \geq Z \mid y']$ is equal to the left-hand side of (5), therefore the claim 1 holds as there exist at most $N-1$ innocent users I.

Secondly, to prove the claim 2, we use Hoeffding's Inequality:

Theorem 3 ([4], Theorem 2). *Let $X_1, X_2, \dots, X_n$ be independent random variables such that $a_i \leq X_i \leq b_i$ for each i. Let $\overline{X}$ be the average value of $X_1, \dots, X_n$. Then for $t > 0$, we have*

$$Pr[\overline{X} - E[\overline{X}] \geq t] \leq \exp\left(\frac{-2n^2t^2}{\sum_{i=1}^{n}(b_i - a_i)^2}\right) . \tag{12}$$

As mentioned above, the left-hand side of (5) is equal to $Pr[S(\mathsf{I}) \geq Z \mid y']$, where I is any specified innocent user. Now by regarding $S(\mathsf{I})$ as the sum of bit-wise scores $X_1, \dots, X_m$, Theorem 3 implies that

$$Pr[S(\mathsf{I}) - \mu \geq mt \mid y'] \leq \exp\left(\frac{-2m^2t^2}{a_{\mathrm{H}}(\log(1/p))^2 + a_{\mathrm{L}}(\log(1/(1-p)))^2}\right) \tag{13}$$

for $t > 0$, where $\mu = a_{\mathrm{H}}p\log(1/p) + a_{\mathrm{L}}(1-p)\log(1/(1-p))$. See the forthcoming full version of this article for the detail. Now by setting $t = \eta/m$ where

$$\eta = \sqrt{(1/2)\left((\log(1/p))^2 a_{\mathrm{H}} + (\log(1/(1-p)))^2 a_{\mathrm{L}}\right)\log(N/\varepsilon_0)} , \tag{14}$$

the right-hand side of (13) is equal to ε_0/N. For the left-hand side of (13), we have $Pr[S(\mathsf{I}) - \mu \geq mt \mid y'] = Pr[S(\mathsf{I}) \geq \mu + \eta \mid y']$, while $Z_0 = \mu + \eta$. Hence the condition (5) is satisfied, concluding the proof of Proposition 1.

B Proof of Proposition 2

To prove Proposition 2, suppose that it is not the case of Type I–IV errors. We show that tracing error does not occur in this case. Recall that $T_{\mathrm{P}} = 123 \in \mathcal{T}(y')$. By virtue of Step 4 of Tr and the absence of Type I error, it suffices to consider the case that $S(i) < Z$ for every $i \in U$. We have $T_{\mathrm{P}} \in \mathcal{T}'$ by the absence of Type II error. Hence every $T \in \mathcal{T}'$ intersects T_{P}, and $\bigcap \mathcal{T}' \subseteq T_{\mathrm{P}}$. By virtue of Step 6, it suffices to consider the case that $\bigcap \mathcal{T}' = \emptyset$. Now there are the following two cases: (A) we have $|T \cap T_{\mathrm{P}}| = 1$ for some $T \in \mathcal{T}'$; (B) we have $|T \cap T_{\mathrm{P}}| = 2$ for every $T \in \mathcal{T}' \setminus \{T_{\mathrm{P}}\}$.

For case (A), let $T_1 \in \mathcal{T}'$ and $|T_1 \cap T_{\mathrm{P}}| = 1$. By symmetry, we may assume that $T_1 \cap T_{\mathrm{P}} = \{1\}$. By the fact $\bigcap \mathcal{T}' = \emptyset$, there is a $T_2 \in \mathcal{T}'$ such that $1 \notin T_2$. We may assume by symmetry that $2 \in T_2$, as $T_2 \cap T_{\mathrm{P}} \neq \emptyset$. We have $T_1 \cap T_2 \neq \emptyset$ as $T_1 \in \mathcal{T}'$, therefore the absence of Type III error implies that $3 \in T_2$. Put $T_2 = 23\mathsf{I}$ with $\mathsf{I} \in U_{\mathrm{I}}$, and $T_1 = 1\mathsf{I}\mathsf{I}'$ with $\mathsf{I}' \in U_{\mathrm{I}}$. Now the fact $\{T_{\mathrm{P}}, T_1, T_2\} \subseteq \mathcal{T}'$

implies that $\mathcal{P} \subseteq \{12, 13, 1\mathsf{I}, 2\mathsf{I}, 2\mathsf{I}', 3\mathsf{I}, 3\mathsf{I}'\}$, therefore the correctness of the output of Tr in this case can be verified by case-by-case analysis (note that it is not the case of Type II or III error). For the detail of the proof, see the forthcoming full version of this article.

For case (B), as $\bigcap \mathcal{T}' = \emptyset$, there are $\mathsf{I}_1, \mathsf{I}_2, \mathsf{I}_3 \in U_{\mathrm{I}}$ such that $12\mathsf{I}_3, 13\mathsf{I}_2, 23\mathsf{I}_1 \in \mathcal{T}'$. By the absence of Type IV error, it does not hold that $\mathsf{I}_1 = \mathsf{I}_2 = \mathsf{I}_3$. By symmetry, we may assume that $\mathsf{I}_1 \neq \mathsf{I}_2$. Now the fact $\{123, 12\mathsf{I}_3, 13\mathsf{I}_2, 23\mathsf{I}_1\} \subseteq \mathcal{T}'$ implies that $\mathcal{P} \subseteq \{12, 13, 23, 1\mathsf{I}_1, 2\mathsf{I}_2, 3\mathsf{I}_3\}$, while $12, 13, 23 \in \mathcal{P}$ by the assumption of the case (B). Therefore the correctness of the output of Tr in this case can be verified by case-by-case analysis as well (note that it is not the case of Type II or III error). For the detail of the proof, see the forthcoming full version of this article. Hence the proof of Proposition 2 is concluded.

C Proof of Proposition 3

To prove Proposition 3, let I_1, I_2 and I_3 be three distinct innocent users. Given y' and $\mathsf{st} = (p_j)_j$, we introduce the following notation for $j \in [m]$:

$$\xi_j^{\mathrm{H}} = \begin{cases} 1 & \text{if } p_j = p \ , \\ 0 & \text{if } p_j = 1 - p \ , \end{cases} \quad \xi_j^{\mathrm{L}} = 1 - \xi_j^{\mathrm{H}} \ . \tag{15}$$

Note that $A_\sigma = \{j \mid y'_j = \xi_j^\sigma\}$ for $\sigma \in \{\mathrm{H}, \mathrm{L}\}$. We write $A_\sigma = A_\sigma(y', \mathsf{st})$ and $a_\sigma = a_\sigma(y', \mathsf{st})$ to emphasize the dependency on y' and st. Then, as the bits of codewords are independently chosen, we have $Pr[\mathsf{I}_1\mathsf{I}_2\mathsf{I}_3 \in \mathcal{T}(y') \mid y', \mathsf{st}] = (1-p^3)^{a_{\mathrm{L}}}(1-(1-p)^3)^{a_{\mathrm{H}}}$, therefore

$$Pr[\mathsf{I}_1\mathsf{I}_2\mathsf{I}_3 \in \mathcal{T}(y')] = \sum_{y', \mathsf{st}} Pr[y', \mathsf{st}](1-p^3)^{a_{\mathrm{L}}(y', \mathsf{st})}(1-(1-p)^3)^{a_{\mathrm{H}}(y', \mathsf{st})} \ . \tag{16}$$

Now we present the following key lemma:

Lemma 1. *Among the possible pirate strategies ρ, the maximum value of the right-hand side of (16) is attained by the majority vote attack, namely the attack word y for codewords w_1, w_2, w_3 of three pirates satisfies that $y_j = 0$ if at least two of $w_{1,j}, w_{2,j}, w_{3,j}$ are 0 and $y_j = 1$ otherwise.*

The formal proof of Lemma 1 will be given in the forthcoming full version of this article. Intuitively, this lemma follows from the facts that the right-hand side of (16) is getting larger as the value of $a_{\mathrm{H}}(y', \mathsf{st})$ is increasing and that the majority of the three bits $w_{1,j}$, $w_{2,j}$, and $w_{3,j}$ is more likely to be ξ_j^{H} than the minority of the three bits. (Note that the difficulty of the proof of Lemma 1 is due to the fact that *the pirate strategy ρ is not necessarily a column-wise strategy.*)

Owing to Lemma 1, we may assume that ρ is the majority vote attack. Then for each $j \in [m]$, we have $j \in A_{\mathrm{H}}(y', \mathsf{st})$ with probability $3p^2(1-p) + p^3 = 3p^2 - 2p^3$ and $j \in A_{\mathrm{L}}(y', \mathsf{st})$ with probability $1 - 3p^2 + 2p^3$. Now a computation shows that

$$Pr[\mathsf{I}_1\mathsf{I}_2\mathsf{I}_3 \in \mathcal{T}(y')] = (1 - 3p^2 + 10p^3 - 15p^4 + 12p^5 - 4p^6)^m \tag{17}$$

(see the forthcoming full version of this article for the detail). Hence Proposition 3 holds, as there are $\binom{N-3}{3}$ choices of the triple $\mathsf{I}_1, \mathsf{I}_2, \mathsf{I}_3$.

D Proof of Proposition 4

To prove Proposition 4, we fix an innocent user $\mathsf{I}_0 \in U_{\mathrm{I}}$ and consider the probability that there are $T_1, T_2 \in \mathcal{T}(y')$ such that $\mathsf{I}_0 \in T_1 \cap T_2 \subseteq U_{\mathrm{I}}$, $T_1 \cap T_{\mathrm{P}} = \{1\}$ and $T_2 \cap T_{\mathrm{P}} = \{2\}$; or equivalently, there are innocent users $\mathsf{I}_1, \mathsf{I}_2 \in U_{\mathrm{I}} \setminus \{\mathsf{I}_0\}$ such that $1\mathsf{I}_0\mathsf{I}_1 \in \mathcal{T}(y')$ and $2\mathsf{I}_0\mathsf{I}_2 \in \mathcal{T}(y')$. We introduce some notations. Given y', w_1, w_2, w_{I_0}, and $\mathsf{st} = (p_j)_j$, we define, for $\alpha, \beta, \gamma, \delta \in \{\mathrm{H}, \mathrm{L}\}$,

$$a_{\alpha\beta\gamma\delta} = |\{j \in [m] \mid y'_j = \xi_j^{\alpha}, w_{1,j} = \xi_j^{\beta}, w_{2,j} = \xi_j^{\gamma}, w_{\mathsf{I}_0,j} = \xi_j^{\delta}\}| \tag{18}$$

(see (15) for notations). Moreover, by using '$*$' as a wild-card, we extend naturally the definition of $a_{\alpha\beta\gamma\delta}$ to the case $\alpha, \beta, \gamma, \delta \in \{\mathrm{H}, \mathrm{L}, *\}$. For example, we have $a_{\alpha**\delta} = a_{\alpha\mathrm{LL}\delta} + a_{\alpha\mathrm{LH}\delta} + a_{\alpha\mathrm{HL}\delta} + a_{\alpha\mathrm{HL}\delta}$. Note that $a_{\sigma***} = a_{\sigma}$ ($\sigma \in \{\mathrm{H}, \mathrm{L}\}$).

Now we have $Pr[1\mathsf{I}_0\mathsf{I}_1 \in \mathcal{T}(y') \mid y', w_1, w_2, w_{\mathsf{I}_0}, \mathsf{st}] = p^{a_{\mathrm{HL*L}}}(1-p)^{a_{\mathrm{LH*H}}}$ for an innocent user $\mathsf{I}_1 \neq \mathsf{I}_0$. As there are $N-4$ innocent users $\mathsf{I} \neq \mathsf{I}_0$, we have

$$\begin{aligned} &Pr[1\mathsf{I}_0\mathsf{I}_1 \in \mathcal{T}(y') \text{ for some } \mathsf{I}_1 \in U_{\mathrm{I}} \mid y', w_1, w_2, w_{\mathsf{I}_0}, \mathsf{st}] \leq (N-4)p^{a_{\mathrm{HL*L}}}(1-p)^{a_{\mathrm{LH*H}}} \enspace, \\ &Pr[2\mathsf{I}_0\mathsf{I}_2 \in \mathcal{T}(y') \text{ for some } \mathsf{I}_2 \in U_{\mathrm{I}} \mid y', w_1, w_2, w_{\mathsf{I}_0}, \mathsf{st}] \leq (N-4)p^{a_{\mathrm{H*LL}}}(1-p)^{a_{\mathrm{L*HH}}} \enspace . \end{aligned} \tag{19}$$

As $Pr[\mathsf{E}_1, \mathsf{E}_2] \leq \min\{Pr[\mathsf{E}_1], Pr[\mathsf{E}_2]\}$ and $\min\{a, b\} \leq \sqrt{ab}$ for $a, b > 0$, we have

$$\begin{aligned} &Pr[1\mathsf{I}_0\mathsf{I}_1, 2\mathsf{I}_0\mathsf{I}_2 \in \mathcal{T}(y') \text{ for some } \mathsf{I}_1, \mathsf{I}_2 \in U_{\mathrm{I}} \mid y', w_1, w_2, w_{\mathsf{I}_0}, \mathsf{st}] \\ &\leq (N-4)\sqrt{p}^{a_{\mathrm{HL*L}} + a_{\mathrm{H*LL}}} \sqrt{1-p}^{a_{\mathrm{LH*H}} + a_{\mathrm{L*HH}}} \enspace . \end{aligned} \tag{20}$$

By Marking Assumption, a computation shows that

$$\begin{aligned} &Pr[1\mathsf{I}_0\mathsf{I}_1, 2\mathsf{I}_0\mathsf{I}_2 \in \mathcal{T}(y') \text{ for some } \mathsf{I}_1, \mathsf{I}_2 \in U_{\mathrm{I}} \mid y', w_1, w_2, \mathsf{st}] \\ &\leq (N-4)(2p-p^2)^{b_{\mathrm{HLLH}}}(1-p^2)^{b_{\mathrm{LHHL}}}\left(p + (1-p)\sqrt{p}\right)^{b_{\mathrm{HLHL}} + b_{\mathrm{HHLH}} + b_{\mathrm{HLHH}} + b_{\mathrm{HHLL}}} \\ &\quad \cdot \left(1 - p + p\sqrt{1-p}\right)^{b_{\mathrm{LHLH}} + b_{\mathrm{LLHL}} + b_{\mathrm{LHLL}} + b_{\mathrm{LLHH}}} \enspace , \end{aligned} \tag{21}$$

where we put $b_{\alpha\beta\gamma\delta} = |\{j \in [m] \mid y'_j = \xi_j^{\alpha}, w_{1,j} = \xi_j^{\beta}, w_{2,j} = \xi_j^{\gamma}, w_{3,j} = \xi_j^{\delta}\}|$ for $\alpha, \beta, \gamma, \delta \in \{\mathrm{H}, \mathrm{L}\}$. By writing the right-hand side of (21) as η, it follows that

$$Pr[1\mathsf{I}_0\mathsf{I}_1, 2\mathsf{I}_0\mathsf{I}_2 \in \mathcal{T}(y') \text{ for some } \mathsf{I}_1, \mathsf{I}_2 \in U_{\mathrm{I}} \mid \boldsymbol{w}_{\mathrm{P}}] \leq \sum_{y', \mathsf{st}} Pr[y', \mathsf{st} \mid \boldsymbol{w}_{\mathrm{P}}]\eta \enspace , \tag{22}$$

where $\boldsymbol{w}_{\mathrm{P}}$ denotes the collection of w_1, w_2, and w_3. Now an argument similar to Lemma 1 implies that the maximum value of the right-hand side of (22) is attained by the majority vote attack, therefore a computation shows that

$$\begin{aligned} &Pr[1\mathsf{I}_0\mathsf{I}_1, 2\mathsf{I}_0\mathsf{I}_2 \in \mathcal{T}(y') \text{ for some } \mathsf{I}_1, \mathsf{I}_2 \in U_{\mathrm{I}}] \\ &\leq (N-4)\left(p^2(1-p)^2(\sqrt{p} + \sqrt{1-p}) + 1 - p - p^2 + 4p^3 - 2p^4\right)^m \end{aligned} \tag{23}$$

(see the forthcoming full version of this article for the details). Hence Proposition 4 follows by considering the number of choices of I_0 and the pair $1, 2$.

E Proof of Proposition 5

To prove Proposition 5, we fix an innocent user I and suppose that $S(i) < Z$ for every $i \in 123$. Given y', w_1, w_2, w_3, and st, we define, for $\alpha, \beta, \gamma, \delta \in \{\mathrm{H}, \mathrm{L}\}$,

$$a_{\alpha\beta\gamma\delta} = |\{j \in [m] \mid y'_j = \xi_j^{\alpha}, w_{1,j} = \xi_j^{\beta}, w_{2,j} = \xi_j^{\gamma}, w_{3,j} = \xi_j^{\delta}\}| \tag{24}$$

(see (15) for notations). Let $\boldsymbol{w}_{\mathrm{P}}$ denote the triple of w_1, w_2, and w_3. Then we have

$$Pr[12\mathsf{I}, 13\mathsf{I}, 23\mathsf{I} \in \mathcal{T}(y') \mid y', \boldsymbol{w}_{\mathrm{P}}, \mathsf{st}] = p^{a_{\mathrm{HLLH}}+a_{\mathrm{HLHL}}+a_{\mathrm{HHLL}}}(1-p)^{a_{\mathrm{LLHH}}+a_{\mathrm{LHLH}}+a_{\mathrm{LHHL}}} . \tag{25}$$

For $\sigma \in \{\mathrm{H}, \mathrm{L}\}$, let a_{σ}^{u} and a_{σ}^{d} be the number of indices $j \in [m]$ of undetectable and detectable columns, respectively, such that $y'_j = \xi_j^{\mathrm{H}}$. Then an upper bound of the logarithm of the right-hand side of the above equality is derived by Marking Assumption and the relations $S(i) < Z \leq Z_0$ for every $i \in 123$ and $p \geq 1-p$, therefore a computation shows that

$$\begin{aligned} &Pr[12\mathsf{I}, 13\mathsf{I}, 23\mathsf{I} \in \mathcal{T}(y') \mid \boldsymbol{w}_{\mathrm{P}}] \\ &< \sum_{\substack{y', \mathsf{st} \\ S(1), S(2), S(3) < Z}} Pr[y', \mathsf{st} \mid \boldsymbol{w}_{\mathrm{P}}]\eta \leq \sum_{y', \mathsf{st}} Pr[y', \mathsf{st} \mid \boldsymbol{w}_{\mathrm{P}}]\eta , \end{aligned} \tag{26}$$

where

$$\begin{aligned} \eta = &(1-p)^{(3p-1)m}(1-p)^{-3\sqrt{(m/2)\log(N/\varepsilon_0)}} \\ &\cdot \left(p^{3-3p}(1-p)^{1-3p}\right)^{a_{\mathrm{H}}^{\mathrm{u}}}(1-p)^{a_{\mathrm{L}}^{\mathrm{u}}}\left(p^{2-3p}(1-p)^{1-3p}\right)^{a_{\mathrm{H}}^{\mathrm{d}}} \end{aligned} \tag{27}$$

(see the forthcoming full version of this article for the details). Now an argument similar to Lemma 1 implies that the maximum value of the right-hand side of (26) is attained by the majority vote attack, therefore a computation shows that

$$\begin{aligned} &Pr[12\mathsf{I}, 13\mathsf{I}, 23\mathsf{I} \in \mathcal{T}(y')] = \sum_{\boldsymbol{w}_{\mathrm{P}}} Pr[\boldsymbol{w}_{\mathrm{P}}]Pr[12\mathsf{I}, 13\mathsf{I}, 23\mathsf{I} \in \mathcal{T}(y') \mid \boldsymbol{w}_{\mathrm{P}}] \\ &= (1-p)^{-3\sqrt{(m/2)\log(N/\varepsilon_0)}}\left(p^{4-3p}(p^2-3p+3)+(1-p)^{3p+1}(p^2+p+1)\right)^m \end{aligned} \tag{28}$$

(see the forthcoming full version of this article for the details). Hence Proposition 5 follows, as there exist $N-3$ choices of the innocent user I.

F Proof of Proposition 6

Given the codewords w_1 and w_2 of the two pirates 1 and 2, let a_{d} denote the number of detectable columns. Then by Marking Assumption and the choice $p = 1/2$, we have $S(1)+S(2) = (2m-a_{\mathrm{d}})\log 2$ regardless of the pirate strategy ρ. This implies that, if $S(1) < Z$ and $S(2) < Z$, then we have

$$2m - a_{\mathrm{d}} < 2Z/\log 2 \leq 2Z_0/\log 2 = m + \sqrt{2m\log(N/\varepsilon_0)} , \tag{29}$$

therefore $a_{\mathrm{d}} - m/2 > m/2 - \sqrt{2m \log(N/\varepsilon_0)}$ (note that the right-hand side is positive under the condition (9)). By Hoeffding's Inequality (Theorem 3) and the condition (9), this occurs with probability not higher than

$$\exp\left(\frac{-2m^2\left(m/2 - \sqrt{2m\log(N/\varepsilon_0)}\right)^2}{m}\right) < \frac{\varepsilon_0}{N} \tag{30}$$

(see the forthcoming full version of this article for the details). Hence the proof of Proposition 6 is concluded.

Tardos's Fingerprinting Code over AWGN Channel

Minoru Kuribayashi

Graduate School of Engineering, Kobe University
1-1 Rokkodai-cho, Nada-ku, Kobe, Hyogo, 657-8501 Japan
kminoru@kobe-u.ac.jp

Abstract. Binary fingerprinting codes with a length of theoretically minimum order were proposed by Tardos, and the traceability has been estimated under the well-known marking assumption. In this paper, we estimate the traceability and the false-positive probability of the fingerprinting code over AWGN channel, and propose a new accusation algorithm to catch more colluders with less innocent users. The design of our algorithm is based on the symmetric accusation algorithm proposed by Škorić et al. that focuses on the characteristic of the p.d.f. of the correlation scores. The proposed algorithm first estimates the strength of noise added to the code, and then calculates the specific correlation scores among candidate codewords using the characteristic of the noisy channel. The scores are finally classified into guilty and innocent by the threshold obtained from the p.d.f. The performance of the proposed tracing algorithm is evaluated by Monte Carlo simulation.

1 Introduction

Digital fingerprinting is used to trace the illegal users, where a unique ID known as a digital fingerprint [10] is embedded into the content before distribution. When a suspicious copy is found, the owner can identify illegal users by extracting the fingerprint. Since each user purchases contents involving his own fingerprint, the fingerprinted copy slightly differs with each other. Therefore, a coalition of users will combine their different marked copies of the same content for the purpose of removing/changing the original fingerprint. One of the solution is to encode the fingerprint information by a binary code, known as collusion secure code.

An early work on designing collusion-resistant binary fingerprinting codes was presented by Boneh and Shaw [1] underlying the principle referred to as the *marking assumption*. In this case, a fingerprint is a set of redundant digits which are distributed in some random positions of an original content. When a coalition of users attempts to discover some of the fingerprint positions by comparing their marked copies for differences, the coalition may modify only those positions where they find a difference in their fingerprinted copies. A c-secure code guarantees the tolerance for the collusion attack with c pirates or less. Tardos [9] has proposed a probabilistic c-secure code with error probability

R. Böhme, P.W.L. Fong, and R. Safavi-Naini (Eds.): IH 2010, LNCS 6387, pp. 103–117, 2010.

ϵ which has a length of theoretically minimal order with respect to the number of colluders. Many researchers have focused on the characteristics of the code to reduce the code length under the marking assumption. One of the interesting reports is presented by Škorić et al. [2] about the symmetric version of the tracing algorithm and Gaussian approximation. Based on the report, the code length is further shortened under a fixed false-positive probability.

Considering about the realistic situation, a fingerprinting codeword is embedded into digital contents using a watermarking technique. Because of the characteristic of the watermark extraction, the codeword is distorted by signal processing operations as well as the collusion attack. Nuida et al. [7] gave a security proof under an assumption weaker than the marking assumption. The code length was evaluated under the binary symmetric channel with a certain error rate. Once a code length is fixed in an application, however, the important factor is how to detect as many colluders as possible with small and constant false-positive probability.

In this paper, we study the tracing algorithm of Tardos's fingerprinting code under the following two assumptions. One is that a codeword is modulated by BPSK at embedding into digital contents. The other is that a pirated codeword produced by collusion attack is further distorted by transmitting over AWGN channel. Different from the conventional assumption that allows bit flips of the pirated codeword, the addition of white Gaussian noise is more realistic even if the robust watermarking method is applied to embed the fingerprint into digital contents. We first attempt to detect colluders directly from the degraded codeword using soft decision method similar to the detection procedure of error correcting codes. Then, we further reduce the probability of false-positive by classifying the elements of the distorted codeword into reliable ones and the others, and detect colluders with two steps. The first step reduces the candidates of suspicious users using only reliable elements, and the second step further narrows down the suspicious users using the whole codeword by the properly designed threshold which is calculated under the Gaussian assumption of the correlation score. The proposed approach can reduce the probability of false-positive and can detect as many colluders as possible.

2 Preliminaries

2.1 Tardos Code

In this section, we first review the original Tardos's binary fingerprinting code [9]. Let N be the allowable number of users in a fingerprinting system. The Tardos fingerprinting scheme distributes a binary codeword of length L to each user. The codewords are arranged as an $N \times L$ matrix $\boldsymbol{X}$, where the j-th row corresponds to the fingerprint given to the j-th user. The generation of the matrix $\boldsymbol{X}$ is composed of two steps.

1. A distributor is supposed to choose the random variables $0 < p_i < 1$ independently for every $1 \leq i \leq L$, according to a given bias distribution $\boldsymbol{P}$, which satisfies the following conditions.

   - $t = 1/300c$
   - $0 < t' < \pi/4$, $\sin^2 t' = t$, $r_i \in [t', \pi/2 - t']$
   - $p_i = \sin^2 r_i$, $t \le p_i \le 1 - t$

   Where r_i is uniformly and randomly selected from the above range.
2. Each entry $X_{j,i}$ of the matrix $\boldsymbol{X}$ is selected independently from the binary alphabet $\{0, 1\}$ with $\Pr(X_{j,i} = 1) = p_i$ and $\Pr(X_{j,i} = 0) = 1 - p_i$ for every $1 \le j \le N$.

Let $\mathcal{C}$ be a set of colluders and c be the number of colluders. Then we denote by $\boldsymbol{X_C}$ the $c \times L$ matrix of codewords assigned to the colluders. Depending on the attack strategy ρ, the fingerprint $\boldsymbol{y} = (y_1, \ldots, y_L), y_i \in \{0, 1\}$ contained in a pirated copy is denoted by $\boldsymbol{y} = \rho(\boldsymbol{X_C})$. In a tracing (accusation) algorithm $\mathcal{A}$, a correlation score S_j of the j-th user is calculated

$$S_j = \sum_{i=1}^{L} y_i U_{j,i} \ , \tag{1}$$

where

$$U_{j,i} = \begin{cases} \sqrt{\frac{1-p_i}{p_i}} & (X_{j,i} = 1) \\ -\sqrt{\frac{p_i}{1-p_i}} & (X_{j,i} = 0) \ . \end{cases} \tag{2}$$

If S_j exceeds a threshold Z, the j-th user is decided as guilty. The algorithm $\mathcal{A}$ outputs a list of colluders.

Škorić et al. [2] proposed a symmetric version of the correlation score:

$$S_j = \sum_{i=1}^{L} (2y_i - 1) U_{j,i} \ . \tag{3}$$

The traceability are usually evaluated in terms of the probability ϵ_1 of accusing innocent users and the probability of missing all colluders ϵ. In order to guarantee that the probability of accusing innocent users is below ϵ_1, the inequality $L > 2\pi^2 c^2 \log(N/\epsilon_1)$ [2] must be satisfied. The number of traceable colluders depends on the design of threshold Z. Referring to the Central Limit Theorem (CLT) in [2] [3], the distribution of the score S_j for innocent users is approximated to be Gaussian distribution $N(0, L)$ because it is a sum of L elements in Eq.(3). Under the Gaussianity assumption, the probability of false-positive that the j-th innocent user is accused is represented as follows.

$$\Pr\left(S_j > Z, j \in \mathcal{I}\right) = \frac{\epsilon_1}{N} = \frac{1}{2} \mathrm{erfc}\left(\frac{Z}{\sqrt{2L}}\right) \ , \tag{4}$$

where $\mathcal{I}$ stands for a set of innocent users and erfc() is the complementary error function. Hence, the threshold Z is written by a given ϵ_1 [5]:

$$Z = \sqrt{2L} \cdot \mathrm{erfc}^{-1}\left(\frac{2\epsilon_1}{N}\right) \ . \tag{5}$$

The validity of such a threshold Z is not assured because the use of the CLT is not recommended in statistics (i.e. integral over the tail of a p.d.f.). Instead of the use of CLT, conventional schemes calculated it based on the Chernoff's bound, union bound, etc. sacrificing the tightness of the upper bound. In this paper, we evaluate the validity of the threshold Z in Eq.(5) using Monte Carlo simulation and address the insight for the recommendation of the use of CLT.

At the collusion attack, c colluders try to find the positions of the embedded codeword from differences of their copies, and then to modify bits of the codeword in these positions. This attack model is called marking assumption formulated as follows.

Let we say that position i is undetectable for colluders in $\mathcal{C}$ if the codewords assigned to c colluders in $\mathcal{C}$ match in i-th position. Then, $y_i = X_{j,i}$ for any $j \in \mathcal{C}$. Under the marking assumption, colluders have no information on the i-th position of innocent users if it is undetectable.

2.2 Relaxation of Marking Assumption

Suppose that a codeword of fingerprint codes is binary and each bit is embedded into one of the segments of digital content without overlapping using a robust watermarking scheme. It is possible for malicious users, called colluders, to compare their fingerprinted copies of the content with each other to find the differences. In the situation, the positions that the bit of their codewords is different are detectable. The marking assumption states that any bit within a detectable position can be selected or even erased, while any bit without the position will be left unchanged in the pirated codeword. A fingerprint code is called totally c-secure if at least one of the colluders is traceable under the marking assumption with the condition that the number of colluders is at most c. Boneh and Shaw, however, proved that when $c > 1$, totally c-secure codes do not exist if the marking assumption is satisfied [1]. Under the weaker condition that one of innocent users will be captured with a tiny probability ϵ, a c-secure code with ϵ-error was constructed. Since then, the study of c-secure code has been investigated under the marking assumption. Although the assumption is reasonable to evaluate the performance of fingerprint codes, there is a big gap from practical cases as follows. Even if a watermarking scheme offers a considerable level of robustness, it is still possible to erase/modify the embedded bits with a non-negligible probability due to the addition of noise to a pirated copy. Therefore, the bits at the undetectable positions as well as the detectable ones may be erased/modified by the attacks for the watermarked signal.

In order to cover more practical cases, various relaxation of marking assumption have been introduced and several c-secure codes under those assumptions, called robust c-secure codes, have been proposed in [7], [4], [8], [6]. Among those assumptions, there are two common conditions: At least one of the colluders is traceable and the number of colluders is at most c. Their goal is mainly to estimate a proper code length L to satisfy that the probability of accusing an innocent user is below ϵ_1, which is dependent on the number of flipped bits at the undetectable position. Although the study of such a code length is meaningful, there is still a difficulty to adapt the fingerprint codes in a system. When the number of colluders is more than c, the false probability may be increased. Even more, the dependency with the number of flipped bits lefts the uncertainty of the code length. Colluders can take a stronger attack strategy for the underlying watermarking scheme in order to affect more the embedded fingerprint code, which is an unavoidable feature of watermarking schemes.

From the different viewpoint, it is an interesting challenge to design a proper threshold Z for a given false probability ϵ_1 under a fixed code length. Based on the CLT, for a given false probability ϵ_1, the threshold Z is calculated by Eq.(5). The study in [5] shows the detectable number of colluders using such a threshold. So, it is also interesting to estimate how many colluders will be able to be caught by a tracing algorithm. To the best of our knowledge, no report about the detectable number of colluders has been presented under a relaxed version of the marking assumption.

Suppose that a fingerprint code is equipped in a fingerprinting system. Then, the code length must be determined under the considerations of system policy and attack strategies such as the number of colluders and the amount of noise. Here, our interest is how to design the good tracing algorithm that can detect more colluders and less innocent users no matter how many colluders get involved in to generate a pirated copy and no matter how much amount of noise is added to the copy.

3 Performance of Tracing Algorithm

In this section, we forget about the limitation of c-secure code such that the number of colluders is less than c. The performance of conventional tracing algorithm based on a threshold Z and its variant are evaluated for arbitrary number $\tilde{c}$ of colluders.

3.1 Channel Model

The conventional analysis considers the case that colluders change several symbols of a pirated codeword in an attempt to attack directly a pirated copy. Regretfully, the attack model is merely a bit flip. If each symbol of codeword is embedded into digital contents assisted by a watermarking technique, the extracted symbol from a pirated copy must contain noise caused by attacks intended to remove/modify the symbol. In such a case, it is a reasonable assumption that the symbol is disturbed by additive white Gaussian noise, namely the codeword is transmitted through AWGN channel.

Let $\boldsymbol{y} = (y_1, \dots, y_L)$ be the fingerprint produced from $\tilde{c}$ colluders' codes under the marking assumption. Namely, the fingerprint is represented by a binary code. Here, we assume that a binary fingerprint code is embedded into digital contents after the BPSK (Binary Phase Shift Keying) modulation, hence, $\hat{\boldsymbol{y}} = (\hat{y}_1, \dots, \hat{y}_L)$, where $\hat{y}_i \in \{-1, 1\}$. The codeword is transmitted through AWGN channel.

1. BPSK modulation
   In the tracing algorithm of Eq.(3), each symbol of the pirated codeword is modulated into two kinds of symbols $\{-1, 1\}$ to calculate the correlation score S_j. Since the modulation can be performed at the embedding, we assume that a binary fingerprint codeword is modulated by BPSK before embedding.
2. AWGN channel
   Even if a robust watermarking method is used to embed the binary fingerprint code into digital contents, it must be degraded by attacks. In our assumption the effects caused by attacks are modeled by additive white Gaussian noise, and the noise is added after collusion attack. The degraded fingerprint codeword is represented by

$$\hat{\boldsymbol{y}}' = \hat{\boldsymbol{y}} + \boldsymbol{e} \ , \tag{6}$$

   where $\boldsymbol{e}$ is the additive white Gaussian noise.

3.2 Hard and Soft Decision

The signal extracted from a pirated copy is represented by analog value $\hat{\boldsymbol{y}}'$. At the tracing algorithm in Eq.(3), however, the codeword y_i of a pirated copy must be binary bit in $\{0, 1\}$. If it is not binary, the design of threshold Z in Eq.(5) is not valid. A simple solution is to quantize $\hat{y}'_i$ into "-1" if $\hat{y}'_i < 0$, otherwise "1". In this solution, an extracted signal is first quantized into digital value, and then the tracing algorithm is performed to identify the colluders. This solution is analogous to the hard decision (HD) method in error correcting code. Here, there is an interesting question whether a soft decision (SD) method is applicable to the tracing algorithm by adaptively designing a proper threshold or not. In general, the performance of SD method is much better than the HD method in error correcting code.

The design of threshold in Eq.(5) is based on the Gaussian approximation of the score S_j. Referring to the CLT, the variance of S_j is L, and hence, the proper threshold Z_{HD} is calculated by the Eq.(5).

$$Z_{HD} = \sqrt{2L} \cdot \mathrm{erfc}^{-1}\left(\frac{2\epsilon_1}{N}\right) \tag{7}$$

Since a pirated codeword is distorted by AWGN channel, the effect on S_j is also approximated to Gaussian. Hence, if the variance σ_{SD}^2 of S_j using the SD method is obtained, the proper threshold Z_{SD} can be designed using the same equation as the case of HD method:

$$Z_{SD} = \sqrt{2\sigma_{SD}^2} \cdot \mathrm{erfc}^{-1}\left(\frac{2\epsilon_1}{N}\right) . \tag{8}$$

Because of the randomness in the generation of codeword, the variance σ_{SD}^2 can be calculated as follows.

1. Generate N' fingerprint codewords $X_{j',i}$ for $j' \notin \{1, \ldots, N\}$.
2. Calculate the correlation scores S'_j.
3. Compute the variance of S'_j, and output it as σ^2_{SD}.

The generated N'codewords $X_{j',i}$ are statistically uncorrelated with the pirated codeword. If N' is sufficiently large, a proper variance can be obtained by the above procedure, and finally, a proper threshold Z_{SD} is derived.

3.3 Numerical Comparison

For the comparison of the performance of HD and SD methods, the number of detected colluders and false-positive probability is plotted in Fig.1 and Fig.2, respectively. The number of users is $N = 10^4$, the code length is $L = 10^4$, SNR is fixed by 8 [dB], and the number of trials for Monte Carlo simulation is 10^5 in this experiment. In addition, the range of bias distribution p_i is given by setting $t = 0.000167$ $(c = 20)$. In the SD method, the number of codewords to calculate σ^2_{SD} is $N' = 10^3$. In this experiment, we check the validity of the use of CLT to set the thresholds Z_{HD} and Z_{SD} from the targeted false-positive probability point of view, which is designed by $\epsilon_1 = 10^{-4}$.

From the Fig.1, we can see that the HD method detects more colluders than the SD method. On the other hand, the false-positive probability of HD method is much higher than that of SD method. It is because several bits are flipped in the HD method by white Gaussian noise. Since the SD method calculates Z_{SD} according to the distribution of S'_j, the false-positive probability is not so degraded. However, such a threshold Z_{SD} is not always valid because the Gaussian assumption of the distribution of S'_j becomes invalid when SNR is decreased. Under the constant number of colluders, the number of detected colluders and the false-positive probability are evaluated by changing the amount of noise. Figures 3 and 4 show the results when the number of colluders is 10. Regardless of the amount of noise, the traceability of HD method is better than that of SD method. It is noticed that the false-positive probability approaches 10^{-5} for both methods when SNR is increased. It is because the probability is

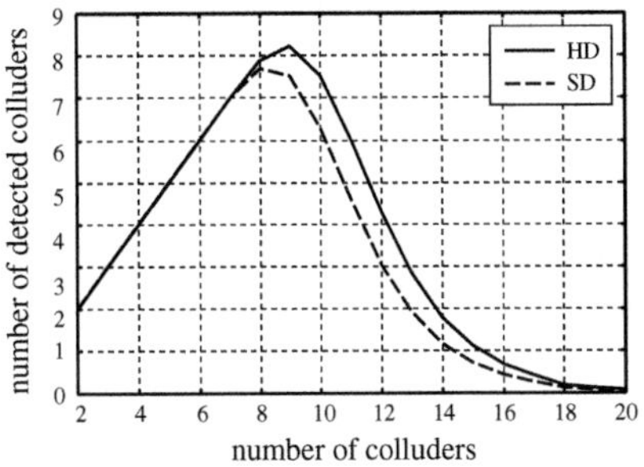

Fig. 1. The number of detected colluders when SNR is 8 [dB]

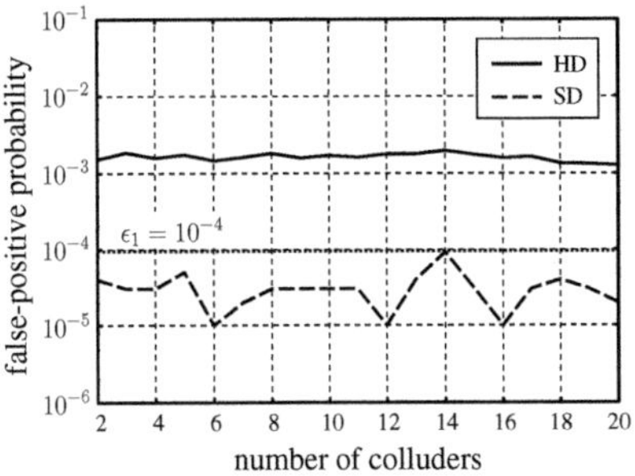

Fig. 2. The false-positive probability when SNR is 8 [dB]

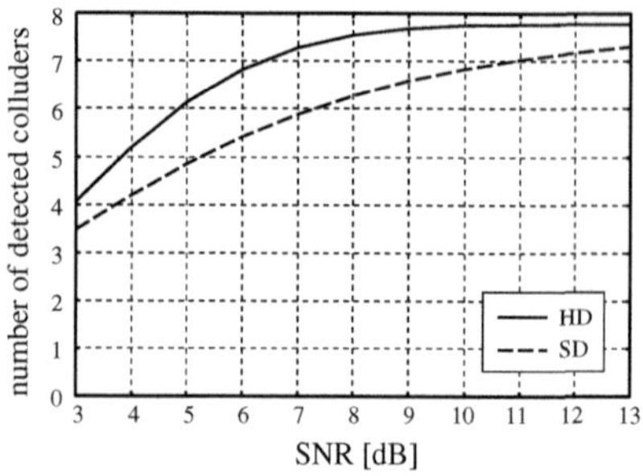

Fig. 3. The number of detected colluders versus SNR when $\tilde{c} = 10$

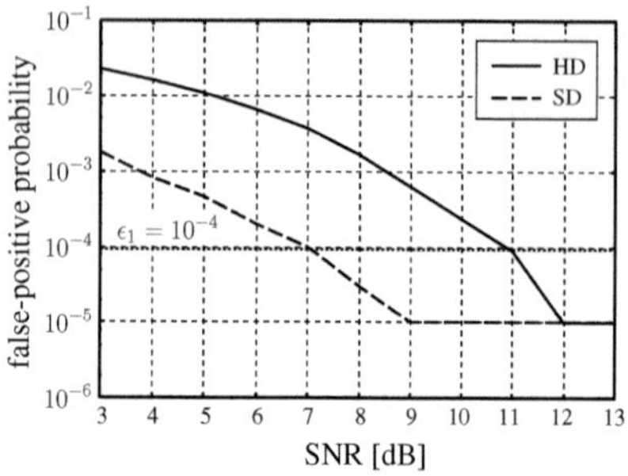

Fig. 4. The false-positive probability versus SNR when $\tilde{c} = 10$

10^{-5} when a pirated copy is not distorted by noise under the above conditions. From these results, the use of CLT seems to be valid only when the amount of noise is very small. Meanwhile, the result in Fig.4 shows the significant property of both methods such that the probability of false-positive is increased with the decrease of SNR. It also means that the increase of the probability is strongly dependent on the number of flipped bits at a pirated codeword. It is remarkable that the increasing rate of the probability of false-positive for the conventional tracing algorithms that use a threshold to determine the guilty must be similar to the above results because such a threshold is independent on the noise and bit flips.

4 Proposed Tracing Algorithm

The number of flipped bits is increased with the noise energy because the probabilities $\Pr(\hat{y}_i = 1 \cap e_i < -1)$ and $\Pr(\hat{y}_i = -1 \cap e_i > 1)$ become non-negligible. In such a case, the designed threshold Z_{HD} is not valid. In the HD method, the degraded signal $\hat{y}'_i$ is classified into only two symbols "-1" and "1" if $\hat{y}'_i$ is more than 0 or not. Considering the variance σ_e^2 of Gaussian noise, the classification should be adaptively modified by a threshold $T_{\sigma_e^2}$ that classifies $\hat{y}'_i$ into three symbols "-1", "1", and "0".

4.1 Channel Estimation

After extracting the fingerprint signal from a pirated copy, we estimate the variance of Gaussian noise using the extracted analog values. Using the variance σ_{SD}^2, the threshold $T_{\sigma_e^2}$ is obtained accordingly.

If $|y'_i| \geq 1$, then the absolute value of the amplitude of noise is estimated as

$$|e_i| = |\hat{y}'_i| - 1 \ , \tag{9}$$

because

$$\Pr(\hat{y}_i = 1 | \hat{y}'_i > 1) \gg \Pr(\hat{y}_i = -1 | \hat{y}'_i > 1) \ , \tag{10}$$

and

$$\Pr(\hat{y}_i = -1 | \hat{y}'_i < -1) \gg \Pr(\hat{y}_i = 1 | \hat{y}'_i < -1) . \tag{11}$$

The probability P_{err} that the estimation of Eq.(9) is failed can be calculated from $\Pr(\hat{y}_i = -1|\hat{y}'_i > 1)$ and $\Pr(\hat{y}_i = 1|\hat{y}'_i < -1)$, and is represented by

$$P_{err} = \frac{1}{2} \Pr\left((e_i < -2) \cup (e_i > 2)\right) = \frac{1}{2} \Pr(|e_i| > 2) . \tag{12}$$

Considering the property of AWGN channel, the probability $\Pr(|e_i| > 2)$ can be calculated by

$$\Pr(|e_i| > 2) = \mathrm{erfc}\left(\frac{2}{\sqrt{2\sigma_e^2}}\right) , \tag{13}$$

where σ_e^2 is the variance of noise. Hence, when σ_e^2 is not very large, P_{err} is negligible and the estimation of Eq.(9) is valid. In such a case, σ_e^2 can be calculated by

$$\sigma_e^2 = \frac{\sum_{i \in \{|\hat{y}'_i| \geq 1\}} (e_i - \overline{e})^2}{L_e} , \tag{14}$$

where $\overline{e}$ is the average value of $e_i, i \in \{|\hat{y}'_i| \geq 1\}$ and L_e is the number of $\hat{y}'_i$ satisfies $|\hat{y}'_i| \geq 1$.

When the variance σ_e^2 of Gaussian noise is very small, the probability $\Pr(\hat{y}'_i > 0|\hat{y}_i = -1)$ is negligible, and then the threshold for the classification is $T_{\sigma_e^2} = 0$. Suppose that the probability is non-negligible. For a given threshold $T_{\sigma_e^2}$, the probability P_{flip} that at least one symbol is flipped is calculated by

$$P_{flip} = \frac{1}{2}\mathrm{erfc}\left(\frac{T_{\sigma_e^2}}{\sqrt{2\sigma_e^2}}\right) . \tag{15}$$

If the code length is L, the average number of flipped bits is $P_{flip}L$. By transforming Eq.(15), the threshold $T_{\sigma_e^2}$ is calculated as follows.

$$T_{\sigma_e^2} = \sqrt{2\sigma_e^2} \cdot \mathrm{erfc}^{-1}(2P_{flip}) \tag{16}$$

Using Eq.(16), we can calculate the threshold $T_{\sigma_e^2}$ for the given probability P_{flip}.

4.2 Adaptive Tracing Algorithm

It is reasonable to apply the HD method for $|\hat{y}'_i| \geq T_{\sigma_e^2}$ because they are judged as the symbols "-1" or "1" with high probability $1 - P_{flip}$. Hence, such elements are reliable to calculate correlation scores, while the other elements, $|\hat{y}'_i| < T_{\sigma_e^2}$, are unreliable. Considering the property derived from the result in Sect.3.3, the unreliable elements should be avoided in order to exclude the bit flips in a pirated codeword. Thus, we replace the elements of pirated codeword $\hat{y}'_i$ with $Y_i^{(1)}$ as follows.

$$Y_i^{(1)} = \begin{cases} 1 & (\hat{y}'_i \geq T_{\sigma_e^2}) \\ -1 & (\hat{y}'_i \leq -T_{\sigma_e^2}) \\ 0 & (|\hat{y}'_i| < T_{\sigma_e^2}) \end{cases} \tag{17}$$

Notice that the above replacement implies the binary erasure channel (BEC).

Let $L^{(1)}$ be the number of $|\hat{y}'_i| \geq T_{\sigma_e^2}$ and $Z_{HD}^{(1)}$ be the proper threshold for the determination of guilty. We first calculate the correlation score $S_j^{(1)}$:

$$S_j^{(1)} = \sum_{i=1}^{L} Y_i^{(1)} U_{j,i} \ . \tag{18}$$

Then, $L^{(1)}$ is derived by counting the number of elements $|\hat{y}'_i| \geq T_{\sigma_e^2}$, which is corresponding to the variance of $S_j^{(1)}$. For the given probability $\epsilon_2^{(1)}$ such that an j-th innocent user gets accused, the threshold $Z_{HD}^{(1)}$ is calculated as follows.

$$Z_{HD}^{(1)} = \sqrt{2L^{(1)}} \cdot \text{erfc}^{-1}\left(2\epsilon_2^{(1)}\right) \tag{19}$$

Because of the randomness of the Gaussian noise, the reliable elements is regarded as the elements of sub-codeword with length $L^{(1)}$. The traceability of the above method is lower than the original HD method because the length of sub-codeword is reduced to $L^{(1)}(\leq L)$, though the increase of the false-positive probability is limited. We denote this method by "method I".

At the method I, only reliable elements $|\hat{y}'_i| \geq T_{\sigma_e^2}$ are selected for the tracing algorithm. It does not mean that the other elements are useless for the judgment. They also contain useful information to improve the traceability though they may increase the probability of false-positive. In order to extract as much information as possible without sacrificing the probability of false-positive, we propose a new tracing algorithm which consists of two stages. First, suspicious users are listed up using the method I by setting $\epsilon_2^{(1)}$ higher to allow the false-positive at this stage. Then, the elements $|\hat{y}'_i| < T_{\sigma_e^2}$ are classified into two symbols "-1" and "1", and the others are changed 0. The replaced elements $Y_i^{(2)}$ are represented as follows;

$$Y_i^{(2)} = \begin{cases} 1 & (0 \leq \hat{y}'_i < T_{\sigma_e^2}) \\ -1 & (-T_{\sigma_e^2} < \hat{y}'_i < 0) \\ 0 & (|\hat{y}'| \geq T_{\sigma_e^2}) \end{cases} \tag{20}$$

Only for the suspicious users whose scores are $S_j^{(1)} > Z_{HD}^{(1)}$, the correlation scores $S_j^{(2)}$ are calculated as follows.

$$S_j^{(2)} = S_j^{(1)} + \sum_{i=1}^{L} Y_i^{(2)} U_{j,i} \tag{21}$$

Finally, j-th user is judged guilty if $S_j^{(2)} > Z_{HD}$, where the threshold Z_{HD} is given by Eq.(7). We denote this method by "method II".

In the method II, we first detect suspicious users from all users using the sub-codeword under the criterion that their sub-codewords retain high correlation with that of pirated codeword. The sub-codeword of pirated codeword is composed of reliable elements $|\hat{y}'_i| \geq T_{\sigma_e^2}$ and the number of bit flips caused by the additive noise is only $P_{flip}L$ in the sub-codeword. If $P_{flip}L$ is small, the number of innocent users involved in the detected suspicious users is expected to be $\epsilon_2^{(1)} N$. In such a case, even if some innocent users are accidentally detected at the first stage, the second stage excludes such innocent users with high probability. In addition, without loss of generality, the following relation is satisfied.

$$\Pr\left(S_j^{(2)} > Z_{HD} | S_j^{(1)} > Z_{HD}^{(1)}, j \in \mathcal{I}\right) > \Pr(S_j > Z_{HD}, j \in \mathcal{I}) , \qquad (22)$$

where $\mathcal{I}$ stands for a set of innocent users. Therefore, the method II can reduce the probability of false-positive effectively.

The performance of method II depends on the selection of $\epsilon_2^{(1)}$ and P_{flip} as the number of suspicious users detected by the tracing algorithm is controlled by these parameters. Remember that it is desirable to keep the number of bit flips $P_{flip}L$ in a sub-codeword as small as possible from the result in Sect.3.3.

5 Experimental Results

We implement the proposed tracing algorithms and evaluate the collusion-resistance. Under the marking assumption, $\tilde{c}$ codewords are randomly chosen and majority attack is performed to produce a pirated codeword $\hat{\boldsymbol{y}}$. After the collusion attack, white Gaussian noise is added to the pirated codeword $\hat{y}'_i$ and try to detect as many colluders as possible from the degraded codeword $\hat{y}'_i$. The length of codeword is $L = 10^4$ and the range of bias distribution p_i is given by setting $t = 0.000167$ ($c = 20$). The number of users is $N = 10^4$, and the false-positive probability is $\epsilon_1 = 10^{-4}$. For the design of the threshold $T_{\sigma_e^2}$, the probability is fixed to $P_{flip} = 10^{-4}$, which average number of flipped bits is $P_{flip}L = 1$. The number of trials for Monte-Carlo simulation is 10^5.

First, the number of detected colluders is evaluated for various SNR of AWGN channel with a fixed number of colluders $\tilde{c} = 10$. As evaluated in Sect.3.3, the number of detected colluders of SD method is lower than that of HD method, so we compare the performance of proposed methods with the HD method, which results are shown in Fig.5 and Fig.6. We can see that the performance of the method I is much lower than the others. It is because of the short code length $L^{(1)} \leq L$. The method II improves the performance compared with the method I. When the threshold $Z_{HD}^{(1)}$, which value is dependent on the given probability $\epsilon_2^{(1)}$, is small, the number of suspicious users is increased, and hence, the number of detectable colluders is improved as shown by the two cases $\epsilon_2^{(1)} = 10^{-3}$ and $\epsilon_2^{(1)} = 10^{-4}$.

For the comparison with the HD method, the probabilities of false-positive are shown in Fig.6. Although the probabilities of method I and method II are slightly growing up with the increase of the amount of noise, the increasing rate is much smaller than that of HD method. Compared with method II, we can see that the false-positive probability of method I is monotonically growing up. It is because of the following reason. P_{err} becomes large with the decrease of SNR, and hence, the variance σ_e^2 estimated by Eq.(14) becomes smaller than the

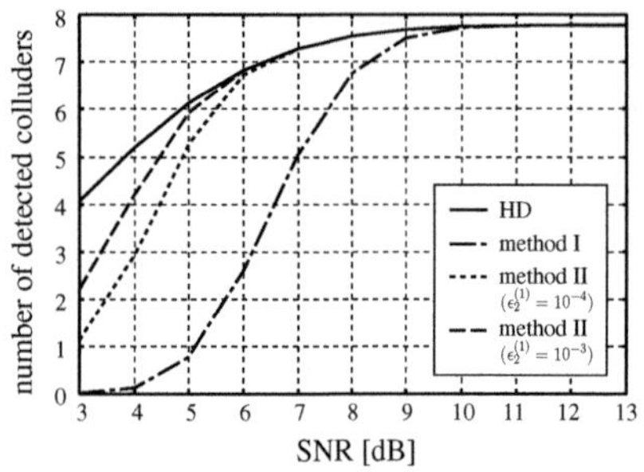

Fig. 5. The number of detected colluders, where $\tilde{c} = 10$ and $\epsilon_1 = 10^{-4}$

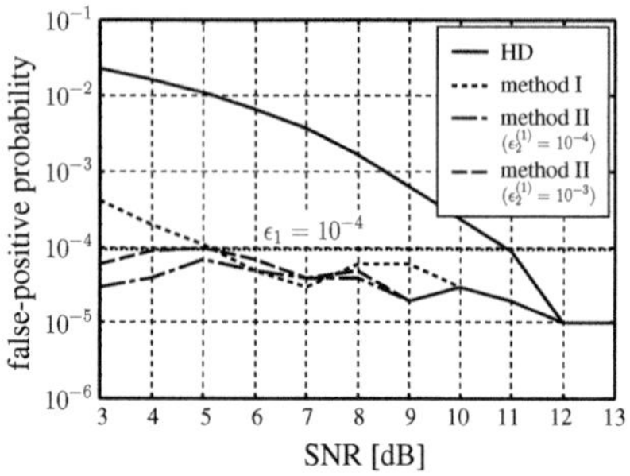

Fig. 6. The probability of false-positive, where $\tilde{c} = 10$ and $\epsilon_1 = 10^{-4}$

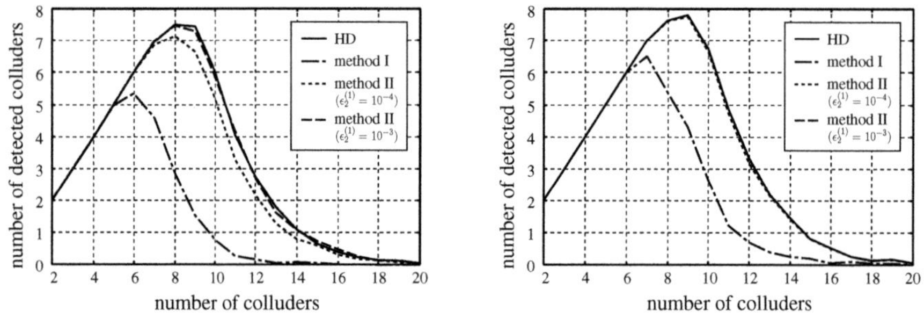

(a) SNR 5 [dB] (b) SNR 6 [dB]

Fig. 7. Comparison of the number of detected colluders for various SNR

actual one. It causes the error on the estimation of the threshold $T_{\sigma_e^2}$. Since the derived threshold is smaller than the actual one, the probability P_{flip} becomes large. As the result, the number of flipped bits are increased and accordingly the probability of false-positive is increased. It is remarkable that the error on the estimation of σ_e^2 is almost canceled at the final determination of guilty in method II, which is conformed by the experimental results.

As the reference data, the comparison of the number of detected colluders is shown in Fig.7 by changing SNR, where the number of trials is 10^2 times. When SNR is 5 [dB], the performance of method II with $\epsilon_2^{(1)} = 10^{-3}$ is better than that of method II with $\epsilon_2^{(1)} = 10^{-4}$ because of the difference in the number of detected suspicious users. When SNR is 6 [dB], no remarkable difference of the performance is appeared, and they are approaching to the lines of HD method. Referring to the results in Fig.6, it is confirmed that the false-positive probability of method II is strongly dependent on the design of $\epsilon_2^{(1)}$.

For the evaluation of stability of false-positive probability, the number of colluders $\tilde{c}$ is increased to produce a pirated codeword, and after the addition of noise, it is input to the proposed tracing algorithm. The probability of false-positive for various number of colluders is plotted in Fig.8. We can see that

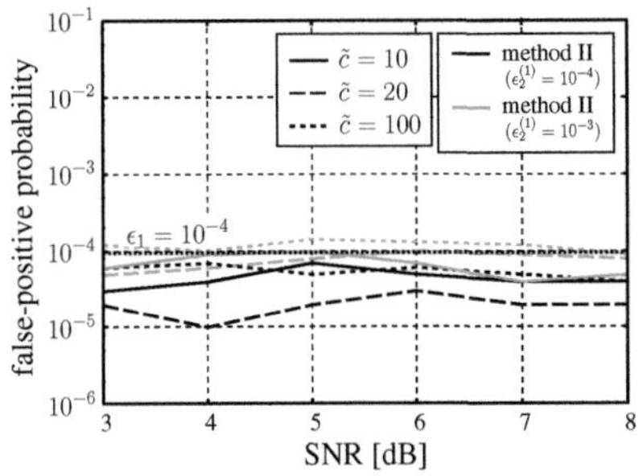

Fig. 8. The probability of false-positive for various number of colluders

the probabilities are almost within a small range even if the number of colluders is changed. We also evaluate the probability of false-positive for various kinds of collusion attacks, which results are shown in Table 1. The table confirms that the probability of false-positive is not dependent on the attack strategy.

The performance of the proposed tracing algorithm is further evaluated for various kinds of parameters. Table 2 and Table 3 show the number of detected colluders and the probability of false-positive under a constant number of colluders $\tilde{c}$ when the allowable numbers of users in a fingerprinting system are $N = 10^5$ and $N = 10^6$, respectively. Due to the limitation of computational resources, the number of trials for Monte Carlo simulation is 10^5 and 10^4 for $N = 10^5$ and 10^6, respectively. In addition, we use the probabilities $\epsilon_1 = 10^{-4}$ and $\epsilon_1 = 10^{-3}$ to keep the precision of the derived probability of false-positive. We set $\epsilon_2^{(1)}$ under the policy that the number of innocent users in the detected suspicious users is 10 in average. From these tables, we can see that the proposed tracing algorithm with the above given parameters detects many colluders with less innocent users, and the probability of false-positive is very close to the designed probability ϵ_1. The comparisons of the number of detected colluders and the probability of false-positive are shown in Fig.9 and Fig.10, respectively. The solid line represents the result of the case that the length is $L = 10000$ and the number of colluders is $\tilde{c} = 10$, the dashed line is the case with $L = 5000$ and $\tilde{c} = 7$, and the dotted

Table 1. Comparison of false-positive probability for various kinds of collusion attacks, where $N = 10^4$, $\tilde{c} = 10$, $L = 10000$, $\epsilon_2^{(1)} = 10^{-3}$, and $\epsilon_1 = 10^{-4}$

SNR [dB]	tracing algorithm	collusion attack				
		majority	minority	random	All-0	All-1
5	HD	111.6×10^{-4}	111.9×10^{-4}	105.2×10^{-4}	113.3×10^{-4}	110.6×10^{-4}
	method II	0.9×10^{-4}	1.2×10^{-4}	0.7×10^{-4}	1.2×10^{-4}	0.7×10^{-4}
8	HD	17.1×10^{-4}	16.0×10^{-4}	17.2×10^{-4}	16.6×10^{-4}	15.6×10^{-4}
	method II	0.5×10^{-4}	1.0×10^{-4}	0.7×10^{-4}	0.8×10^{-4}	0.4×10^{-4}
13	HD	0.1×10^{-4}	0.7×10^{-4}	0.2×10^{-4}	0.2×10^{-4}	0.4×10^{-4}
	method II	0.1×10^{-4}	0.7×10^{-4}	0.2×10^{-4}	0.2×10^{-4}	0.4×10^{-4}

Table 2. The number of detected colluders under a constant number of colluders $\tilde{c} = 10$ when the allowable number of users is expanded, where the code length is $L = 10^4$

(a) $N = 10^5$, $\epsilon_2^{(1)} = 10^{-4}$, and $\epsilon_1 = 10^{-4}$

SNR [dB]	HD	method II
5	4.62	4.17
8	6.21	6.21
13	6.50	6.50

(b) $N = 10^6$, $\epsilon_2^{(1)} = 10^{-5}$, and $\epsilon_1 = 10^{-3}$

SNR [dB]	HD	method II
5	4.57	4.12
8	6.16	6.16
13	6.45	6.45

Table 3. The probability of false-positive under a constant number of colluders $\tilde{c} = 10$ when the allowable number of users is expanded, where the code length is $L = 10^4$

(a) $N = 10^5$, $\epsilon_2^{(1)} = 10^{-4}$, and $\epsilon_1 = 10^{-4}$

SNR [dB]	HD	method II
5	871.6×10^{-4}	3.5×10^{-4}
8	135.7×10^{-4}	3.7×10^{-4}
13	1.1×10^{-4}	1.1×10^{-4}

(b) $N = 10^6$, $\epsilon_2^{(1)} = 10^{-5}$, and $\epsilon_1 = 10^{-3}$

SNR [dB]	HD	method II
5	910.8×10^{-3}	2.6×10^{-3}
8	146.5×10^{-3}	2.2×10^{-3}
13	0.5×10^{-3}	0.5×10^{-3}

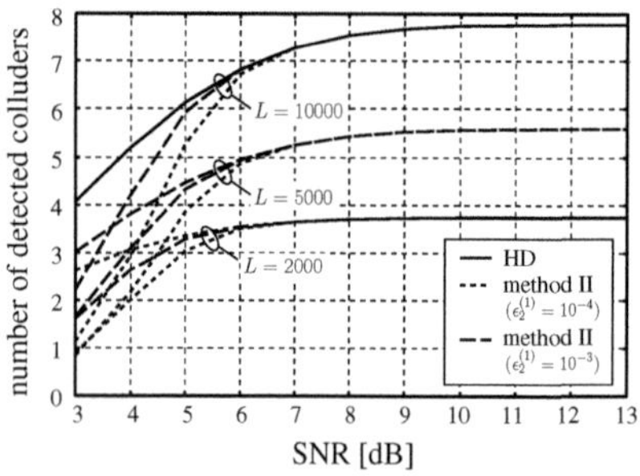

Fig. 9. Comparison of the number of detected colluders, where $N = 10^4$ and $\tilde{c} = 10, 7, 4$ for the codes with length $L = 10000, 5000, 2000$, respectively

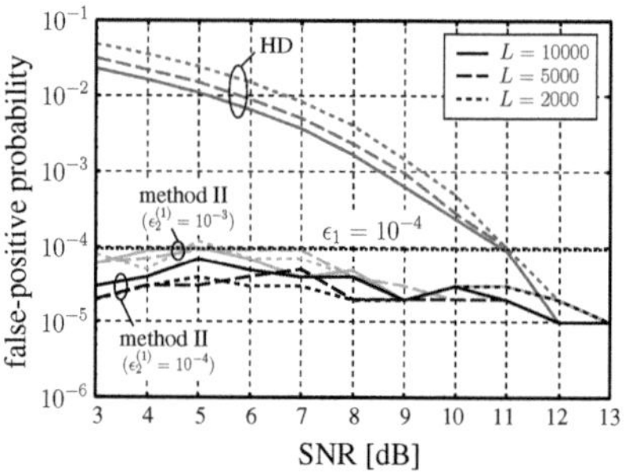

Fig. 10. Comparison of the false-positive probability, where $N = 10^4$ and $\tilde{c} = 10, 7, 4$ for the codes with length $L = 10000, 5000, 2000$, respectively

line is the case with $L = 2000$ and $\tilde{c} = 4$. From Fig.9, it is confirmed that the performance of method II is degraded from that of HD method when SNR drops to less than 6 [dB]. The probability of false-positive is, however, much smaller than that of HD method, and it almost keeps within a small range even if the length is changed.

6 Conclusion

In this paper, we relaxed the marking assumption to consider more realistic situation. In our attack model, a pirated codeword is modulated by BPSK, and is degraded by additive white Gaussian noise after performing collusion attack.

Considering the watermarking technique, the extracted codeword from the pirated copy is represented by analog values. To accommodate with the degradation caused by the noise, the proposed tracing algorithm first estimates the amount of noise injected to a channel, and then, detects as many colluders as possible. In order not to increase the probability of false-positive, the proposed algorithm classify the elements of the codeword into reliable ones and the others and detect suspicious users using the former ones with the threshold calculated under the Gaussian assumption of the correlation score. Then, among the suspicious users, the proposed algorithm narrow down the suspicious users using the whole codeword with the corresponding threshold. From the simulation results, it is confirmed that the proposed tracing algorithm can detect many colluders with less innocent users.

Acknowledgment

This research was partially supported by the Ministry of Education, Culture, Sports Science and Technology, Grant-in-Aid for Young Scientists (B) (21760291), 2010.

References

1. Boneh, D., Shaw, J.: Collusion-secure fingerprinting for digital data. IEEE Trans. Inform. Theory 44(5), 1897–1905 (1998)
2. Škorić, B., Vladimirova, T.U., Celik, M., Talstra, J.C.: Tardos fingerprinting is better than we thought. IEEE Trans. Inform. Theory 54(8), 3663–3676 (2008)
3. Furon, T., Guyader, A., Cérou, F.: On the design and optimization of Tardos probabilistic fingerprinting codes. In: Solanki, K., Sullivan, K., Madhow, U. (eds.) IH 2008. LNCS, vol. 5284, pp. 341–356. Springer, Heidelberg (2008)
4. Guth, H.J., Pfitzmann, B.: Error- and collusion-secure fingerprinting for digital data. In: Pfitzmann, A. (ed.) IH 1999. LNCS, vol. 1768, pp. 134–145. Springer, Heidelberg (2000)
5. Kuribayashi, M., Morii, M.: Systematic generation of Tardos's fingerprinting codes. IEICE Trans. Fundamentals E93-A(2), 508–515 (2009)
6. Nuida, K.: Making collusion-secure codes (more) robust against bit erasure. In: eprint. 2009-549 (2009)
7. Nuida, K., Fujitu, S., Hagiwara, M., Kitagawa, T., Watanabe, H., Ogawa, K., Imai, H.: An improvement of discrete Tardos fingerprinting codes. Designs, Codes and Cryptography 52(3), 339–362 (2009)
8. Safavi-Naini, R., Wang, Y.: Collusion secure q-ary fingerprinting for perceptual content. In: Sander, T. (ed.) DRM 2001. LNCS, vol. 2320, pp. 57–75. Springer, Heidelberg (2002)
9. Tardos, G.: Optimal probabilistic fingerprint codes. J. ACM 55(2), 1–24 (2008)
10. Wu, M., Trappe, W., Wang, Z.J., Liu, K.J.R.: Collusion resistant fingerprinting for multimedia. IEEE Signal Processing Mag., 15–27 (2004)

Steganalysis Using Partially Ordered Markov Models

Jennifer Davidson[1] and Jaikishan Jalan[2]

[1] Department of Mathematics, Iowa State University, Ames 50011, USA
davidson@iastate.edu

[2] Department of Computer Science, Iowa State University, Ames 50011, USA
jjalan@iastate.edu

Abstract. The field of steganalysis has blossomed prolifically in the past few years, providing the community with a number of very good blind steganalyzers. Features for blind steganalysis are generated in many different ways, typically using statistical measures. This paper presents a new image modeling technique for steganalysis that uses as features the conditional probabilities described by a stochastic model called a *partially ordered Markov model* (POMM). The POMM allows concise modeling of pixel dependencies among quantized discrete cosine transform coefficients. We develop a steganalyzer based on support vector machines that distinguishes between cover and stego JPEG images using 98 POMM features. We show that the proposed steganalyzer outperforms two comparative Markov-based steganalyzers [25,6] and outperforms a third steganalyzer [23] on half of the tested classes, by testing our approach with many different image databases on five embedding algorithms, with a total of 20,000 images.

1 Introduction

Steganography is the field of covert communication: sending a message in such a way that only the sender and receiver are aware of the existence of the message. While steganography has been practiced since ancient times, the availability of digital media for hiding secret messages has exploded the use of this means of communication. The goal of steganography is to embed a *payload* into a *cover* object to obtain a *stego* object so that the presence of hidden information cannot be detected by either perceptual or statistical analysis of the stego object. The development of stego-detection schemes, called steganalysis, began immediately after popular embedding freeware became available on the Internet. The main goal of steganalysis is to identify whether a given object has a payload embedded in it. Other information about the payload is often sought, including identification of the steganography algorithm, estimation of payload length, recovery of the payload, or obliteration of the payload. If there exists an algorithm that can determine whether or not a given image contains a secret message with a success rate better than random guessing, the steganographic scheme is considered to be broken. A more detailed introduction to steganography and steganalysis can be found in [13].

R. Böhme, P.W.L. Fong, and R. Safavi-Naini (Eds.): IH 2010, LNCS 6387, pp. 118–132, 2010.

This paper focuses on steganalysis of still image data in the Joint Photographic Experts Group (JPEG) format. Image data is a good choice for hiding payload, as this type of media file is readily available in copious amounts and their use in steganography can be difficult to detect. JPEG images also have the advantage of low bandwidth for storage and transmission, unlike raw or other uncompressed formats, and there is a wide variety of freeware available for hiding secret information; see the site stegoarchive.com [27].

Attacks on steganography have two broad divisions: *targeted* and *blind* steganalysis. In targeted steganalysis, known embedding signatures, such as characteristic histogram shapes, are exploited to create specific feature values that can distinguish between stego and cover images. Blind steganalysis systems use a set of generic feature values that model image statistics so as to distinguish between cover and stego images, and a pattern classifier for identification of classes. Blind methods can be used on a variety of steganographic algorithms and offer the potential to identify unknown but similar embedding algorithms. In this paper, we approach the blind steganalysis problem as a two-class problem that identifies an image as cover or stego. Features based on statistical measures of the data have been most successful, as they contain information that quantify differences between cover and stego images [6,23,12,18]. Our features arise from a stochastic model that is able to differentiate "noise" characteristics between cover and stego images. The pattern classifier we use is a two-class support vector machine (SVM), used in prior successful steganalysis systems [6,23].

Previous works using Markov-based features include the works by Shi et al. [25] who used Markov transition matrices applied to differences of quantized DCT coefficient values. The steganalyzer calculated four directional differences in neighboring values in the DCT coefficients. Their steganalyzer used 324 feature values, an SVM classifier, and approximately 7500 training images. In Pevný et al.'s more recent work [23], the authors perform calibration of the image data first and then average the four Markov feature direction sets into one set, producing 81 instead of 324 feature values. The 81 features are added to an extended set of 193 DCT features, producing a total of 274 features. This set produces an improvement in classification accuracy over Shi's model. Later in 2008, Chen and Shi [6] extended the steganalyzer in [25] to 486 features by computing transition probability matrices for each difference JPEG 2-D array to utilize the intrablock correlation, and "averaged" transition probability matrices for those difference mode 2-D arrays to utilize interblock correlation. Other Markov modeling for steganalysis on spatial domain data has also been performed [28,22].

The rest of the paper is organized as follows. In Section 2, we give an introduction to POMMs and then develop a particular POMM model for the purpose of steganalysis in Section 3. In Section 4 we give details of our experiments followed by a discussion of the results in Section 5. We conclude with remarks in Section 6.

2 Partially Ordered Markov Models

The probabilistic model called *partially ordered Markov models* (POMM) can be used to describe statistical characteristics of an image [9]. A subclass of Markov random fields, it is a model that was used previously for texture analysis and synthesis [9,8,15] . In this paper, we show that it can provide a rigorous foundation for describing steganalysis features in a theoretical probabilistic way. There is dependency between DCT coefficients which the steganalyst can exploit to detect hidden payload, as steganogrpahy embedding in the DCT domain typically changes the statistical dependency among DCT coefficients. The steganalyst's job is to discover measures that quantify these changes. A Markov based process has used to capture this dependency with success in [25,6]. In this section, we present the partially ordered Markov model as a stochastic model used to exploit inherent dependencies between DCT coefficients for steganalysis.

Markov random fields (MRFs) are a well-known modeling tool and have many successes in image analysis. However, the use of MRFs continues to be problematic when problems require computing an explicit joint probability, such as for texture classification and parameter estimation. Development of POMMs grew from investigating computationally efficient models to implement spatial stochastic models for images. POMMs allow, under minimal and reasonable assumptions, an explicit closed form for the joint probability of the random variables (r.v.s) at hand, expressed in terms of a conditional probability. The conditional probabilities express the spatial dependency of the data, via a directional neighborhood, unlike the undirected neighborhood of a MRF model. POMMs are a generalization of Markov Mesh Models (MMMs) [2], which are themselves a generalization of Markov chains. A partially ordered Markov model generalizes the concept of local neighborhood directionality by using an acyclic directed graph underlying the pixel locations. An acyclic digraph, in turn, has a well-known relation to a partially ordered set [3]. This characteristic results in a computational advantage of POMMS over MRFs: whenever the normalizing constant needs to be calculated, such as in determining the joint probability distribution function (pdf), the joint pdf of a POMM is available in closed form, and the normalizing constant for a POMM is always known and equal to the value one [9]. We next present graph-theoretic concepts used to define a POMM.

Let V be a set of vertices, and E be a set of directed edges between vertices in V. Thus, $V = (V_1, V_2, ..., V_k)$ and $E = \{(i,j) : V_i, V_j \in V$, and (i,j) is an edge with tail on i and head on $j\}$. For our purposes, we also assume that there are no *cycles* in E, that is, there does not exist a sequence of edges $(i_1, j_1), (i_2, j_2), ..., (i_u, j_u)$ where $j_n = i_{n+1}$, $n = 1, ..., u-1$ and $j_u = i_1$. Thus, we assume that (V, E) is an *acyclic* directed graph, or acyclic digraph.

The example we are interested in occurs on an M x N array of r.v.s that represents the image data. Other examples can be found in [9]. The notation $P(A)$ denotes the (discrete) probability measure A, based on the r.v. A. We use the common notation that an upper-case letter denotes the r.v. A, while a realization of A is denoted by a. $P(A|B)$ is used to denote the conditional probability measure of A given another r.v. B. Let $A = \{A_{i,j} : 1 \leq i \leq M, 1 \leq j \leq N\}$ be an array of r.v.s representing the image data on an $M \times N$ array. Let S be a collection of ordered subsets described by $S = \{S_1, ..., S_t\}$, where each S_i is a ordered subset of r.v.s in array A. For example, we can let

$S_1 = \{A_{11}, A_{12}\}$, $S_2 = \{A_{12}, A_{13}\}$, etc. for all horizontal pairs of pixels in A. Let f be a function from S to $\mathbb{R}$ where $\mathbb{R}$ is the set of real numbers. In our example, $f(S_i) = f(A_{jk}, A_{j,k+1})$ for some indices i, j and k. Describe an edge E_i between an element of the range of f and an element of S by $E_i = (f(S_i), S_i)$. Note that the image under f of S_i is $f(S_i)$, and the pre-image of $f(S_i)$ is S_i. The edge has tail on image $f(S_i)$ and head on pre-image S_i. Let E be the set of all edges created on V in this manner. Let the set of vertices be the r.v.s $V = S \cup f(S)$. It is straightforward to show that (V, E) is an acyclic digraph. We discuss reasons for this particular choice of acyclic digraph below. We call this example the $function - subset$ acyclic digraph, denoted $f - S$, and use it in Section 3 to create features.

An acyclic digraph has an associated *partial order* $\prec$ in the sense of Birkhoff [3]. A partial order is defined as follows.

Definition 1. *Let V be a set and let $\prec$ be a subset of $V \times V$, that is, $\prec$ is a binary relation on V. A set of elements V with binary relation $\prec$ is said to have a partial order with respect to $\prec$ if the following properties hold:*
1. $w \prec w$ for all $w \in V$ (reflexivity)
2. $w \prec x$, $x \prec y \Rightarrow w \prec y$ (transitivity)
3. If $w \prec x$ and $x \prec w$ then $w = x$ (anti-symmetry)

In this case, $(V, \prec)$ is called a *partially ordered set*, or a *poset*. By convention, we write $w \succ x$ to mean $w \prec x$. A partial order can be loosely described as a relationship between some pairs of elements in V, where not all elements are necessarily related. A common example of a poset is the set of all subsets of a set, ordered or related by set inclusion.

The relationship between an acyclic directed graph and a partial order is: two elements V_i and V_j in the vertex set V are related under the partial order $\prec$ if (i, j) is a directed edge in E. In this case we write $V_i \prec V_j$.

We next give a few additional definitions that enable us to define a partially ordered Markov model.

Definition 2. *Let V be a set of vertices. For any $B \in V$, the cone of B is the set cone $B = \{C \in V : C \prec B, C \neq B\}$.*

Definition 3. *For any $B \in V$, the adjacent lower neighbors of B are those elements $C \in V$ such that (C, B) is a directed edge in the graph (V, E). Formally, $adj_\prec B = \{C : (C, B)$ is a directed edge in $E\}$.*

The set $adj_\prec B$ is the set of vertices that have directed edges into B.

Definition 4. *An element B in V is a minimal element if there is no element C in V such that $C \prec B$.*

Equivalently, B is a minimal element if there is no directed edge (C, B) in E for any $C \in V$. Let L_0 be the set of minimal elements in the poset. For the remainder of the paper, we assume that the elements of set V are random variables.

Definition 5. *The partially ordered Markov model (POMM) is defined as follows: Let $B \in V$ where (V, E) is a finite acyclic digraph of r.v.s and $(V, \prec)$ is its* *corresponding poset. Describe the set of r.v.s not related to B by $Y_B = \{C : B$ and C are not related $\}$. Then $(V, \prec)$ is called a partially ordered Markov model (POMM) if for any $B \in V\backslash L^0$ and any subset $U_B \subset Y_B$ we have*

$$P(B|cone\ B, U_B) = P(B|adj_{\prec}B). \tag{1}$$

This material is sufficient to discuss the features described next. The interested reader is directed to [9].

3 POMMS for Steganalysis

With the notation introduced in the previous section, we now discuss the application of POMMs to steganalysis. We describe a POMM whose random variables use the quantized DCT coefficients array. Assume that the quantized coefficients are described by random variables on a rectangular pixel set, and are given by $A = \{A_{i,j} : 1 \leq i \leq M, 1 \leq j \leq N\}$. Let f be a function chosen by the steganalyzer that exploits the dependency among DCT coefficients. If f is a useful function for the steganalyst, then the quantity $P(S_k|f(S_k))$, which is a measure of the frequency of occurrence of the pre-image of $f(S_k)$, can be used to distinguish between cover and stego images. This is the motivation for using the partial order as described earlier. Experimental results shown below indicate that at least for the choice of functions f we applied, these conditional probabilities, when used as features, can distinguish between cover and stego very well.

For our purposes, the function

$$f(v_1, v_2) = v_1 - v_2 \tag{2}$$

has been shown to be a useful feature for steganalysis [25], [6], where v_1 and v_2 are two adjacent pixels. Using the function f in Eq. 2, we create a series of individual POMMs whose underlying acyclic digraphs are constructed according to the $f-S$ example from Section 2. The feature values are conditional probabilities, which are averaged over a collection of POMMs. Calibration is also applied, and the final set of features is the difference between the input and calibrated image features.

Given a rectangular array of r.v.s, we use the four directions horizontal, vertical, diagonal, and minor diagonal to define the four sets S^h, S^v, S^d, S^m based on directional subsets $S^h_{i,j} = \{A_{i,j}, A_{i,j+1}\}$, $S^v_{i,j} = \{A_{i,j}, A_{i+1,j}\}$, $S^d_{i,j} = \{A_{i,j}, A_{i+1,j+1}\}$, and $S^m_{i,j} = \{A_{i+1,j}, A_{i,j+1}\}$. From these sets we construct the four acyclic digraphs (V^*, E^*) with $V^* = S^* \cup f(S^*)$ and $E^* = \{E^*_i : E^*_i = (f(S^*_i), S^*_i)\}$, $^* \in \{h, v, d, m\}$. For each (V^*, E^*), we construct the corresponding POMM P^* defined by its conditional probabilities

$$P^*(S^*|f(S^*)) = \frac{P^*(S^*, f(S^*))}{P^*(f(S^*))} \tag{3}$$

These conditional probabilities are calculated by histogram binning of the data. For computational purposes, we threshold values in the rectangular array A to between $-T$ and $+T$ for some integer T:

$$A_{i,j} = \begin{cases} A_{i,j} & \text{if } -T \leq A_{i,j} \leq T \\ -T & \text{if } A_{i,j} < -T \\ +T & \text{if } A_{i,j} > +T \end{cases}$$

This limits the number of values for $P^*(S^*|f(S^*))$ to $(2T+1)^2$ for a given direction *. If we define a $(2T+1) \times (2T+1)$ matrix F^* by

$$F^*(w,z) = P^*(S^*|f(S^*)) = P^*(w,z|f(w-z)),$$

then a set of $(2T+1)^2$ features can be defined by

$$F(w,z) = \frac{1}{4} \sum_{*\in\{h,v,d,m\}} F^*(w,z) = \frac{1}{4} \sum_{*\in\{h,v,d,m\}} P^*(S^*|f(S^*)) = \\ = \frac{1}{4} \sum_{*\in\{h,v,d,m\}} P^*(w,z|f(w,z)). \quad (4)$$

Thresholding reduces the number of features necessary for classification. Indeed, larger values of T do not necessarily give better classification rates, as discussed in Section 4. We next discuss the application of this POMM to two different sets of arrays, one the global quantized DCT array values that captures the intrablock dependencies, and the other the mode arrays that capture interblock dependencies. Mode arrays are created by collecting the mode frequency from different blocks.

3.1 Intrablock Features

Certainly, steganographic embedding can cause disturbances on the smoothness, regularity, continuity, consistency, and/or periodicity of quantized DCT coefficients, and therefore affect correlations among DCT coefficients. To quantify this change, we create the above defined POMM on the global quantized DCT coefficients array. We use the averaged conditional probabilities given in Eq. 4 as intrablock features for a given image. Since this POMM models the dependency among DCT coefficients within a 8x8 DCT block, we refer to these as intrablock features. This results in $(2T+1)^2$ features.

3.2 Interblock Features

Interblock dependency is determined by first collecting JPEG mode values into an array, that is, the DCT coefficients located at the same relative position within each 8x8 block. JPEG steganographic embedding typically disturbs this kind of interblock dependency. Let the r.v. array A represent the quantized DCT coefficient array with size MxN. Then there are $N_r * N_c$ number of 8x8 DCT blocks where $N_r = \lceil \frac{M}{8} \rceil$ and $N_c = \lceil \frac{N}{8} \rceil$. Let $X^{i,j}$ be the r.v. array formed by collecting DCT coefficients located at i,j from every 8x8 block. Equivalently,

$X_{u,v}^{i,j}$ is the DCT coefficient located at position (i, j) in the (u, v)-block where $1 \leq u \leq N_r$, $1 \leq v \leq N_c$. The array $X^{i,j}$ is called a *mode array* as it represents mode or specific frequencies from every 8x8 block. There are 64 mode arrays $X^{i,j}$.

To capture interblock dependency, we calculate conditional probabilities as in Eq. 4 on every mode array $X^{i,j}$. We use the conditional probabilities averaged over all 64 mode arrays as interblock features. This results in an additional $(2T + 1)^2$ features.

We apply calibration to reduce the dependency of the feature values on the image content itself. Calibration is a well known technique in steganalysis that can improve detection accuracy by making features dependent on the changes incurred by data hiding rather on the image content itself [12]. More information on calibration can be found in [21]. We cropped by 4 pixels on each side to perform calibration.

Let I^o be the given image, and let its calibrated image be I^{cal}. We calculate intra- and inter-block features for I^{cal} also, and then use the difference between the two features $(F^o - F^{cal})(w, z)$ as our final set of features. There are $(2T+1)^2$ number of intrablock features and $(2T + 1)^2$ number of interblock features, for a total of 2*$(2T + 1)^2$ number of features for a fixed value of T.

4 Experiments

It has been shown recently in [20] that the performance of a steganalyzer depends on the database used for training and testing the steganalyzer. Therefore, for our experiments, we use four different databases to compare our steganalyzer based on proposed feature set with other steganalyzers.

- **Bows2:** This database contains 10,000 images of size 512x512 in pgm format.
- **Camera:** This database consists of 3164 images captured using 24 different digital cameras (Canon, Kodak, Nikon, Olympus and Sony) previously used in [14]. They include photographs of natural landscapes, buildings and object details. All images are of size 512x512 and stored in a raw format (tif) i.e. the images have never undergone lossy compression.
- **Corel:** This database consists of 8185 images from the Corel database [7]. They include images of natural landscapes, people, animals, instruments, buildings, artwork, etc. Although there is no indication of how these images have been acquired, they are very likely to have been scanned from a variety of photos and slides. This database has been previously used in [30]. All images are of size 512x512 and stored in a raw format (tif).
- **NRCS:** This database consists of 2,375 images from the NRCS Photo Gallery [1].The photos are of natural scenery, e.g. landscape, cornfields, etc. These images were acquired by scanning photographic negatives at three wavelengths, introducing a relatively high level of noise. This database has been previously used in [19]. All images in this database too are of size 512x512 and stored in a raw format (tif). Results from this database are omitted from the paper due to space limitiations.

The last three databases were downloaded from [11]. We generate the training and testing set for each database separately as follows. The images were divided randomly into two disjoint groups of equal size. The first group was used to create the training examples with both cover and stego data. The second group contained the remaining images that were used for testing. We did not use those stegoimages for which the steganography embedding algorithm exited unsuccessfully. Thus, no image or its different variations were simultaneously seen by the SVMs for testing and training, and there were an equal number of images used for cover and stego. This strict division of images enabled us to estimate the performance on never seen images. We intentionally chose to generate our cover and stego images for training and classification in a way to avoid double compression, reformatting the raw pgm files into JPEG files saved with 75% quality factor. Recall that a JPEG image is double compressed when it is first compressed using quantization matrix Q^1, then uncompressed and re-compressed using quantization matrix Q^2, where $Q^2 \neq Q^1$. When an image undergoes double compression, the statistics of DCT coefficients can change, and may resulting in misclassification if the steganalyzer is not designed to handle detection of double compression. Our detection scheme assumes that the data has not been double-compressed.

We generate stego images by embedding data with different message lengths and different embedding algorithms. The five different steganography embedding methods we consider are: OutGuess [24], F5 [29], JPHide& Seek [16], StegHide [26], and JSteg [17]. Each of these methods embeds bits of value 0 or 1 directly into the quantized DCT coefficient array. The payload is assumed to be an encrypted bitstream. We use *bits per nonzero ac coefficient*, or *bpnz* to describe message length, with $bpnz = 0.05, 0.1, 0.2, 0.4$. Images embedded using Outguess have only $bpnz = 0.05, 0.1, 0.2$. Once the stego images are generated, we extract feature values from the cover and stego images according to the steganalyzer. For each database, feature set and for each algorithm we train a soft margin support vector machine with gaussian kernel [4] (using LIBSVM [5]).

We determined the training parameters of the C-SVMs by grid-search performed on the following multiplicative grid

$$(C, \gamma) \in \left\{ \left(2^i, 2^j\right) | i \in Z, j \in Z \right\}.$$

We compare our results with steganalyzer from feature set proposed in [25], [6] and [23] abbreviating them as Markov324, Markov486 and Merged, respectively.

5 Discussion of Results

Tables 1-4 display the classifier results for values of $T = 1, 2, 3, 4$ and 5, presenting the accuracy of detection in percentage values for each binary classifier: cover vs. stego, for different embedding algorithms, for varying values of T, and for each of the four databases. We first understand the effect of threshold parameter T on the detection accuracy. Note that different values of T will produce different POMMs and hence we get different feature sets. For purposes of comparison,

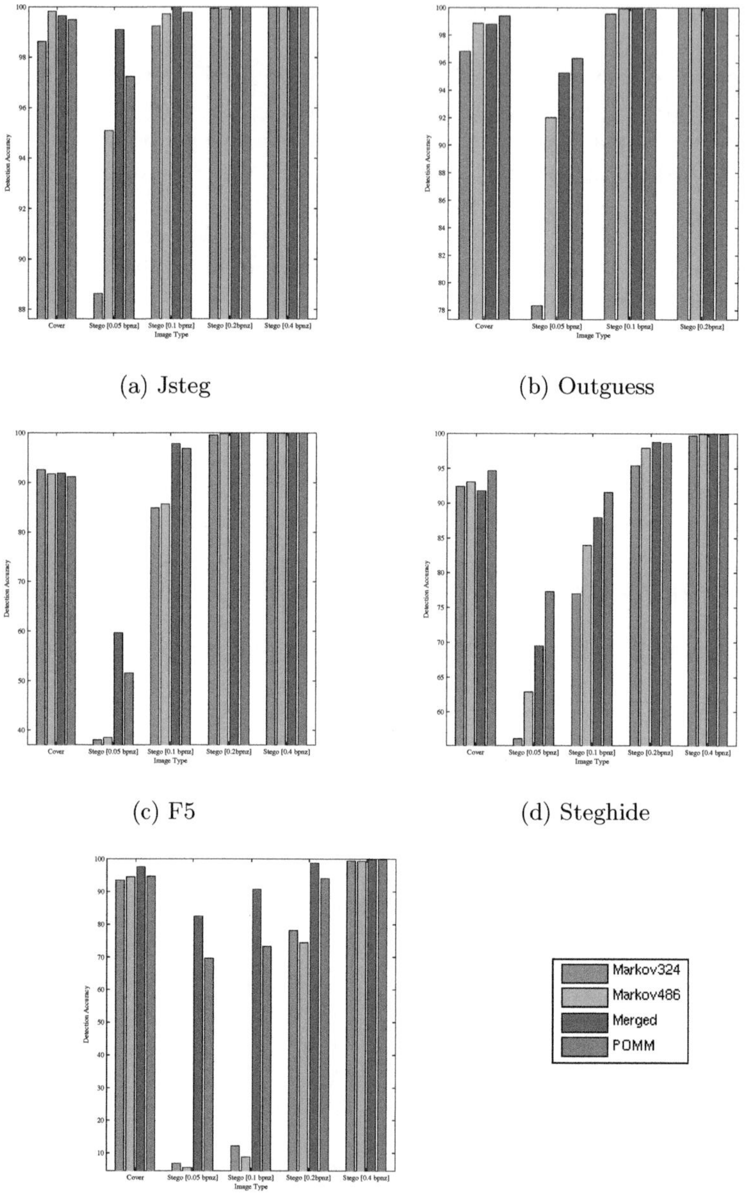

(a) Jsteg

(b) Outguess

(c) F5

(d) Steghide

(e) JPHide

Fig. 1. Detection accuracy results for different steganalyzers on BOWS2 database. Note different scales on y axis.

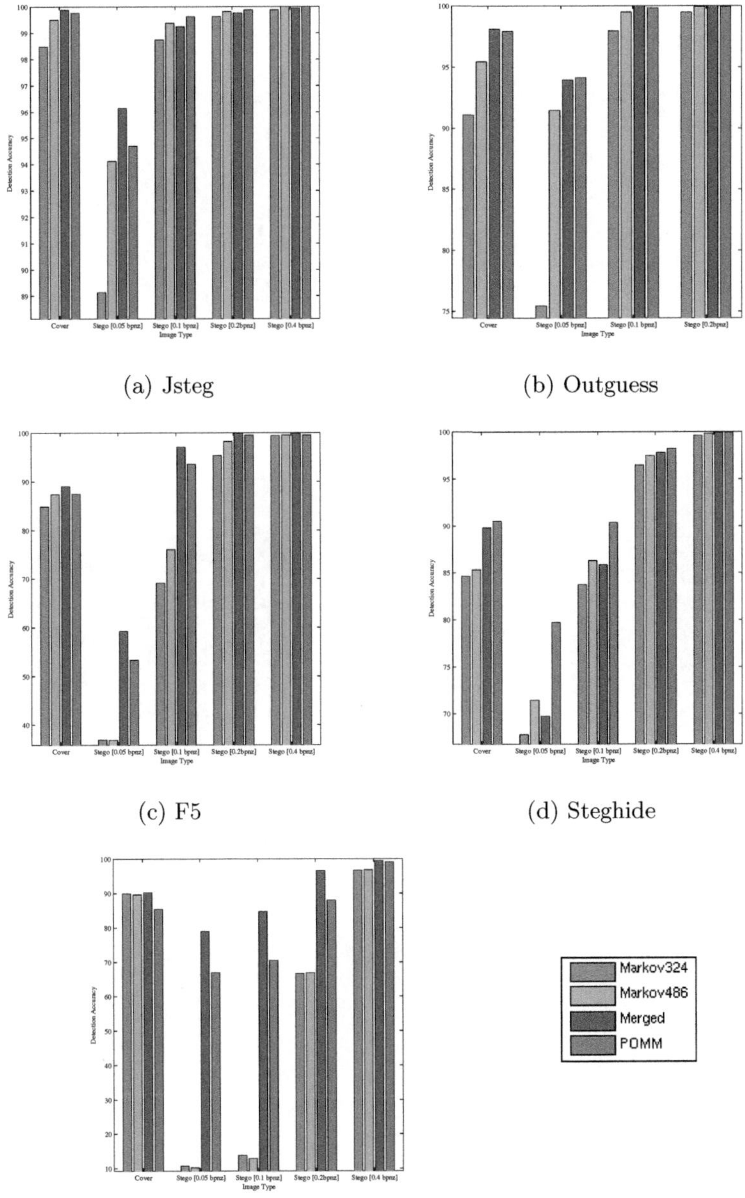

(a) Jsteg

(b) Outguess

(c) F5

(d) Steghide

(e) JPHide

Fig. 2. Detection accuracy results for different steganalyzers on Camera database. Note different scales on y axis.

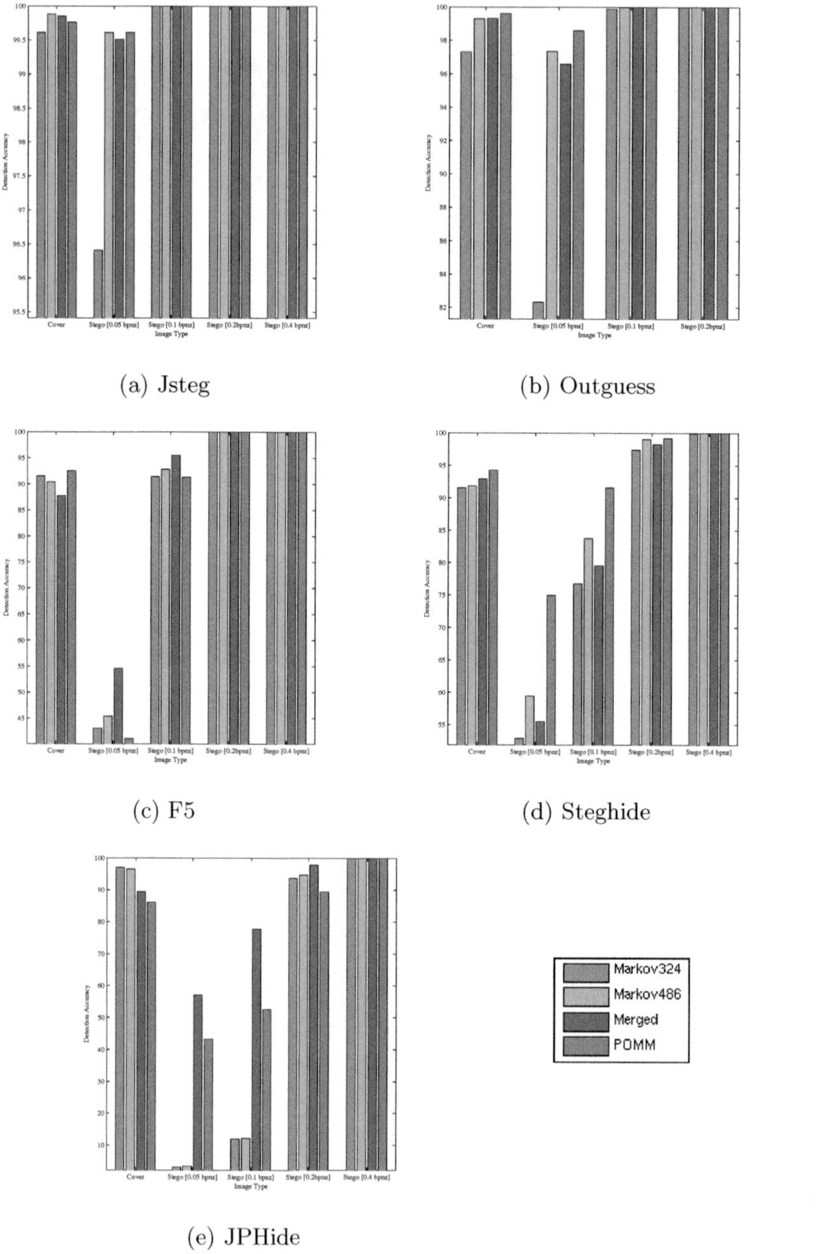

Fig. 3. Detection accuracy results for different steganalyzers on Corel database. Note different scales on y axis.

if more than 5000 images were used for training, we assume that differences in classification of less than one percent are relatively equivalent. It is clear from Tables 1-4 that the POMMs with $T = 1$ (18 features) and $T = 2$ (50 features) do not capture steganographic changes at lower embedding rate, and this is true across the different databases. It is also clear that the POMMs with $T = 3$ perform better than the other POMMs in smaller database (NRCS and Camera) for JPHide whereas the POMM with $T = 5$ performs better on larger databases. This trend is also true for F5. For example, POMM with $T = 3$ gives a detection accuracy of 56% for 0.05 bpnz embedding for F5 algorithm whereas POMM with $T = 5$ gives an accuracy of 50%. But the trend is reversed once you go to a larger database. In Bows2 database which has 10000 images, the POMM with $T = 5$ gives a detection accuracy rate of 54% for F5 at 0.05 bpnz whereas POMM with $T = 3$ gives 51%. It can also be seen that POMM with $T = 3$ gives an overall better detection for OutGuess irrespective of database size. Performances of the POMM steganalyzers with $T \geq 3$ are approximately same for Jsteg and Steghide across different databases. It is clear that based on the database size, a POMM can be build by selecting an appropriate value of T. Based on the above observations, we chose POMM with $T = 3$ to compare our approach with other steganalyzers proposed in the literature.

Table 1. Detection accuracy results for POMM based features on NRCS database

		T=1	T=2	T=3	T=4	T=5
Jsteg	Cover	92.67	99.75	99.75	99.83	100.00
	0.05 bpc	24.43	98.15	99.58	98.99	98.99
	0.10 bpc	41.62	100.00	100.00	100.00	100.00
	0.20 bpc	77.93	100.00	100.00	100.00	100.00
	0.40 bpc	98.74	100.00	100.00	100.00	100.00
Outguess	Cover	81.89	99.58	99.92	99.66	99.66
	0.05 bpc	37.18	97.39	98.31	99.16	98.90
	0.10 bpc	61.64	99.92	100.00	100.00	100.00
	0.20 bpc	90.37	100.00	100.00	100.00	100.00
F5	Cover	90.31	93.01	93.18	93.43	92.00
	0.05 bpc	29.65	48.19	56.53	50.72	49.71
	0.10 bpc	74.47	95.79	97.39	96.46	95.20
	0.20 bpc	99.92	99.92	100.00	100.00	100.00
	0.40 bpc	100.00	99.75	100.00	100.00	100.00
Steghide	Cover	71.61	95.37	96.04	96.29	96.12
	0.05 bpc	36.82	74.05	74.56	72.11	73.38
	0.10 bpc	40.86	89.81	92.25	91.24	90.73
	0.20 bpc	49.03	97.73	98.99	98.99	99.16
	0.40 bpc	62.59	99.58	99.75	100.00	100.00
JPHide	Cover	83.40	84.92	89.22	91.49	90.90
	0.05 bpc	29.16	37.45	30.68	28.23	25.44
	0.10 bpc	32.94	44.45	39.63	39.46	37.85
	0.20 bpc	51.86	69.71	84.43	88.83	92.22
	0.40 bpc	90.76	96.02	98.39	99.07	99.66

Table 2. Detection accuracy results for POMM based features on Camera database

		T=1	T=2	T=3	T=4	T=5
Jsteg	Cover	86.22	99.18	99.75	99.56	99.24
	0.05 bpc	35.08	91.66	94.69	94.56	94.75
	0.10 bpc	50.51	99.30	99.62	99.62	99.62
	0.20 bpc	78.07	99.81	99.87	99.94	99.94
	0.40 bpc	98.48	99.94	100.00	99.87	100.00
Outguess	Cover	74.53	97.47	97.91	97.47	96.65
	0.05 bpc	52.39	89.80	94.14	94.39	93.69
	0.10 bpc	72.71	99.35	99.81	99.81	99.81
	0.20 bpc	94.03	99.93	99.93	100.00	100.00
F5	Cover	85.21	87.29	87.42	85.40	90.14
	0.05 bpc	44.82	52.97	53.29	54.99	45.70
	0.10 bpc	86.09	93.11	93.55	93.99	89.82
	0.20 bpc	99.05	99.43	99.49	99.56	99.37
	0.40 bpc	99.18	99.43	99.49	99.62	99.49
Steghide	Cover	58.85	90.01	90.52	90.14	89.51
	0.05 bpc	56.01	76.80	79.71	80.28	79.14
	0.10 bpc	60.05	87.55	90.39	90.39	89.82
	0.20 bpc	67.13	96.97	98.23	97.98	97.98
	0.40 bpc	77.24	99.56	99.94	99.81	99.75
JPHide	Cover	85.40	85.71	85.40	90.52	91.28
	0.05 bpc	61.90	69.18	66.90	60.95	60.57
	0.10 bpc	63.12	71.28	70.53	64.45	64.71
	0.20 bpc	72.51	80.30	87.97	87.84	89.42
	0.40 bpc	95.89	97.91	99.05	99.24	99.37

In Figures 1, 2, and 3, the order of the detection accuracy for each level of embedding is, from left to right: Markov324, Markov486, Merged, and POMM. It is clear that the steganalyzer based on our proposed feature set clearly beats Markov324 and Markov486 in all the databases at all the embedding rates for almost every steganography algorithm. Even though the performances of the steganalyzers are very close to 100% for higher embedding rates, their performances vary much more at lower embedding rates. Note that Markov486 is an

Table 3. Detection accuracy results for POMM based features on Corel database

		T=1	T=2	T=3	T=4	T=5
Jsteg	Cover	87.02	97.80	99.76	99.80	99.95
	0.05 bpc	29.79	97.39	99.61	99.58	99.39
	0.10 bpc	54.72	99.98	100.00	100.00	100.00
	0.20 bpc	86.78	100.00	100.00	100.00	100.00
	0.40 bpc	98.53	100.00	100.00	100.00	100.00
Outguess	Cover	75.81	99.00	99.61	99.80	99.80
	0.05 bpc	42.79	97.48	98.58	98.34	98.34
	0.10 bpc	62.07	100.00	100.00	100.00	100.00
	0.20 bpc	89.89	100.00	100.00	100.00	100.00
F5	Cover	86.19	93.55	92.55	92.18	91.79
	0.05 bpc	27.32	37.51	41.10	40.81	40.74
	0.10 bpc	50.42	88.76	91.30	90.69	90.27
	0.20 bpc	99.07	100.00	100.00	100.00	100.00
	0.40 bpc	99.88	100.00	100.00	100.00	100.00
Steghide	Cover	60.68	92.67	94.26	93.94	94.79
	0.05 bpc	46.29	73.34	74.98	75.66	73.78
	0.10 bpc	49.39	89.10	91.59	91.94	90.32
	0.20 bpc	55.33	98.44	99.17	99.19	99.10
	0.40 bpc	66.01	99.95	100.00	99.98	100.00
JPHide	Cover	79.06	83.68	86.17	85.46	85.24
	0.05 bpc	42.18	42.59	43.26	44.33	44.41
	0.10 bpc	44.52	48.80	52.54	54.94	56.23
	0.20 bpc	54.37	72.63	89.44	92.66	92.88
	0.40 bpc	81.80	99.14	99.93	99.90	99.95

Table 4. Detection accuracy results for POMM based features on Bows2 database

		T=1	T=2	T=3	T=4	T=5
Jsteg	Cover	93.18	98.84	99.48	99.46	99.48
	0.05 bpc	33.48	92.72	97.24	96.80	96.38
	0.10 bpc	58.30	99.52	99.78	99.78	99.72
	0.20 bpc	83.48	99.98	100.00	100.00	99.96
	0.40 bpc	97.14	100.00	100.00	100.00	100.00
Outguess	Cover	81.84	98.70	99.38	99.08	99.38
	0.05 bpc	38.63	94.88	96.32	96.90	96.90
	0.10 bpc	62.75	99.84	99.88	99.90	99.92
	0.20 bpc	92.29	100.00	100.00	100.00	100.00
F5	Cover	87.66	90.86	91.12	90.66	89.42
	0.05 bpc	29.16	49.56	51.56	52.66	54.58
	0.10 bpc	66.90	95.90	96.78	96.90	97.18
	0.20 bpc	99.46	99.98	99.98	100.00	99.98
	0.40 bpc	99.78	99.98	99.98	100.00	100.00
Steghide	Cover	66.34	92.48	94.66	93.96	94.34
	0.05 bpc	44.36	74.40	77.28	78.04	77.82
	0.10 bpc	48.60	88.40	91.56	92.14	91.56
	0.20 bpc	55.88	97.36	98.62	98.82	98.74
	0.40 bpc	70.28	99.82	99.88	99.92	99.94
JPHide	Cover	90.34	92.64	94.66	95.14	94.98
	0.05 bpc	64.95	71.73	69.60	70.90	70.80
	0.10 bpc	66.66	74.13	73.25	74.67	75.32
	0.20 bpc	74.18	85.12	94.01	94.25	94.79
	0.40 bpc	92.88	99.14	99.78	99.60	99.58

extension of Markov324 and the added inter block features have helped in boosting the performance. However, our POMM based steganalyzer shows significant improvement in performance at the lower embedding rates compared to either Markov scheme. For example, for Bows2 database, Markov324 and Markov486 has a detection accuracy close to 6% whereas POMM based steganalyzer gave a detection accuracy of 71%. Our proposed steganalyzer has also performed better than Merged for Outguess and Steghide across all the databases whereas Merged performs better at lower embedding rates for F5 and JPHide. Both steganalyzers perform equivalently at higher embedding rates and for detecting Jsteg. For example, the Merged steganalyzer gave a detection accuracy of 55% for Steghide at 0.05 bpnz for Corel database whereas our POMM based steganalyzer gave a detection accuracy of 74%. On the other hand, for the same database, Merged performed better for JPHide at 0.05 bpnz with a detection accuracy of 57% whereas our POMM based steganalyzer detected 44% of the stego images correctly. And this trend holds across different databases.

6 Conclusion

Clearly the POMM outperforms the Markov324 and Markov486 steganalyzers on these databases. The POMM also outperforms the Merged on Steghide and Outguess on the Corel and NCRS databases, as well as on F5 cover (Corel, NCRS) and Jsteg (0.05, NCRS). The creation of features using the POMM is a completely different approach than Merged, and it can offer a theoretical basis foundation for describing, and possibly analzying, steganalysis features.

The use of a partially ordered Markov model to describe pixel dependencies for steganalysis is presented in this paper. The conditional probabilities given by the POMM are used as features to describe inter and intra block pixel relations in JPEG image data. Other choices for the function f given in Equation

2 can be devised to describe other pixel dependencies, giving rise to features whose potential for steganalysis can be inspected. The POMM with 98 features performed better than either Markov model with 324 or 486 features, and as well as or better than the Merged model in half of the classes. Other POMM feature models are being investigated for future steganalysis, and a multiclass detector using this POMM has been developed for use in a police forensic lab [10]. A double-compression detector front-end is being developed as well using these POMM features. Questions that remain open include investigating the use of the joint pdf for steganalysis, and also the use of POMMs for spatial domain steganalysis. Also, can the spatial dependency itself be determined from the data, and used to provide a POMM whose features can be used to produce an accurate steganalyzer?

Acknowledgment

This work was partially funded by the National Institute of Justice, through the Midwest Forensics Resource Center at Ames Laboratory under Interagency Agreement number 2008-DN-R-038. The Ames Laboratory is operated for the U.S. Department of Energy by Iowa State University, under contract No. DE-AC02-07CH11358. The authors would also like to thank Dr. Jessica Fridrich and Dr. Yun Q. Shi for providing code for the steganalyzers used in this research. Special thanks to Dr. Gwenael Doerr for providing a rich database of JPEG images which helped in this research.

References

1. Natural resources conservation service photo gallery, `http://photogallery.nrcs.usda.gov`
2. Abend, K., Harley, T., Kanal, L.: Classification of binary random patterns. IEEE Transactions on Information Theory 11(4), 538–544 (1965)
3. Birkhoff, G.: Lattice Theory, vol. 25. American Mathematical Society, Providence (1940)
4. Burges, C.J.: A tutorial on support vector machines for pattern recognition. Data Mining and Knowledge Discovery 2, 121–167 (1998)
5. Chang, C.C., Lin, C.J.: LIBSVM: a library for support vector machines (2001), software available at, `http://www.csie.ntu.edu.tw/~cjlin/libsvm`
6. Chen, C., Shi, Y.Q.: JPEG image steganalysis utilizing both intrablock and interblock correlations. In: IEEE International Symposium on Circuits and Systems, pp. 3029–3032 (May 2008)
7. Corporation, C.: Corel stock photo library 3, Ontario, Canada
8. Cressie, N., Davidson, J.: Image analysis with partially ordered Markov models. Computational Statistics and Data Analysis 29, 1–26 (1998)
9. Davidson, J.L., Cressie, N., Hua, X.: Texture synthesis and pattern recognition for partially ordered Markov models. Pattern Recognition 32(9), 1475–1505 (1999)
10. Davidson, J., Jalan, J.: Canvass: A steganographic forensic tool for JPEG images. Submitted to the ADFSL 2010 Conference on Digital Forensics, Security and Law (2010)

11. Doerr, G.: Image database for steganalysis studies (2007), `http://www.cs.ucl.ac.uk/staff/ingemar/Content/Downloads.html`
12. Fridrich, J.: Feature-based steganalysis for JPEG images and its implications for future design of steganographic schemes. In: Fridrich, J. (ed.) IH 2004. LNCS, vol. 3200, pp. 67–81. Springer, Heidelberg (2005)
13. Fridrich, J.: Steganography in Digital Media: Principles, Algorithms, and Applications. Cambridge University Press, Cambridge (2010)
14. Goljan, M., Fridrich, J., Holotyak, T.: New blind steganalysis and its implications. In: Delp III, E.J., Wong, P.W. (eds.) Security, Steganography, and Watermarking of Multimedia Contents VIII, vol. 6072, p. 607201. SPIE, Bellingham (2006)
15. Helterbrand, J., Cressie, M., Davidson, J.: Optimal closed boundary identification in gray-scale imagery. Journal of Mathematical Imaging and Vision 5(3), 179–206 (1995)
16. JPHide&Seek, `http://linux01.gwdg.de/~alatham/stego.html`
17. JSteg, `http://zooid.org/paul/crypto/jsteg/`
18. Ker, A.: Quantitive evaluation of pairs and RS steganalysis. In: Security, Steganography, and Watermarking of Multimedia Contents III, vol. 3677, pp. 273–274. SPIE, Bellingham (2004)
19. Ker, A.: Steganalysis of LSB matching in grayscale images. IEEE Signal Processing Letters 12(6), 441–444 (2005)
20. Ker, A.D., Lubenko, I.: Feature reduction and payload location with WAM steganalysis. In: Media Forensics and Security, vol. 7254, p. 72540A. SPIE, Bellingham (2009)
21. Kodovský, J., Fridrich, J.: Calibration revisited. In: Proceedings of the 11th ACM workshop on Multimedia and security, pp. 63–74. ACM, New York (2009)
22. Pevný, T., Bas, P., Fridrich, J.: Steganalysis by subtractive pixel adjacency matrix. In: Proceedings of ACM Multimedia and Security Workshop, pp. 75–84 (September 2009)
23. Pevný, T., Fridrich, J.: Merging Markov and DCT features for multi-class JPEG steganalysis. In: Delp III, E.J., Wong, P.W. (eds.) Security, Steganography, and Watermarking of Multimedia Contents IX, vol. 6505. SPIE, Bellingham (2007)
24. Provos, N.: Defending against statistical steganalysis. In: Proceedings of the 10th conference on USENIX Security Symposium, pp. 24–24. USENIX Association, Berkeley (2001)
25. Shi, Y.Q., Chen, C., Chen, W.: A Markov process based approach to effective attacking JPEG steganography. In: Camenisch, J.L., Collberg, C.S., Johnson, N.F., Sallee, P. (eds.) IH 2006. LNCS, vol. 4437, pp. 249–264. Springer, Heidelberg (2007)
26. Steghide, `http://steghide.sourceforge.net/`
27. StegoArchive, `http://www.stegoarchive.com/`
28. Sullivan, K., Madhow, U., Manjunath, B., Chandrasekaran, S.: Steganalysis for Markov cover data with applications to images. IEEE Transactions on Information Security and Forensics 1(2), 275–287 (2006)
29. Westfeld, A.: F5: A steganographic algorithm. In: Moskowitz, I.S. (ed.) IH 2001. LNCS, vol. 2137, pp. 289–302. Springer, Heidelberg (2001)
30. Zhang, J., Cox, I., Doerr, G.: Steganalysis for LSB matching in images with high-frequency noise. In: IEEE 9th Workshop on Multimedia Signal Processing, pp. 385–388 (October 2007)

The Influence of the Image Basis on Modeling and Steganalysis Performance

Valentin Schwamberger[1], Pham Hai Dang Le[2], Bernhard Schölkopf[1], and Matthias O. Franz[2]

[1] Max Planck Institute for Biological Cybernetics, Spemannstr. 38, 72076 Tübingen, Germany
vschwamb@gmail.com, bs@tuebingen.mpg.de
[2] HTWG Konstanz, Institute for Optical Systems, Brauneggerstr. 55, 78462 Konstanz, Germany
{mfranz,dangle}@htwg-konstanz.de

Abstract. We compare two image bases with respect to their capabilities for image modeling and steganalysis. The first basis consists of wavelets, the second is a Laplacian pyramid. Both bases are used to decompose the image into subbands where the local dependency structure is modeled with a linear Bayesian estimator. Similar to existing approaches, the image model is used to predict coefficient values from their neighborhoods, and the final classification step uses statistical descriptors of the residual. Our findings are counter-intuitive on first sight: Although Laplacian pyramids have better image modeling capabilities than wavelets, steganalysis based on wavelets is much more successful. We present a number of experiments that suggest possible explanations for this result.

1 Introduction

Most steganalytic methods are not capable of detecting general steganographic manipulations in images (universal steganalysis), since they are tuned to specific steganographic algorithms. The few currently available universal steganalytic algorithms [13,8,11,2] are relatively insensitive towards small embeddings. This is due to the problem of detecting a tiny manipulation (the embedded data) in a large amplitude signal (the carrier image).

The large amplitude of the carrier signal can be largely reduced by applying an image model to a suitably transformed image. The image model is capable of predicting transform coefficients from their local neighborhoods [4] based on the coefficient statistics of the image. Since an embedded message cannot be predicted from the neighborhood statistics of the image, it must be part of the prediction error of the model [13]. Thus, by analyzing the prediction error instead of the whole image, we effectively remove most of the carrier signal. The residual is much more affected by the embedding manipulation than the full image which results in a better detectability.

The initial image transform determines the basis in which the image is modeled and in which the residual is characterized by a suitable set of statistical

R. Böhme, P.W.L. Fong, and R. Safavi-Naini (Eds.): IH 2010, LNCS 6387, pp. 133–144, 2010.

descriptors which constitute the input to a final classifier stage. In such a steganalyzer architecture, a plausible hypothesis can be stated as follows: "The best image basis (or the best associated subband transform) is that which leads to the image model with the highest predictability since this most effectively removes the carrier from a potential stego image". Here, we show that this is not the case, and provide some hints on the possible reasons for this counter-intuitive result.

Our study is based on a modified version of the well-known steganalyzer of Lyu and Farid [13] which we describe in the next section. The investigated image bases are QMF wavelets [18] and Laplacian pyramids [1]. In Sect. 3, we present our results on image modeling and steganalysis performance. Additional experiments for explaining these results are discussed in Sect. 4. We conclude with a brief summary in Sect. 5.

2 Lyu and Farid's Algorithm and Modifications

The input of Lyu and Farid's algorithm is an image in its pixel representation. Originally, a wavelet pyramid is built for each color channel (as shown in the upper path in Fig. 1). Alternatively, the image can be decomposed into a Laplacian pyramid described later (lower path in Fig. 1). Quadrature mirror filters are used for building the wavelet pyramid [18] with a quadrature mirror filter of width 9. We get $3(3s+1)$ subbands for an RGB image and a pyramid with s scales and three orientation subbands, i. e. diagonal, vertical, and horizontal orientation. In the case of the alternative Laplacian pyramid representation [1], we

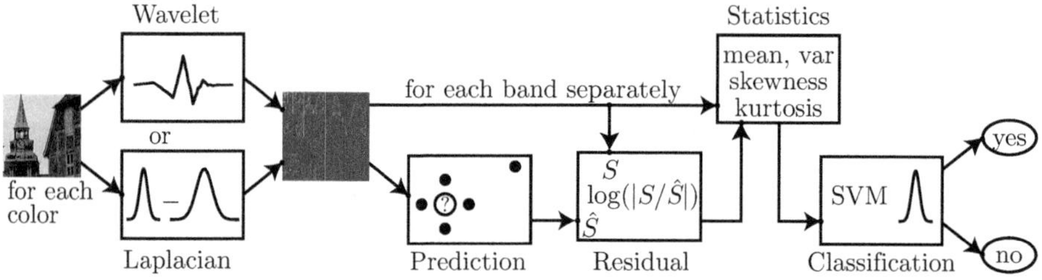

Fig. 1. The algorithm—schematics

use a standard binomial filter of width 5 to obtain the lowband approximation of the image. The number of pyramid levels was chosen to be the same as in the wavelet pyramid, but—since the Laplacian pyramid does not decompose the image according to orientation—we have only one subband per pyramid level which results in overall $3s$ subbands for color images.

For predicting coefficients from their neighborhood, we need to specify a neighborhood structure for each image representation which is shown in Fig. 2. The neighborhood structure for wavelets is the same as in [13] (cf. Fig. 2a), the Laplacian neighborhood is constructed analogously (cf. Fig. 2b), but without

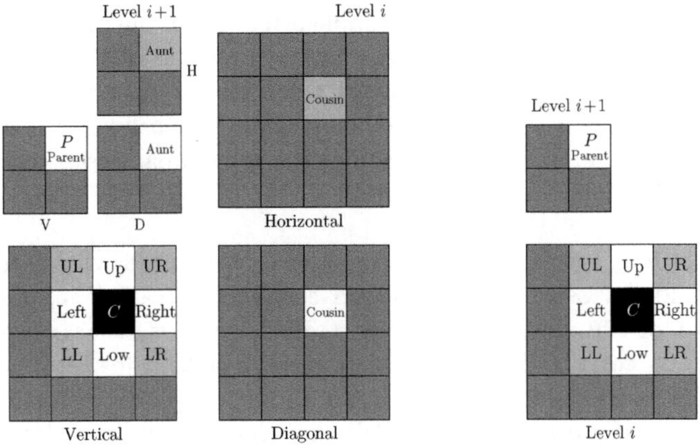

(a) Wavelet (b) Laplacian

Fig. 2. Neighborhood structure for image modeling (color neighbors not included). The central coefficient to be predicted is C, the light gray neighbors can be optionally included but did not lead to significantly different results.

orientation neighbors. Both representations contain the corresponding central coefficient from the other color channels in their neighborhoods (not shown in Fig. 2). Due to only including the neighboring coefficients from closest orientations on the same scale (hence including horizontal and vertical coefficients for predicting the diagonal subband, but only diagonal coefficients for both the horizontal and vertical subbands), and correspondingly only one (diagonal) or two neighbors (horizontal and vertical) from the coarser scales, neighborhoods in the wavelet representation contain 9 coefficients, in the Laplacian representation 7 coefficients.

The predictions are computed with linear regression applied to each subband separately, i. e., the magnitude of the central coefficient is obtained as a weighted sum of the magnitudes of its neighboring coefficients greater than a given threshold: It has been shown empirically that only the magnitudes of coefficients are correlated, and the correlation decreases for smaller magnitudes [4]. The weight sets over all subbands thus constitute the image model. In their original approach, Lyu & Farid used standard least-squares regression for this purpose. In our implementation, we use Gaussian process (GP) regression [14,15] instead after normalizing all subband coefficients to the interval $[0, 1]$. This approach leads to slightly more robust, but essentially comparable results for the purpose of this study. GP regression needs an additional model selection step for estimating the noise content in the image. For that, we use Geisser's surrogate predictive probability [7]. It is computed on a subset of of the coefficients: The finest scales are subsampled by a factor of 5 and the coarser by a factor of 3, each in both directions. Details on this regression technique can be found in [15].

Each estimator is trained and used for prediction on the same subband. Thus, training and test set coincide for this application. From the predicted coefficients $\widehat{S}$, small coefficients with amplitude below a threshold of $t = 1/255$ are set to zero. For reconstructing complete images, the algebraic signs are transferred from the original to the predicted subband coefficients. The residual r is computed by taking the logarithm of the coefficients of the input image transform S and the predicted coefficients $\widehat{S}$ and subtracting them subsequently, hence $r = \log S - \log \widehat{S}$.

Next, the four lowest statistical moments—i. e. mean, standard deviation, skewness, and kurtosis—of the subband coefficients (called *marginal statistics* in [13]) and of the subband residuals (called *error statistics*) are computed, again for each color and subband separately. Finally, all these independently normalized statistics serve as feature inputs for a support vector machine [17]. In this study, we use $s = 3$ pyramid levels which results in a 120-dimensional feature vector for the wavelet representation and in a 48-dimensional vector for the Laplacian decomposition. The final classification was done with a 1-norm soft margin non-linear C-SVM using a Gaussian kernel. The choice of the parameter C of the SVM and the width σ of the Gaussian kernel was based on a paired cross-validation procedure [16]. The SVM is tunable in order to adapt the rate of false alarms and the detection rate.

3 Comparison between Wavelet and Laplacian Basis

3.1 Image Modeling Performance

The prediction quality of the image model is measured in terms of the *explained variance* in pixel space

$$\mathcal{V}_{\text{expl}} \equiv \frac{\mathcal{V}_{\text{img}} - \mathcal{V}_{\text{err}}}{\mathcal{V}_{\text{img}}}$$

with the variance $\mathcal{V}_{\text{img}} \equiv (1/n) \sum_{i,j} \left(S(x_i, y_j) - \bar{S}\right)^2$ of the n image pixels $S(x_i, y_j)$ (with mean $\bar{S}$) and the mean square error $\mathcal{V}_{\text{err}} \equiv (1/n) \sum_{i,j} \left(S(x_i, y_j) - \widehat{S}(x_i, y_j)\right)^2$ where $\widehat{S}(x_i, y_j)$ are the predicted pixel values and the (i, j) run over all pixels in the image[1]. In addition, we provide explained variances for each analyzed image scale separately to highlight the relative contribution of each image scale to the overall error. In this case, variances, errors and predictions are computed in transform coefficient space instead of pixel space, and the (i, j) run over all coefficients belonging to a given scale.

We compared the explained variances of the two image bases on the familiar Brodatz texture database [3] which contains 111 640 × 640-sized grayscale images scanned off black and white prints. In addition, we tested both bases on an image database containing more than 1600 never compressed RGB color images

[1] We prefer explained variance over the frequently used mean square error since it provides a relative measure of image modeling performance and thus is independent of the actual scaling of the pixel values.

provided by the German Federal Office for Information Security. Although textures are not representative for natural images, they constitute a good testbed for local Markov random field (MRF) type image models such as ours since they are statistically uniform at a limited range of scales and orientations and thus help to reveal potential weaknesses of a model which otherwise could remain invisible in natural images with their variable mixture of local textures. Typically, a higher performance of a local MRF-type model on a texture database leads to a higher performance on natural images which was also the case in our tests.

Fig. 3 shows that the Laplacian image basis outperforms the wavelet basis significantly in terms of explained variance, even if training and test region of the images were not the same. This happened consistently, both in pixel space (first bar group, "pixel space reconstruction") and across the different scales of the Laplace or wavelet decomposition (bar groups numbered 1–4). For RGB images, the advantages of the Laplacian basis are less pronounced but still significant, since the high correlations between the color channels are exploited by the models as well and thus lead to smaller differences in their prediction performance, see Fig. 4. The better prediction performance of the Laplace basis can be attributed to two factors: (1) Laplace coefficients are higher correlated with their neighbourhood than wavelet coefficient magnitudes and thus are easier to predict; (2) The Laplace pyramid is overcomplete by a factor of 4/3 which allows for a more finely grained modeling of the local dependency structure.

3.2 Steganalysis Performance

The wavelet and the Laplacian image models were used for determining both marginal and residual statistics for the above-mentioned image database of never compressed color images. This is known to be the most difficult setting for steganalysis, as the entropies of the images remain high. For instance, JPEG artifacts contained in the images from previous compression simplify steganalysis [11]. Different embedding algorithms and rates were used for creating sets of stego images from these clean images.

The comparison of the distributions of the residuals for a clean color image and its corresponding stego version can be seen in Fig. 5. Here, for the sake of easily recognizable differences, the complete least significant bit plane was replaced by white noise, serving as a representative of a very simple steganogram. In Fig. 6, the corresponding distributions for the same color image are shown for the Laplacian image model. The differences are very small, compared to the distributions computed with the wavelet model.

From these distributions, we computed the statistical moments for every color image that serve as associated feature vectors. The dimensionality of these vectors was 120 for wavelet decomposition, and 48 for Laplacian decomposition. After normalizing the components of these vectors independently, we carefully selected the parameters of the support vector machine and its Gaussian kernel (SVM, cf. [5]) C and γ by employing a 5-fold cross-validiation scheme specifically adapted to the steganalysis scenario [16] on 1000 clean and 1000 stego images. Subsequently, we trained the SVM using the values of C and γ determined in the

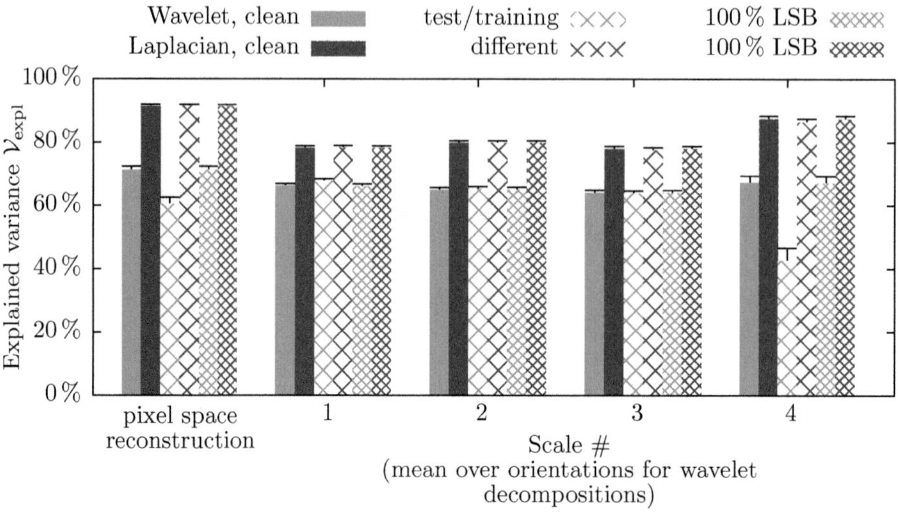

Fig. 3. Explained variances $\mathcal{V}_{\mathrm{expl}}$ for Wavelet and Laplacian decompositions averaged over the Brodatz database, for each subband scale and pixel space reconstruction (leftmost bar group). The black bars indicate the standard error of the mean over the database. Results based on the wavelet representation are shown in orange, results based on the Laplacian representation in blue. The title "test/training different" in the legend belongs to both representations and indicates that training and prediction of the image model were carried out on different sections of the same image.

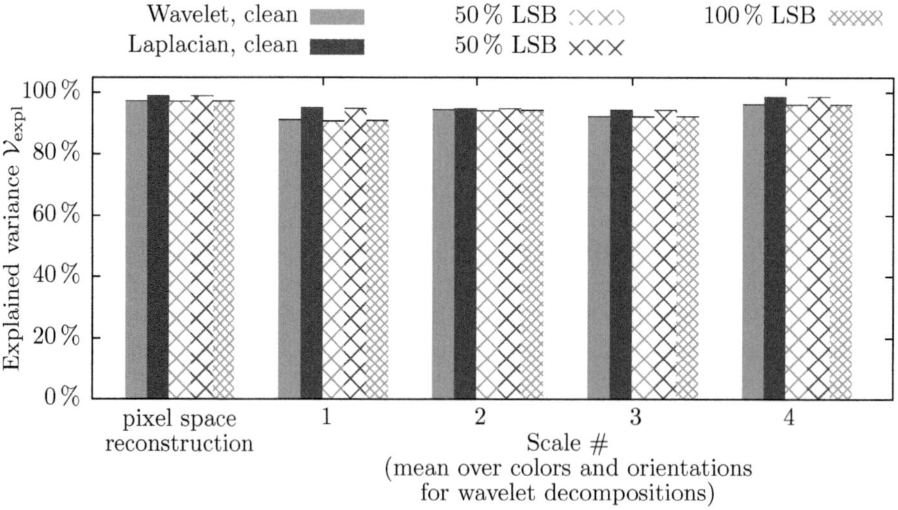

Fig. 4. Just as Fig. 3, but for color images and different embedding rates. The strong correlations between the RGB color channels improve the prediction, thus differences between the methods are considerably smaller than in grayscale images.

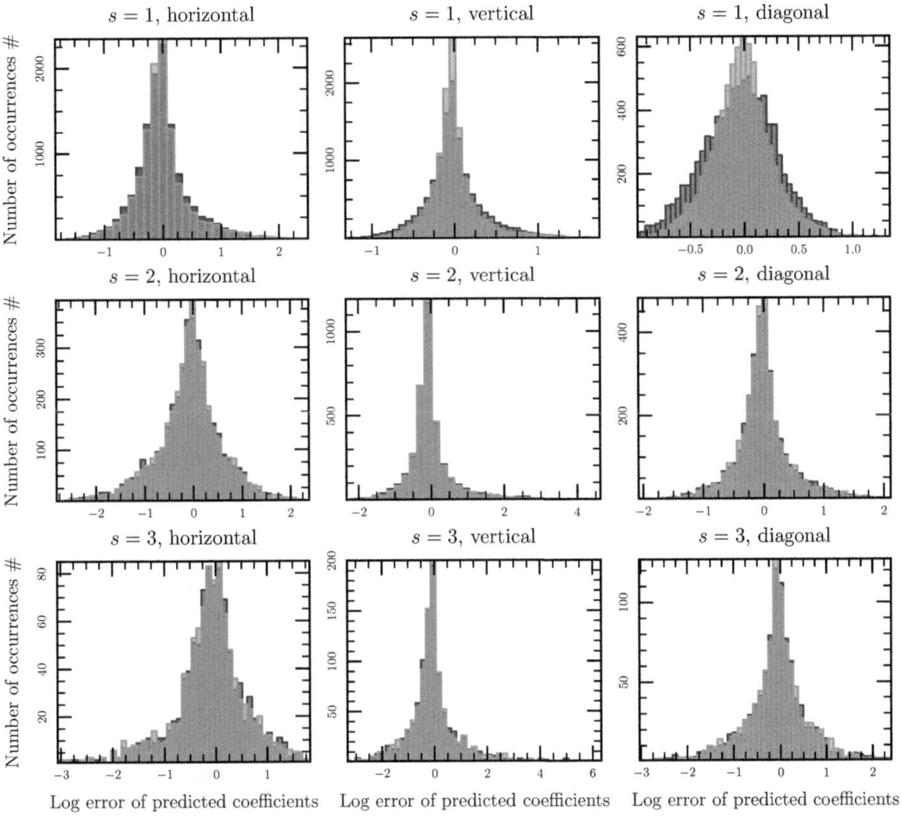

Fig. 5. Log error for each scale and orientation of a wavelet pyramid for the green channel of a clean image (light grey) showing a church vs its stego version (black). Uniform random noise was embedded into the least significant bit plane, with a rate of 100 %.

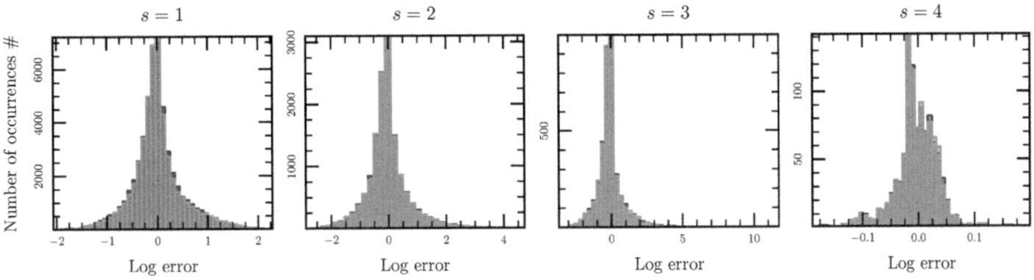

Fig. 6. Just as Fig. 5, but for the Laplacian image model

last step on 2000 images (the training data, 1000 clean and 1000 stego images), and then tested with a set of 1200 examples (600 clean and 600 stego images). We randomly divided the entire set into training and test sets and averaged over 100 splittings, which enabled us to estimate the error of the detection rate on the test set. The results showed a considerable variance for the standard test scenario of steganalysis with a fixed false positive rate of 1 %, but averaging over the 100 splittings turned out to be sufficient for finding statistically significant differences between the Laplace and wavelet basis.

We tested with two distinct normalization methods: Either each component of the vectors is scaled independently to $[0, 1]$; or only the quantile range $Q_{0.95} - Q_{0.05}$ of a component is standardized to $[0, 1]$, while clipping values above and below. We call the former standard and the latter interquantile normalization. With normalization, features of high magnitude exert a less dominant influence on the results. Additionally, for interquantile normalization, the detrimental effects of outliers are reduced.

If adapting the wavelet-based classifier such that the false positive rate falls below 1 %, then (12.6 ± 0.2) %, (30.8 ± 0.7) %, and (69.5 ± 0.9) % of the stego images can be detected for standard normalization and embedding rates of 10 %, 25 %, and 50 %, respectively. When normalizing the interquantile range of the features to $[0, 1]$, the detection rates are (8.5 ± 0.2) %, (14.4 ± 0.7) %, and (65.7 ± 1.4) %. This is a higher performance than found by Lyu & Farid in [13]. The accuracies are averaged over 100 test runs. These runs differ in that we randomly divided the entire set into training and test sets repeatedly. The full set of the results is shown graphically in Fig. 7, plus the prediction rate on ternary embeddings [10,12] and on ±1 embeddings. ±1 embeddings conserve parity properties of the images as it is described in [12]. In this case, for embedding rates of 50 % and 25 %, (76.8 ± 1.4) % and (37.1 ± 0.6) % of the stego images can be revealed for standard normalization, and (81.9 ± 0.7) % and (20.7 ± 1.0) % for quantile normalization. The error bars denote the standard error of the mean $\overline{\sigma} = \sigma/\sqrt{n}$ over the $n = 100$ test runs, where $\sigma = \sqrt{\mathrm{Var}(x)}$ and x is the true positive rate found.

The prediction accuracies for higher embedding rates are frequently higher for standard normalization. However, for a reduced feature vector, containing only error residuals or even a subset thereof, interquantile normalization allows for higher detection rates.

In Fig. 7, Laplacian predictions are only shown for the highest embedding rate of 50 %, because their accuracies fall rapidly to values close to that achieved by random guesses: The true positive accuracies are (4.5 ± 0.1) % and (3.9 ± 0.1) %, for standard and interquantile normalization, respectively. Obviously, the detection rates of the Laplacian steganalyzer are inferior to those of the wavelet steganalyzer.

4 Discussion

The hypothesis we investigate is that a better image model should yield a better detection rate. According to Figures 3 and 4, the Laplacian representation is

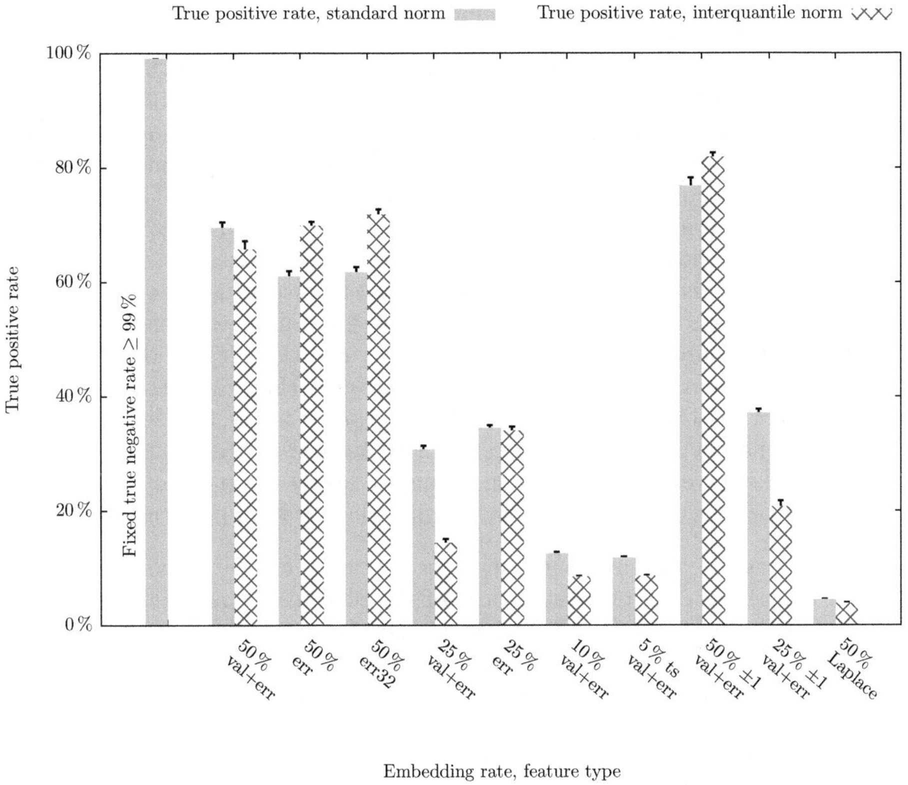

Fig. 7. Comparison of classification accuracies for four distinct embedding rates (50 %, 25 %, 10 %, 5 %), for two types of predictions (wavelet, Laplacian) and two normalization methods (interquantile, standard). The used embedding scheme is least significant bit replacement, except for "ts", which indicates ternary, scanner-adaptive embedding, and "±1", which indicates ±1 embedding. "err+val" denotes the usage of marginal and log error statistics, "err" denotes the utilization of log error statistics only, "err32" that of the log error statistics of the $3 \cdot 32$ moments originating from the finest scales.

better than the wavelet representation in terms of explained variance. However, the steganalysis experiments show that the detection performance of the wavelet model is significantly higher than that of the Laplacian so that our plausible hypothesis turned out to be wrong. As a consequence the wavelet representation must have additional properties that allow for a better steganalysis performance in spite of its inferior modeling capability.

We think that the improved discriminability in the wavelet domain can be mainly attributed to two reasons:

1. In comparison to the Laplacian, the dimensionality of the feature vectors in the wavelet representation is tripled since there are more subband statistics. It is a well-known fact (see, e. g., [9]) that the probability that two classes are separable increases with the dimensionality of their representation.

2. As described in Sect. 2, the steganalyzer uses a threshold on the coefficients before estimating the statistical moments of the subband coefficients. This turns out to be a critical step since this prevents small coefficients from influencing the feature vectors used for classification. This type of thresholding estimator is a well-known concept in signal processing where such estimators are used for denoising. We can think of the prediction step in the steganalyzer as a denoising step since we reconstruct the "true" or denoised image from the contaminated or "noisy" stego image. Here comes into play a theorem by Donoho & Johnstone [6] which states that a threshold estimator has a higher denoising performance if the signal representation is sparse. Sparse in this context means a representation of a signal that needs only a few coefficients to approximate a signal with low error. Although we did not formally test this on our data, it is a common observation that wavelets are a sparser representation of natural images when compared to the Laplacian pyramid. As a consequence, we can expect a more accurate estimation of the residual in the wavelet representation leading to better estimates of the subband statistics which finally results in a higher classification performance.

The advantage of a high-dimensional representation can be demonstrated in a simple experiment: On our test dataset of the 1600 color image pairs in uncompressed format with a 50 % LSB embedding (see Sect. 3.2), we first computed the model predictions in the Laplacian domain and transformed the predicted images back into their pixel representation. In the next step, both original image and stego image were wavelet-transformed and subjected to the same analysis as before. In this way, the modeling step took place in the Laplacian domain, whereas the classification step was based on feature vectors with the three-fold dimensionality of the wavelet domain. As a result, detection performance of this hybrid steganalyzer improved considerably from a true positive rate of (4.5 ± 0.1) % of the pure Laplacian steganalyzer to (37.3 ± 1.0)%, although the performance of the pure wavelet steganalyzer of (69.5 ± 0.9) % could not be achieved. The results are given at a fixed true negative rate of 99 %.

A second experiment highlights the critical influence of the thresholding step: Disabling the thresholding process in the wavelet steganalyzer reduces the detection performance from (69.5 ± 0.9) % to (27.1 ± 0.6) %. This demonstrates that wavelet thresholding plays a role of similar importance to the higher dimensionality of the resulting feature vectors.

Finally one might ask why the hybrid steganalyzer from the first experiment did not reach the performance of the pure wavelet steganalyzer. The reason for this can be seen in a third experiment where we analyzed the explained variance in the wavelet domain of both the Laplacian model predictions and the wavelet model predictions (cf. Table 1). The results show that, although the Laplacian model is more accurate in its own domain, it does not reach the accuracy of the wavelet model in the wavelet domain. In our opinion, this accounts for the observed performance difference between the hybrid and the pure wavelet steganalyzer.

Table 1. Mean over orientations for wavelet decompositions in % explained variance of the hybrid and the pure wavelet steganalyzer

	Clean images (wavelet)	Clean images (hybrid)	Stego images (wavelet)	Stego images (hybrid)
Pyramid level 1	89.6 %	88.7 %	88.5 %	87.9 %
Pyramid level 2	91.8 %	82.5 %	91.2 %	81.3 %
Pyramid level 3	91.9 %	94.3 %	91.4 %	94.2 %

5 Conclusion

In this study we analyzed the relationship between image modeling and detection performance in a universal Lyu & Farid type steganalyzer. Our results show that a high performance in image modeling does not directly transfer to a higher steganalysis performance. Although the Laplacian representation leads to a better image model, it shows an inferior detection performance. From the steganalysis point of view, the characteristics of the wavelet representation (sparse representation and higher dimensionality of the feature vector) turned out to be more important as it evidently allows the final classification stage to discriminate between the resulting feature vectors more easily. Furthermore, it seems important to stay in the same transformation space in all steganalysis steps from image modeling, thresholding, computation of feature vectors to classification. This study shows a connection between sparsity of the image basis, denoising and steganalysis performance. In future work, we plan to make this link more explicit.

Acknowledgments

P.H.D. Le was supported by the grant "Detection of Steganography in Images with Statistical Models" of the German Federal Ministry of Research and Education, and partially—together with M. O. Franz—by the grant "Communication and Automation" of the Baden-Württemberg Foundation. This work was also partially supported by the German National Academic Foundation. We would like to thank M. Messmer, O. Schönemann, Medav GmbH (Ilmenau) and J. Keppler for their help in the project, and H. Schwigon of the German Federal Office for Information Security for providing us useful data and discussions.

References

1. Adelson, E.H., Burt, P.J.: Image data compression with the Laplacian pyramid. In: Proceedings of the 1981 Conference on Pattern Recognition and Information Processing, pp. 218–223. IEEE Computer Society Press, Los Alamitos (1981)
2. Avcibas, I., Memon, N.D., Sankur, B.: Steganalysis using image quality metrics. IEEE Transactions on Image Processing 12(2), 221–229 (2003)

3. Brodatz, P.: Textures: A Photographic Album for Artists and Designers. Dover Publications, New York (1966)
4. Buccigrossi, R.W., Simoncelli, E.P.: Image compression via joint statistical characterization in the wavelet domain. IEEE Transactions on Image Processing 8(12), 1688–1701 (1999)
5. Chang, C.C., Lin, C.J.: LIBSVM: A library for support vector machines (2001), software available at http://www.csie.ntu.edu.tw/~cjlin/libsvm/
6. Donoho, D., Johnstone, I., Johnstone, I.M.: Ideal spatial adaptation by wavelet shrinkage. Biometrika 81, 425–455 (1993)
7. Geisser, S., Eddy, W.F.: A predictive approach to model selection. Journal of the American Statistical Association 74(365), 153–160 (1979)
8. Goljan, M., Fridrich, J.J., Holotyak, T.: New blind steganalysis and its implications. In: Security, Steganography, and Watermarking of Multimedia Contents VIII 6072(1), pp. 1–13 (February 2006)
9. Haykin, S.: Neural Networks, 2nd edn. Prentice-Hall, Upper Saddle River (1999)
10. Holotyak, T., Fridrich, J.J., Soukal, D.: Stochastic approach to secret message length estimation in ±k embedding steganography. In: Delp, E.J., Wong, P.W. (eds.) Security, Steganography, and Watermarking of Multimedia Contents. Proceedings of SPIE, vol. 5681, pp. 673–684. International Society for Optical Engineering, SPIE, San Jose (2005)
11. Holotyak, T., Fridrich, J.J., Voloshynovskiy, S.: Blind statistical steganalysis of additive steganography using wavelet higher order statistics. In: Dittmann, J., Katzenbeisser, S., Uhl, A. (eds.) CMS 2005. LNCS, vol. 3677, pp. 273–274. Springer, Heidelberg (2005)
12. Ker, A.D.: Improved detection of LSB steganography in grayscale images. In: Fridrich, J. (ed.) IH 2004. LNCS, vol. 3200, pp. 97–115. Springer, Heidelberg (2004)
13. Lyu, S., Farid, H.: Steganalysis using higher-order image statistics. IEEE Transactions on Information Forensics and Security 1(1), 111–119 (2006)
14. Rasmussen, C.E.: Gaussian processes in machine learning. In: Bousquet, O., von Luxburg, U., Rätsch, G. (eds.) Machine Learning 2003. LNCS (LNAI), vol. 3176, pp. 63–71. Springer, Heidelberg (2004)
15. Rasmussen, C.E., Williams, C.K.I.: Gaussian Processes for Machine Learning. MIT Press, Cambridge (2006)
16. Schwamberger, V., Franz, M.O.: Simple algorithmic modifications for improving blind steganalysis performance. In: Proceedings of the 2010 Workshop on Multimedia and Security, MM&Sec 2010. ACM Press, New York (2010)
17. Schölkopf, B., Smola, A.J.: Learning with Kernels. In: Support Vector Machines, Regularization, Optimization, and Beyond, MIT Press, Cambridge (2002)
18. Simoncelli, E.P., Adelson, E.H.: Subband transforms. In: Woods, J.W. (ed.) Subband Image Coding. Kluwer Academic Publishers, Norwell (1990)

The Square Root Law in Stegosystems with Imperfect Information

Andrew D. Ker

Oxford University Computing Laboratory, Parks Road, Oxford OX1 3QD, England
adk@comlab.ox.ac.uk

Abstract. Theoretical results about the capacity of stegosystems typically assume that one or both of the adversaries has perfect knowledge of the cover source. So-called perfect steganography is possible if the embedder has this perfect knowledge, and the Square Root Law of capacity applies when the embedder has imperfect knowledge but the detector has perfect knowledge. The epistemology of stegosystems is underdeveloped and these assumptions are sometimes unstated. In this work we consider stegosystems where the detector has imperfect information about the cover source: once the problem is suitably formalized, we show a parallel to the Square Root Law. This answers a question raised by Böhme.

1 Introduction

The most important ingredient of a stegosystem is the cover source, and the relevant epistemology – who knows what about the covers – is sometimes neglected in the literature. But knowledge about the covers is fundamental to the problem. It is known that perfect steganography is possible, conveying an amount of payload linear in the cover size with zero risk of detection, as long as the embedder has perfect knowledge of the cover distribution [18]. This is so because the embedder can ensure that all the statistics of the cover source are preserved by their embedding function[1]. On the other hand, we have a completely different situation when the embedder does not know, and fails to preserve, all the statistics of the covers: this leads to the Square Root Law [15,6,13] stating that the information conveyed can grow at most of order $\sqrt{n}$, where n represents the size of the cover, or face eventual certain detection.

The latter result assumes that the *detector* has perfect knowledge of the distribution of covers, so to compare potential stegotexts against it. This attack model is defensible: taking the role of the embedder, it is conservative to assume that the opponent has perfect knowledge, since we cannot then make the error of underestimating them. However, it is arguable that for steganography in real objects, be they digital images, audio, or even plain text, no perfect model of the source exists [1] and therefore the opponent cannot have such perfect knowledge

[1] In fact, perfectly secure steganography remains theoretically possible even if the embedder does not know about the cover source, as long as they have an inexhaustible cover supply from which to sample [9], but such embedding is practically infeasible.

R. Böhme, P.W.L. Fong, and R. Safavi-Naini (Eds.): IH 2010, LNCS 6387, pp. 145–160, 2010.

any more than can the embedder. In practice, steganalysts create models of cover objects by examining finitely many examples of genuine covers, which leads to an imperfect level of knowledge about the source. What is the consequence of this imperfect knowledge for a detector, and can the square root capacity rule be exceeded? This question was raised by Böhme [2] who pointed out

> "The square root law is supported with evidence for fixed cover models of the adversary... so far it does not anticipate adversaries who refine their cover models adaptively."

Modifying the square root law to this circumstance, in the simplest and most abstract setting, is our goal in this paper.

In Sect. 2 we describe and justify our framework for a stegosystem with imperfect information, and in Sect. 3 state and prove a modified square root law for this situation. There are new statistical challenges in this setting and we require a discussion of the statistical property of *unbiasedness*, along with some analysis. The significance of the theorem is discussed in Sect. 4: we shall see that the embedder remains limited to payloads of order $\sqrt{n}$ as long as the detector has access to a linear number of examples of covers from which to learn about their distribution. A superlinear number of covers conveys no asymptotic advantage but a sublinear number of covers allows the embedder to improve on the order of capacity $\sqrt{n}$. We briefly extend the result to the case when the embedder also learns about the cover source from examples, and adapts their embedding accordingly. We also discuss many avenues for further research.

We will use the following notation: $f(n) = O(g(n))$ indicates that f grows asymptotically no faster than g, i.e. $|f(n)| \leq c|g(n)|$ for $n \geq N$, for some c and N; $f(n) \sim g(n)$ means that f and g are asymptotically equal, i.e. $f(n)/g(n) \to 1$ as $n \to \infty$; $f(n) = o(g(n))$ means that f grows strictly slower than g, i.e. $f(n)/g(n) \to 0$ as $n \to \infty$. Ω is the converse to O: $f(n) = \Omega(g(n))$ if $g(n) = O(f(n))$. We will use Knuth's notation [8] for the falling Pochhammer symbol $n^{\underline{k}} = n(n-1)\cdots(n-k+1)$, so that the binomial coefficients are $\binom{n}{k} = n^{\underline{k}}/k!$. $\mathrm{E}[X]$ and $\mathrm{Var}[X]$ will denote the mean and variance of the random variable X, and $X \sim \mathrm{Bi}(n,p)$ that X takes the binomial distribution with size parameter n and probability parameter p.

2 Stegosystems with Imperfect Information

In a stegosystem with perfect information, both parties know the exact distribution of the covers. In this scenario it has already been shown that the embedder can communicate completely undetectably at a nonzero (linear) rate by preserving all the statistics of the covers [18]. However, no such system, for embedding in real-world cover media, has ever been constructed. That is because of the distinction, made clear by Böhme in [2,1], between *artificial* covers – mathematical structures such as random variables or Markov chains – and *empirical* covers which are

> "digital representations of parts of reality... They have to be obtained through observation of reality... All kinds of digitised media data, such as images, audio or video files, belong to this class."

Böhme argues that the true distribution of real-world covers is incognisable. Given this, we should focus on *stegosystems with imperfect information* where the distribution can only be obtained approximately, by observation of the source.

It is consideration of imperfect embedding, where some statistic of the cover source is not preserved by embedding, that leads to the Square Root Law of capacity [15,6,13]. As the cover size increases the embedding rate must diminish, lest evidence in abnormal statistics build to eventual certain detection. But these results still require the detector to have perfect knowledge of the cover source, arguably an equally unrealistic situation to that of perfect embedding. Böhme comments that

> "security in empirical covers depends on the steganalyst's knowledge of the cover source and the amount of evidence available to distinguish any abnormality"

and we aim to quantify how secure capacity – the maximum payload which cannot be detected with nontrivial error rates – depends on this steganalyst's knowledge and the size of the stego object.

Proving a theoretical result about capacity requires us to remain in the world of artificial covers, and in Sect. 3 we will use the very simplest mathematical model, of i.i.d. binary sequences, but we can reflect something of the incognisability of cover distributions by saying that the exact value of the parameter(s) of the model are not known to either party. This is our model for a stegosystem with imperfect information. To an extent, this contradicts Kerckhoffs' Principle: we are assuming that the "enemy" (the steganalyst) does not have full knowledge of the "system", if we take the system to include the covers; in view of the preceding discussion, this demonstrates that Kerckhoffs' Principle should not be applied blindly to the steganography problem. The covers are, in reality, external to the system.

For our result, most of the assumptions apart from knowledge of the covers will not be relevant: it will not matter whether the detector knows the embedding algorithm or the size of embedded payload, though it is important that they not know which locations in the cover have been used for payload. The latter information is part of the secret embedding key, but for more on this assumption see [13].

There remains the question of how we measure the embedder's and detector's uncertainty about the parameter(s) of the artificial cover model. As in the classical square root law setting, we will assume that the embedder does not evolve and proceeds with some fixed method which replaces symbols in the cover by symbols with a non-identical distribution. For the detector, we will be guided by the practice of steganalysis, in which researchers most commonly train a detector using sets of genuine covers. We will assume that the detector has access to some genuine covers from an independent but identically-distributed source to that of the embedder, i.e. limited access to a cover oracle. It will turn out (Subsect. 4.1) that any finite limit on the size or number of covers leads to a trivial situation, so we will suppose a limit of m_n accesses to the oracle when the cover size is n. (Briefly, in Subsect. 4.3, we will consider the case when the embedder also makes use of a cover oracle. The result is rather different.)

We should ask whether it is plausible that the detector has access to the embedder's cover source. Of course it depends on the application, but for example one could imagine that a well-motivated steganalyst can at least determine forensically the model of camera used to take cover images, and then purchase their own. Or, once the steganographer comes under suspicion, seize the camera itself. A similar detector model, using limited access to a cover oracle, is found in [10], although no capacity result is proved. The limitation in that work, inspired by computational complexity bounds, is of polynomially-many accesses to the oracle; in our model we shall see (Subsect. 4.1) that linearly-many accesses suffice to recover a square root law.

Taking the role of the embedder, it is conservative to assume that the opponent has perfect knowledge of the cover source. It is also rather pessimistic, and the work in this paper is motivated by the desire to relax the condition. We, the embedder, may feel happier about transmitting information at a rate *above* the capacity predicted by the square root law if we could, for example, prove that our opponent needed to spend an exponential amount of time learning about the cover source in order to catch us. Sadly for steganographers, it will turn out that this is not the case.

3 The Square Root Law for Imperfect Detectors

To explain the difficulties in constructing a modified square root law for imperfect detectors, and to contrast with new results, we begin by restating the simplest square root law for the classical system when the detector has perfect knowledge of the cover source.

Theorem 1. *Suppose that cover objects consist of sequences of symbols which are Bernoulli random variables with parameter p; we exclude the pathological cases $p = 0, 1$. Suppose that the embedder replaces the cover symbols, independently with probability γ, by stego symbols which are Bernoulli with parameter $q \neq p$. The steganalyst wishes to distinguish the cases $\gamma = 0$ and $\gamma > 0$. Let n be size of the cover, i.e. the number of symbols.*

(1) If $\gamma\sqrt{n} \to \infty$ as $n \to \infty$ then, for sufficiently large n, the steganalyst can create an arbitrarily accurate detector.

(2) If $\gamma\sqrt{n} \to 0$ as $n \to \infty$ then, for sufficiently large n, every detector has arbitrarily low accuracy.

We interpret the theorem to mean that $\gamma = O(1/\sqrt{n})$ is the critical rate: unless γ diminishes at least this fast, large enough n will result in certain detection. If the embedding involves a simple substitution (no source coding by the embedder) then the payload size is γn and this must grow slower than $\sqrt{n}$, hence the name Square Root Law.

Theorem 1 has been known for a few years but the first published proof is found in [13]. We briefly sketch the techniques used, for comparison with the result about imperfect stegosystems below. For (1), the simple detector which compares the proportion of observed "1" symbols with p, and rejects the null hypothesis $\gamma = 0$ if the difference is significant, can be analyzed using tail inequalities for the binomial distribution. It can be shown that this detector has false positive and negative rates which tend to zero as $n \to \infty$ as long as $\gamma\sqrt{n} \to \infty$. For (2), the Kullback-Leibler divergence between the distribution of the observations when $\gamma = 0$ and $\gamma = \gamma_1 > 0$ is computed: it can be shown that this tends to zero as long as $\gamma_1\sqrt{n} \to 0$, which means that any hypothesis test for $\gamma = 0$ against $\gamma > 0$ must have power tending to size (i.e. the error rate tends to that of a purely random decision).

Some of the apparent limitations in this result can be avoided. First, we have stated it here in the context of binary alphabets, but this is not essential and the same is true for arbitrary finite alphabets [13], for a reason we shall discuss in Subsect. 4.2. Second, the symbols in the cover must be independent. This is a severe condition not likely to be satisfied by real cover media. However, the result still holds when there is non-pathological dependence between the symbols [6]. A square root relationship between cover size and capacity has been verified empirically, for contemporary steganography and steganalysis methods, in [15]. Finally, the theorem does not address the critical case when $\gamma = c/\sqrt{n}$, in which case the value of c determines a maximum possible detection accuracy. This is arguably the most important situation because it gives the embedder a genuine "secure" capacity for their embedding; this is related to Steganographic Fisher Information which has recently been examined in [12, 5]. One other minor limitation is that formalising the embedding as affecting each symbol independently with probability γ is not quite right for embedding a fixed-length message; this is addressed in [13].

However there is one limitation that remains: the opponent is assumed to have knowledge of p. They do not need to know q or γ: the detector constructed for (1) does not depend on these quantities, and the bound in (2) applies even if we grant the detector this knowledge anyway. But if we wish to prove a similar result for a stegosystem with imperfect information, as described in Sect. 2, the detector must not know p, instead learning something about its value through a limited number of observations of true cover bits from an oracle.

There is no great difficulty in adapting part (1) of Th. 1 to such a situation. But part (2) is simply untrue, because there *does* exist a detector which distinguishes cover and stego objects, with the same condition on γ as in Th. 1 (2) and without using the oracle at all: it is simply the detector which makes use of a fixed value of p "hardwired" into it, in the case when it just happens to have the correct p. It is difficult to use Kullback-Leibler divergence to bound the performance of a detector subject to the constraint that the detector does not know a certain piece of information: an accurate detector *does* exist, even if it requires a very lucky guess to pick the right value of p to begin with.

Inspired by the idea of a "lucky guess," we tried to model the true value of p as random, uniformly on $[0, 1]$: placing a uniform prior on unknown parameters seems a reasonable way to reflect absence of information about them. Unfortunately this model does not function as desired because the probability of observing a "1" in stego sequences is $p + \gamma(q - p)$ which is *not* uniformly distributed: it is biased towards either q (if q is known) or $1/2$ (if q is also given a uniform prior). Computing the KL divergence between the cover and stego sequences when p has a uniform prior is algebraically challenging because, unconditionally, the bits are no longer independent; the calculations are far too long to include here but the author has outlined them in a technical report [14]. It turns out that this KL divergence is always positive, even when the cover oracle is completely absent – but the detector was supposed to have no knowledge of the cover source! This means that the stego signal leaks information about the presence of payload even when the cover oracle is disregarded. Sadly, placing a uniform prior on p did not properly reflect a lack of knowledge of p, and, crucially, did not allow us to use Kullback-Leibler divergence to bound the accuracy of ignorant detectors. For more details of this argument, see [14].

Instead, we seek to rule out those detectors which have knowledge of p, using the statistical concept of unbiasedness. A test for the null hypothesis class $H_0 : \theta \in \Theta_0$ against an alternative class $H_1 : \theta \in \Theta_1$ is called *unbiased* if, whenever the true value of θ is in Θ_1, the probability of rejection of H_0 is never less than the probability of rejection of H_0 when $\theta \in \Theta_0$. In the language of detectors, this is to say that the probability of a true positive is always at least the probability of a false positive.

There is comprehensive information about the theory and application of unbiased hypothesis tests in [16, Chs. 4-5]: it is popular to restrict attention to unbiased tests for many reasons, including the existence of uniformly most powerful (UMP) tests within this class when general UMP tests do not exist. In the application we consider in this paper, we can justify restricting attention to unbiased detectors for two reasons. First, any detector with more false positives than true positives is not going to be much use in practice. Second, forbidding a test biased towards any particular value of p reflects a lack of knowledge of p, without even placing a prior distribution on it, which was exactly our aim. Having restricted to unbiased tests, the detector with a "hardwired" value of p is inadmissible because it is too likely to give a false negative when the true value of q is actually the hardwired value of p. We can then make use of literature on UMP unbiased tests for exponential families to prove a modified square root law. It links secure embedding rates with both the cover size and the level of imperfect cover information at the detector, defined in terms of a number of true cover samples which the detector receives from an oracle.

We will retain the simplicity of Theorem 1 in that the covers will still be independent bit strings; when the detector learns about the cover source through

m_n bits from a cover oracle (we would expect that m_n is an increasing function of n), and if restricted to unbiased detectors, we then have the following result.

Theorem 2. *Suppose that cover objects consist of sequences of symbols which are Bernoulli random variables with parameter p; we exclude the pathological cases $p = 0, 1$. Suppose that the embedder replaces a randomly-selected proportion γ of the cover symbols by stego symbols which are Bernoulli with parameter $q \neq p$. Let n be size of the cover, i.e. the number of symbols.*

The steganalyst wishes to distinguish the cases $\gamma = 0$ and $\gamma > 0$, but they have no knowledge of p or q, instead they have access to m_n independently-generated cover symbols from which they may learn about the cover source.

(1) If

$$\frac{\gamma}{\sqrt{\frac{1}{m_n} + \frac{1}{n}}} \to \infty$$

as $n \to \infty$ then, for sufficiently large n, the steganalyst can create an arbitrarily accurate detector.

(2) If

$$\frac{\gamma}{\sqrt{\frac{1}{m_n} + \frac{1}{n}}} \to 0$$

as $n \to \infty$ then, for sufficiently large n, every unbiased *detector has arbitrarily low accuracy.*

The proof follows. We will interpret the result in Subsect. 4.1, and consider its limitations, paralleling those of Th. 1, in Subsect. 4.2.

3.1 Proof of Theorem 2 (1)

This half of the proof is the easier, using techniques similar to that of Th. 1 (1). Write X for the number of 1 bits in the cover stream, and Y for the number in the object to be classified. Then $X \sim \mathrm{Bi}(m_n, p)$ and $Y \sim \mathrm{Bi}(n, p + \gamma(q - p))$; under the null hypothesis that the object is a cover, $\gamma = 0$, otherwise $\gamma > 0$.

A suitably asymptotically powerful detector can be constructed, without knowledge of p, q, or γ, using the obvious statistic which measures the difference in proportion of ones between the unknown signal and the cover bits:

$$T = \frac{X}{m_n} - \frac{Y}{n} \tag{1}$$

and we will reject the null hypothesis, giving a positive detection of steganography, if $|T|$ exceeds a critical threshold $c\sqrt{\frac{1}{m_n} + \frac{1}{n}}$ for some positive constant c.

The variance of a binomial distribution $\mathrm{Bi}(k, p)$ is bounded by $\frac{1}{4}k$, regardless of p, so

$$\mathrm{Var}[T] \leq \frac{1}{4}\left(\frac{1}{m_n} + \frac{1}{n}\right) .$$

Under the null hypothesis, when the unknown object is a cover, $\mathrm{E}[T] = 0$, so the probability of a false positive α satisfies

$$\alpha = \Pr\Big(\big|T - \mathrm{E}[T]\big| > c\sqrt{\tfrac{1}{m_n} + \tfrac{1}{n}}\Big) \le \frac{\mathrm{Var}[T]}{c^2\big(\frac{1}{m_n} + \frac{1}{n}\big)} \le \frac{1}{4c^2}$$

(the first inequality is Chebyshev's). This can be made arbitrarily small by suitable choice of c.

Under the alternative hypothesis, when the unknown object is a stego object, $\mathrm{E}[T] = \gamma r$ where $r = p - q \neq 0$, so the probability of missed detection β satisfies

$$\begin{aligned}\beta &= \Pr\Big(\big|T - \mathrm{E}[T] + \gamma r\big| \le c\sqrt{\tfrac{1}{m_n} + \tfrac{1}{n}}\Big)\\ &\le \Pr\Big(\big|T - \mathrm{E}[T]\big| > \Big|\gamma|r| - c\sqrt{\tfrac{1}{m_n} + \tfrac{1}{n}}\Big|\Big)\\ &\le \frac{\frac{1}{4}\big(\frac{1}{m_n} + \frac{1}{n}\big)}{\gamma^2 r^2 - 2\gamma|r|c\sqrt{\frac{1}{m_n} + \frac{1}{n}} + c^2\big(\frac{1}{m_n} + \frac{1}{n}\big)}\\ &\to 0\end{aligned}$$

as $n \to \infty$, for any positive c, because of the assumption that $\frac{\gamma}{\sqrt{\frac{1}{m_n}+\frac{1}{n}}} \to \infty$ and again using Chebyshev's inequality. We have shown that, for sufficiently large n, T distinguishes cover and stego objects with arbitrarily high accuracy.

3.2 Proof of Theorem 2 (2)

For simplicity we will omit the subscript in m_n. The condition

$$\frac{\gamma}{\sqrt{\frac{1}{m_n} + \frac{1}{n}}} \to 0$$

is equivalent to

$$\gamma^2 \frac{mn}{m+n} \to 0 \; . \qquad (2)$$

Let X and Y be as in the previous subsection. A test of whether $\gamma = 0$ or $\gamma > 0$ amounts to a hypothesis test for whether the probability parameter, in the two binomial distributions $X \sim \mathrm{Bi}(m, p)$ and $Y \sim \mathrm{Bi}(n, p + \gamma(q - p))$, is equal or not. The problem of comparing two binomials has been studied in the statistics literature and an optimal (UMP) unbiased test can be found in [16]. To summarise the result in [16, §4.5], the optimal unbiased critical region (i.e. values for which we reject the null hypothesis and give a positive detection) is given by $Y \notin [C_1(X+Y), C_2(X+Y)]$, where C_1 and C_2 are functions which, for each value of $X+Y$, give the desired size (lead to the desired false positive rate). We can imagine a suite of detectors, one for each observed value of $X+Y$, each optimal in the sense of the Neyman-Pearson Lemma and giving a positive result for too-large or too-small values of Y. As long as these detectors all have the same false positive probability then collectively they form an optimal unbiased detector: for any given false positive rate, the false negative rate will be minimal amongst all unbiased detectors.

It is simple to verify that $Y \notin [c_1, c_2]$, for constants c_1 and c_2, is the optimal unbiased detector, given that $X + Y = t$ for some fixed t, by using the result in [16, §4.2]. But the apparently simple conclusion is quite deep because, in general, one cannot expect that optimal conditional tests will together form an optimal unconditional test. It holds in this case because the joint distribution of X and Y forms an exponential family with p as the nuisance parameter, and by applying Th. 4.4.1 from [16]. Moreover, this optimal unbiased detector is not easy to construct in practice because all the conditional tests must have *exactly* the same false positive rate for their combination to be optimal, yet the underlying distribution is discrete and unlikely to oblige with such exact probabilities; we may require randomisation at the detector.

Thankfully, none of these details need concern us. We now know that, for the lowest possible false negative rate at any given false positive rate, the optimal unbiased detector depends only on the value of Y, conditional on $X+Y$. So we can analyze its behaviour using Kullback-Leibler divergence. All we need to do is to show that the KL divergence, between the cases $\gamma = 0$ and otherwise, tends to zero under assumption (2), although the analysis is rather tricky.

Let $P(y;\gamma)$ be the conditional distribution of Y, given $X + Y = t$. We have

$$P(y;\gamma) = \frac{\Pr(Y = y \wedge X = t - y)}{\sum_{y'=0}^{t} \Pr(Y = y' \wedge X = t - y')} = \frac{\rho(\gamma)^y \binom{m}{t-y}\binom{n}{y}}{\sum_{y'=0}^{t} \rho(\gamma)^{y'} \binom{m}{t-y'}\binom{n}{y'}}$$

where $\rho(\gamma)$ is the odds ratio between the events in Y and X,

$$\rho(\gamma) = \frac{(p + \gamma(q-p))/(1-p-\gamma(q-p))}{p/(1-p)} = \frac{1 + \gamma(\frac{q-p}{p})}{1 - \gamma(\frac{q-p}{1-p})} .$$

Without loss of generality we may assume that $q > p$, so that $\rho(\gamma) \geq 1$ with equality at $\gamma = 0$. Furthermore, $\rho'(\gamma) = (q-p)(1-p)/p\big(1-p-\gamma(q-p)\big)^2$, a bounded function, so by the mean value theorem

$$0 \leq \rho(\gamma) - 1 \leq C\gamma , \tag{3}$$

where C is a positive constant. When $\gamma = 0$, the conditional distribution of Y is hypergeometric, therefore

$$\frac{P(y;0)}{P(y;\gamma)} = \frac{\sum_{y'=0}^{t} \rho(\gamma)^{y'} \binom{m}{t-y'}\binom{n}{y'} / \binom{m+n}{t}}{\rho(\gamma)^y} ;$$

where the numerator is the expectation of $\rho(\gamma)^Y$ when Y has a hypergeometric distribution. Taking the expectation over a random variable Y with this distribution, we have

$$\begin{aligned}
0 &\le D_{\mathrm{KL}}(P(y;0)\,\|\,P(y;\gamma)) \\
&= \mathrm{E}_Y\left[\log\left(\frac{P(Y;0)}{P(Y;\gamma)}\right)\right] \\
&= \mathrm{E}\Big[\log\Big(\mathrm{E}\big[\rho(\gamma)^Y\big]\Big) - \log\rho(\gamma)^Y\Big] \\
&= \log\Big(\mathrm{E}\big[(1+\rho(\gamma){-}1)^Y\big]\Big) - \mathrm{E}[Y]\log\rho(\gamma) \\
&\overset{(a)}{=} \log\left(\sum_{k=0}^{n}(\rho(\gamma){-}1)^k\frac{\mathrm{E}[Y^{\underline{k}}]}{k!}\right) - \mathrm{E}[Y]\log\rho(\gamma) \\
&\overset{(b)}{=} \log\left(\sum_{k=0}^{n}(\rho(\gamma){-}1)^k\frac{n^{\underline{k}}\,t^{\underline{k}}}{(m+n)^{\underline{k}}\,k!}\right) - \frac{nt}{m+n}\log\rho(\gamma) \\
&\overset{(c)}{\le} \log\left(\sum_{k=0}^{t}\left(\tfrac{n}{m+n}(\rho(\gamma){-}1)\right)^k\frac{t^{\underline{k}}}{k!}\right) - \frac{nt}{m+n}\log\rho(\gamma) \\
&\overset{(d)}{=} t\left(\log\left(1+\tfrac{n}{m+n}(\rho(\gamma){-}1)\right) - \frac{n}{m+n}\log\rho(\gamma)\right) \\
&\overset{(e)}{\le} \frac{tnm}{2(n+m)^2}(\rho(\gamma)-1)^2 \\
&\overset{(f)}{\le} \frac{nm}{2(n+m)}C^2\gamma^2 \\
&\to 0\,.
\end{aligned}$$

We justify the steps as follows. (a) is the binomial expansion; note that we may take the sum to n because Y is an integer, so any summand with $k > Y$ will be zero. (b) is from the following property of the hypergeometric distribution:

$$\begin{aligned}
\mathrm{E}[Y^{\underline{k}}] &= \sum_{i=k}^{n} i^{\underline{k}}\tbinom{n}{i}\tbinom{m}{t-i}\Big/\tbinom{m+n}{t} = \sum_{i=k}^{n} n^{\underline{k}}\tbinom{n-k}{i-k}\tbinom{m}{(t-k)-(i-k)}t^{\underline{k}}\Big/(m+n)^{\underline{k}}\tbinom{m+n-k}{t-k} \\
&= \frac{n^{\underline{k}}\,t^{\underline{k}}}{(m+n)^{\underline{k}}}\,.
\end{aligned}$$

(c) is because $n^{\underline{k}}/(m+n)^{\underline{k}} \le (n/(m+n))^k$; we may take the sum to t because the summands are zero when $k > n$ or $k > t$. (d) is another binomial expansion. (e) follows from

$$\log(1+ax) - a\log(1+x) \le \tfrac{1}{2}a(1-a)x^2$$

for $a \in [0,1]$ and $x \ge 0$, which may easily be verified by differentiating the difference. (f) is because $t \le m+n$, and using (3). The final limit is just (2).

Since the Kullback-Leibler divergence tends to zero, the information processing theorem [4] implies that the performance of any decision based on Y given $X + Y = t$ must tend to that of a purely random decision. This is true whatever the value of t so the same holds for the collection of detectors, one for each value of $X + Y$, which together form the optimal unbiased detector: no other unbiased detector can do better. Therefore every unbiased detector has asymptotically random output: in the language of Cachin [3], the system is ϵ-secure, for arbitrarily small ϵ, if n is large enough; in terms of detectors, the false positive probability α tends to $1 - \beta$, the true positive probability.

4 Discussion

We conclude with a discussion of the significance of the result as regards secure payload size, its limitations, how it might be extended, and briefly consider the case when the embedder also learns about the cover source through an oracle.

4.1 Interpretation

We consider the consequences of Th. 2 as regards secure embedding capacities. Let us first assume that the embedder requires a fixed average number of changes per bit of payload conveyed (for example in the simplest case of overwriting pseudorandomly-selected locations in the cover with payload bits, whether or not the payload is compressed prior to embedding). Then the payload transmitted M is proportional to γn and we can deduce the following corollaries to Th. 2.

Corollary 1. *If $m_n = \Omega(n)$ then the situation, up to asymptotic order, is the same as in the perfect knowledge case: $M = o(\sqrt{n})$ for asymptotically perfect security.*

The classical square root exponent cannot be reduced, no matter how large m_n. We know this because, of course, the classical Square Root Law tells us that the square root cannot be beaten even when the detector has perfect knowledge of p. We also now know that a linear number of accesses to the cover oracle is sufficient for the detector: more accesses might reduce the secure capacity by a constant multiple, but do not affect its order of growth.

Corollary 2. *If $m_n = o(n)$ then $M = o(n/\sqrt{m_n})$ and the classical square root rate can be beaten. For example, if $m_n \sim n^e$ with $0 \leq e < 1$, then $M = o(n^{1-e/2})$ for asymptotically perfect security.*

This tells us that a linear number of cover examples is necessary for the detector, otherwise their knowledge about the cover source grows too slowly and their opponent can exceed the square root law. In particular,

Corollary 3. *If $m_n = O(1)$ then the embedder can achieve any sublinear payload rate $M = o(n)$ with asymptotically perfect security.*

In the case when the detector only has finitely much information about the cover source, i.e. their information does not grow with the cover size, they will always have some uncertainty about the true value of p. By decreasing the embedding rate, no matter how slowly, the embedder can ensure that the perturbation to the frequency of 0s and 1s falls within this uncertainty.

In fact, of course, all that is required is for γ to be sufficiently small to ensure that the risk of detection is sufficiently low. Quantifying this, over and above the asymptotic relationship, is a problem related to Steganographic Fisher Information (SFI) [12,5], and a direction for future research. It should not be infeasible to estimate the SFI and hence find a concrete capacity bound, in terms of m_n and p, given a KL divergence limit on the risk of detection.

Finally, we turn to the case when the steganographer uses not a simple substitution but an adaptive source-coding at the embedding stage, e.g. matrix embedding [7]. Effectively, the location of the changes, as well as their content, conveys information to the recipient, allowing the payload transmitted to be (slightly) superlinear in γ.

Corollary 4. *With optimal source coding, the payload size M is bounded by $M = O(\gamma n \log(1/\gamma))$, which is achievable (asymptotically) using simple syndromes from the Hamming code.*

If $m_n = \Omega(n)$ then $M = o(\sqrt{n} \log n)$, otherwise $M = o(\log m_n n / \sqrt{m_n})$, for asymptotic perfect security.

Care must be taken to ensure that the embedding locations remain unpredictable by the detector, and that the code does not introduce predictable dependencies between the symbols embedded. This could be achieved by using a random codebook, but the secret key parameterising it might need to be large (note a parallel result in [13]). The equivalent problem for the perfect knowledge detector is examined in [11], where it is shown that the dependencies introduced by a certain type of matrix embedding do not grant, asymptotically, any extra evidence. It is for further work to consider the imperfect information case.

4.2 Limitations and Extensions

Theorem 2 shares the limitations of Th. 1: the version proved applies only to an abstract mathematical structure which is not a good model for any realistic digital medium.

One limitation can be removed immediately: we can extend the binary i.i.d. case to any finite alphabet i.i.d. case by the following argument. If the alphabet is $\Sigma = \{z_1, \ldots, z_N\}$ then we can consider the N binary hypothesis tests which, for each i, count only z_i against all other symbols. If stego objects perturb the frequency of any one of those symbols, one of the hypothesis tests has asymptotically negligible error, and the combination of N hypothesis tests (with the rule that rejecting one leads to rejecting the ensemble) will have asymptotic error a factor of N higher, but still tending to zero. For the converse, it can be shown that whenever the KL divergence between all pairs tends to zero, so does the KL divergence of the entire set.

The most obvious avenue for future research is to remove the i.i.d. condition, for example extending to Markov chains as in [6]. But an involved analysis was required to prove that result, and the difficulty will be compounded by existence of a cover oracle. And we would prefer to go even further, to properly two dimensional objects with arbitrary correlation structures. Perhaps the method just described, which lifts results from the binary to arbitrary finite symbol case, can provide a simplification. Whether the technique we used to prove the main result of this paper can be adapted to more complex cover models depends on whether a sufficiently simple UMP unbiased detector exists. Some of our current research indicates that a square root law can hold even in the case of nonstationary cover sources (perhaps contrary to intuition), and there may be further extensions for an imperfect knowledge detector.

Another direction for further research is to compute concrete capacities in the case when $\gamma \sim c\sqrt{\frac{1}{m_n} + \frac{1}{n}}$. The definition of capacity, in the context of a restriction to unbiased detectors, needs some care, but in principle it should not be too difficult to compute a Steganographic Fisher Information quantity for the model of Th. 2.

It is, of course, important that the steganalyst's cover oracle matches exactly the distribution of the covers used by the embedder. Even the smallest deviation means that the detector will give false positive results with asymptotic probability one. And notice that the detector constructed in Subsect. 3.1 does not need to know q. This is analogous to a steganalyst who is not sure of the embedding method used by their adversary. They require that $p \neq q$ – the embedding method does change the distribution of covers – and if not then their detector will produce asymptotically no false positives, but no true positives either. A further extension to the information model is considered next.

4.3 Embedding with Learning

Our imperfect information stegosystem was not "fair" to the steganographer: we assumed that the embedding method was fixed, causing an alteration to the bit probabilities in a stego object, whereas we allowed the detector access to a cover oracle to learn about the distribution of covers. Briefly, we examine the situation when both embedder and detector have access to cover oracles, and both adapt their behaviour according to the information they learn.

First, how does the embedder make use of the oracle? Assuming that the embedder is still constrained to overwrite symbols in the cover, their optimal behaviour is to estimate the symbol distribution and then adjust their embedding so that the overwritten symbols have the same distribution. We will not concern ourselves with how this is achieved, but note that it is fairly simple, at least in the i.i.d. cover bit scenario, to do using a randomised arithmetic coding (a similar idea appears in [17]).

We have yet to prove a general counterpart to Th. 2, but can make a strong conjecture. Suppose that the embedder has l_n accesses to cover oracle bits, of which Z turn out to take value 1. They estimate p from the ratio Z/l_n – here $Z \sim \mathrm{Bi}(l_n, p)$ – so that the conditional distribution of Y given Z is $\mathrm{Bi}(n, p +$

$\gamma(Z/l_n - p))$. If T is as in (1), one can compute $\mathrm{Var}[T] \sim l_n(\frac{1}{m_n} + \frac{1}{n})$. On the other hand, the constant C in (3) is proportional to $Z/l_n - p$, so C^2 is of average order $1/l_n$, so the KL divergence in Subsect. 3.2 probably tends to zero if $\frac{nm}{(n+m)l_n}\gamma^2 \to 0$ (this is a long way from a proper proof!). Together these suggest:

Conjecture 3. *In the i.i.d. binary sequence model of Th. 2, in the case when the embedder estimates the distribution of covers from l_n cover symbols generated independently of the detector's information, the critical rate of γ is $\sqrt{l_n(\frac{1}{m_n} + \frac{1}{n})}$.*

We mean "critical rate" in the sense that if γ exceeds it asymptotically this gives eventual certain detection, and below it gives asymptotic perfect security.

Although the conjecture has yet to be proved, we can deal with a special case, although lack of space prevents the inclusion of the proof here. Since the embedder already has a source of covers – the one they embed in – they can use this for linearly many examples. (Note that, if $l_n = \Omega(n)$, m_n would be irrelevant.) Even accounting for the dependencies thus introduced,

Theorem 4. *In the i.i.d. binary sequence model of Th. 2, if the embedder learns from the cover they embed in and replaces cover symbols with bits distributed according to their estimate of p, even if the detector has exact knowledge of p, the critical rate of γ is 1 in the sense that, if $\gamma \to 0$ as $n \to \infty$, the embedding is asymptotically perfectly secure.*

It appears that the embedder "wins" this contest, since they can learn enough information about the covers to keep the detector uncertain, no matter how slowly their embedding rate tends to zero. (Of more interest is the case when γ is constant, and the consequential bounds on detection accuracy. This is for further research.) It makes the square root law redundant. But, although it seems fair to allow both embedder and detector to learn about the cover source, this is not what happens in practice: steganalysts constantly refine their cover models, but current steganography algorithms are not sophisticated enough to learn about the cover source[2]. It remains to be seen whether the same results hold in more realistic cover models where dependence between symbols is permitted.

5 Conclusions

We have proved a result which shows, amongst other things, that linearly many accesses to a cover oracle suffice for the detector to restrict the embedder to a square root capacity law; this is a significantly weaker condition than the perfect information required for the standard square root law. We have illustrated the difficulty in reasoning about a *lack* of knowledge on the part of the detector: placing a uniform prior on the unknown parameters is not the same as having

[2] Although some embedding methods try to preserve statistical properties of the individual cover object used for embedding, they do not use information from any larger sample of covers and so cannot reduce learning error asymptotically.

no knowledge. The statistical property of unbiasedness has provided the solution in this case, and the literature on UMP unbiased hypothesis tests has been a useful resource.

We also briefly considered the case when the embedder learns from a cover oracle. Although we omitted the proof of the technical result, we can show that an adaptive embedder who learns from the cover source can come arbitrarily close to a linear law of capacity, even if their opponent is granted complete knowledge of the cover source. All results, however, are in the context of a highly simplified cover model with i.i.d. bits and there remains much further research to expand them to, at least, correlated cover symbols.

What both these results emphasise, however, is that the epistemology of steganography needs more study. Indeed, the situation is more complex than the simple classification of stegosystems into perfect and imperfect information: to say that "the detector needs linearly many cover examples to restrict their opponent to a square root law" is not quite the whole story. What matters is that *the embedder knows* that the detector has, or might have, linearly many cover examples. In a similar vein, for the classical square root law *the detector knows that the embedder does not know* all the statistics of the cover source, and once *the embedder knows that the detector knows this* they are forced to a square root capacity law. In a similar vein, if the detector's cover oracle has a slightly different distribution to the true covers used by the embedder then it is practically useless and a linear law is recovered, but it requires *the embedder to know that the detector has faulty information* about the covers.

These considerations once again illustrate that steganography requires a more subtle information asymmetry than cryptography, and Kerckhoffs' Principle is not suitable to provide the whole framework. We expect that there will be a number of different applications which drive problems all currently named "steganography", with different capacities and solutions, depending on different levels of knowledge amongst the actors.

Acknowledgements

The author is a Royal Society University Research Fellow. This line of research was prompted by conversations with Rainer Böhme at the last Information Hiding conference.

References

1. Böhme, R.: Improved Statistical Steganalysis using Models of Heterogeneous Cover Signals. Ph.D. thesis, Technische Universität Dresden (2008)
2. Böhme, R.: An epistemological approach to steganography. In: Katzenbeisser, S., Sadeghi, A.-R. (eds.) IH 2009. LNCS, vol. 5806, pp. 15–30. Springer, Heidelberg (2009)
3. Cachin, C.: An information-theoretic model for steganography. Information and Computation 192(1), 41–56 (2004)

4. Cover, T., Thomas, J.: Elements of Information Theory. Wiley, USA (1991)
5. Filler, T., Fridrich, J.: Fisher Information determines capacity of ϵ-secure steganography. In: Katzenbeisser, S., Sadeghi, A.-R. (eds.) IH 2009. LNCS, vol. 5806, pp. 31–47. Springer, Heidelberg (2009)
6. Filler, T., Ker, A., Fridrich, J.: The square root law of steganographic capacity for Markov covers. In: Delp III, E., Dittmann, J., Memon, N., Wong, P. (eds.) Media Forensics and Security XI. Proc. SPIE, vol. 7254, pp. 0801–0811 (2009)
7. Fridrich, J., Lisoněk, P., Soukal, D.: On steganographic embedding efficiency. In: Camenisch, J.L., Collberg, C.S., Johnson, N.F., Sallee, P. (eds.) IH 2006. LNCS, vol. 4437, pp. 282–296. Springer, Heidelberg (2007)
8. Graham, R., Knuth, D., Patashnik, O.: Concrete Mathematics: A Foundation for Computer Science. Addison-Wesley, Boston (1994)
9. Hopper, N., Langford, J., von Ahn, L.: Provable secure steganography. In: Yung, M. (ed.) CRYPTO 2002. LNCS, vol. 2442, pp. 77–92. Springer, Heidelberg (2002)
10. Katzenbeisser, S., Petitcolas, F.: Defining security in steganographic systems. In: Delp III, E., Wong, P. (eds.) Security and Watermarking of Multimedia Contents IV. Proc. SPIE, vol. 4675, pp. 50–56 (2002)
11. Ker, A.: The Square Root Law does not require a linear key. To appear in MM&Sec 2010: Proceedings of the 12th ACM workshop on Multimedia and Security (2010)
12. Ker, A.: Estimating steganographic Fisher Information in real images. In: Katzenbeisser, S., Sadeghi, A.-R. (eds.) IH 2009. LNCS, vol. 5806, pp. 73–88. Springer, Heidelberg (2009)
13. Ker, A.: The Square Root Law requires a linear key. In: Felten, E., Dittmann, J., Fridrich, J., Craver, S. (eds.) Proceedings of the 11th ACM Workshop on Multimedia and Security. MM&Sec 2009, pp. 85–92. ACM Press, New York (2009)
14. Ker, A.: The Uniform Prior and Zero Information: A Technical Note, Oxford University Computing Laboratory Research Report CS-RR-10-06 (2010)
15. Ker, A., Pevný, T., Kodovský, J., Fridrich, J.: The square root law of steganographic capacity. In: Ker, A., Dittmann, J., Fridrich, J. (eds.) MM&Sec 2008: Proceedings of the 10th ACM Workshop on Multimedia and Security. MM&Sec 2008, pp. 107–116. ACM Press, New York (2008)
16. Lehmann, E., Romano, J.: Testing Statistical Hypotheses, 3rd edn. Springer, Heidelberg (2005)
17. Sallee, P.: Model-based methods for steganography and steganalysis. International J. Image and Graphics 5(1), 167–189 (2005)
18. Wang, Y., Moulin, P.: Perfectly secure steganography: Capacity, error exponents, and code constructions. IEEE Trans. Inf. Th. 54(6), 2706–2722 (2008)

Using High-Dimensional Image Models to Perform Highly Undetectable Steganography

Tomáš Pevný[1], Tomáš Filler[2], and Patrick Bas[3]

[1] Czech Technical University in Prague, Czech Republic
pevnak@gmail.com
[2] State University of New York in Binghamton, NY, USA
tomas.filler@gmail.com
[3] CNRS - LAGIS, Lille, France
patrick.bas@ec-lille.fr

Abstract. This paper presents a complete methodology for designing practical and highly-undetectable stegosystems for real digital media. The main design principle is to minimize a suitably-defined distortion by means of efficient coding algorithm. The distortion is defined as a weighted difference of extended state-of-the-art feature vectors already used in steganalysis. This allows us to "preserve" the model used by steganalyst and thus be undetectable even for large payloads. This framework can be efficiently implemented even when the dimensionality of the feature set used by the embedder is larger than 10^7. The high dimensional model is necessary to avoid known security weaknesses. Although high-dimensional models might be problem in steganalysis, we explain, why they are acceptable in steganography. As an example, we introduce HUGO, a new embedding algorithm for spatial-domain digital images and we contrast its performance with LSB matching. On the BOWS2 image database and in contrast with LSB matching, HUGO allows the embedder to hide 7× longer message with the same level of security level.

1 Introduction

The main goal of a passive-warden steganographic channel [10] (stegosystem) between Alice and Bob is to transmit a secret message hidden in an innocuously looking object without any possibility for the warden Eve to detect such communication. A stegosystem is called *perfectly secure* [2] if the cover distribution exactly matches the stego distribution. Although this problem has been solved by the so-called "cover generation" [1,29,24], this solution requires *exact* knowledge of the probability distribution on cover objects, which is hard (if possible at all) to obtain for real digital media in practice. The most common practical solution is to hide the message by making small perturbations with the hope that these perturbations will be covered by image noise.

One of the most popular embedding methods used with digital images is the Least Significant Bit (LSB) replacement, where the LSBs of individual cover elements are replaced with message bits. It has been quickly realized that the

R. Böhme, P.W.L. Fong, and R. Safavi-Naini (Eds.): IH 2010, LNCS 6387, pp. 161–177, 2010.

asymmetry in the embedding operation[1] is a potential weakness opening doors to the development of highly accurate targeted steganalyzers (see [17] and references therein) pushing the secure payload almost to zero.

A trivial modification of the LSB replacement method is LSB matching (often called ± 1 embedding). This algorithm randomly modulates pixel values by ± 1 so that the LSBs of pixels match the communicated message. Despite the similarity to LSB replacement, LSB matching is much harder to detect, because the embedding operation is no longer unbalanced. In fact, LSB matching has been shown to be near optimal [5] when only information from a single pixel can be utilized. The biggest weakness of LSB matching is the assumption that image noise is independent from pixel to pixel. It has been shown that this is not true in natural images, which was in different ways exploited by LSB matching detectors [16,14,22].

From the short overview of spatial domain steganography above, it is clearly seen that the embedding algorithms are not secure. This is mainly because their image model is not general enough and some marginal or joint image statistics are not preserved. In this paper, we propose a novel method for designing new steganographic algorithms allowing to use very general and high-dimensional models covering various dependencies in natural images in order to create more secure steganographic algorithms. The method follows and extends the best principles known in steganography and steganalysis so far.

The proposed method relies on the principle of minimal impact embedding [11], which is revisited in Section 2. This principle allows decomposition of the design of steganographic algorithms into the design of the image model and the coder. By virtue of this principle, steganographic algorithms can be improved either by using a better coder, or by using a better model. Thus, the image model becomes one of the most important parts of the design. Section 3 is devoted to this problem. We explain why steganalytic features can be used as a good start to design a steganographic model, if they are extended to avoid overfitting to a particular steganalyzer. Although such steganographic models can be very large (we give an example of a model with dimension 10^7), we argue that for steganographic purposes such large dimension does not pose a problem. In Section 4, we practically demonstrate the presented method by constructing a new steganographic algorithm for the spatial domain based on the SPAM (Subtractive Pixel Adjacency Matrix) features [22]. The security of the proposed scheme and the effect of individual design elements on the security is experimentally verified. The paper is concluded in Section 5.

The ideas presented in this paper can been seen in prior art. (a) Virtually all steganographic algorithms aim to minimize distortion to preserve some image model. The image model is derived either from the image itself (e.g., F5 algorithm [30] and its improvement [13], Model Based Steganography [26], etc.), or the distortion is defined by means of error introduced by quantization. The latter class of algorithms (MMX [19] and its improvement [25], PQ [12], etc.) uses "side information" in the form of a higher quality image, which is not available to the recipient (and Eve). (b) Many algorithms (F5 [30], nsF5 [13], MMX [19],

[1] Even cover elements are never decreased whereas odd ones are never increased.

and [25]) already utilized various coding schemes (matrix embedding) to minimize the distortion. While early schemes (e.g., F5 or LSB matching) used coding to minimize the number of embedding changes, a significant departure was proposed in MMX, which allowed more embedding changes than optimal (with given coding), in order to decrease the overall distortion. Thus, MMX can be interpreted as making local content-adaptive embedding by means of coding, which is close to the proposed scheme.

With respect to the above prior work, the main contributions of this work are as follows. (a) We promote and advocate the use of high-dimensional image models in steganography that cannot be used in steganalysis (yet). (b) We separate the image model from coding, which allows simulating optimal coding and thus comparing image models without the effect of coding. Moreover, the message can be hidden in parts of the image difficult for steganalysis while *considering all pixels simultaneously during the embedding.*

Although the proposed steganographic scheme might be considered as an adaptive, it is not adaptive in the usuall approach, when first good pixels are selected [14,9,8] (e.g. pixels in noisy and textured areas) and than the message is inserted in the image while modifying only the selected pixels (e.g by using wet paper codes). Our scheme always uses all pixels for the embedding, but it changes them with probability inversely proportional to the detectability of their change.

In the rest, we use the following notation. Small-case boldface symbols are used for vectors and capital-case boldface symbols for matrices and possibly tensors. Symbols $\mathbf{X} = (x_{ij}) \in \mathcal{X} = \{0, \ldots, 255\}^{n_1 \times n_2}$ and $\mathbf{Y} = (y_{ij}) \in \mathcal{X}$ are exclusively used to represent intensities of $n = n_1 n_2$-pixel cover and stego image. For the sake of simplicity, we sometimes index the pixels with a single number, $\mathbf{X} = (x_i)_{i=1}^n$ and similarly for stego image $\mathbf{Y} = (y_i)_{i=1}^n$.

2 Minimizing Embedding Impact

Virtually all practical steganographic algorithms for digital media strive to minimize an ad hoc *embedding impact* [11,6], which, if properly defined, is correlated with detectability. In its simplest form, embedding impact is simply the number of changes (known as matrix embedding). However, more general ways, as already suggested by Crandal [4], should be considered. In general, the embedding impact is captured by a non-negative distortion measure $D : \mathcal{X} \times \mathcal{X} \rightarrow [0, \infty]$. During embedding, the algorithm should find a stego image $\mathbf{Y}$, which (a) communicates a given message and (b) achieves minimal value of $D(\mathbf{X}, \mathbf{Y})$. Unfortunately, this problem is generally very difficult in practice.

From this reason, we constrain ourselves to a well-studied special (but still powerful enough) case assuming (a) binary embedding changes[2], i.e., $|x_i - y_i| \leq 1$, $i \in \{1, \ldots, n\}$, and (b) additive distortion measure in the form

$$D(\mathbf{X}, \mathbf{Y}) = \sum_{i=1}^{n} \rho_i |x_i - y_i|. \qquad (1)$$

[2] Extensions to ternary case can be done by the "$e + 1$" construction described in [31].

The constants $0 \leq \rho_i \leq \infty$ are fixed parameters expressing costs of (or distortion caused by) pixel changes. The case $\rho_i = \infty$ corresponds to the so-called *wet pixel* not allowed to be modified during embedding. Notice that the additivity of the distortion function D implies that that the embedding changes do not interact between each other. This is a reasonable assumption, especially if we assume low embedding rates and embedding changes being far from each other. Unfortunately, there are cases of important distortion measures which cannot be written in this form. One such case will be introduced in Section 4.

For additive distortion functions (1), the following theorem taken from [11] gives the minimal expected distortion obtained by hiding m bits in an n-pixel cover object.

Theorem 1. *Let $\rho = (\rho_i)_{i=1}^n$, $0 \leq \rho_i < \infty$, be the set of constants defining the additive distortion measure (1) for $i \in \{1, \ldots, n\}$. Let $0 \leq m \leq n$ be the number of bits we want to communicate by using a binary embedding operation. The minimal expected distortion has the following form*

$$D_{min}(m, n, \rho) = \sum_{i=1}^{n} p_i \rho_i,$$

where

$$p_i = \frac{e^{-\lambda \rho_i}}{1 + e^{-\lambda \rho_i}} \tag{2}$$

is the probability of changing the ith pixel. The parameter λ is obtained by solving

$$-\sum_{i=1}^{n} \Big(p_i \log_2 p_i + (1 - p_i) \log_2 (1 - p_i) \Big) = m. \tag{3}$$

The importance of Theorem 1 is in the separation of the image model (needed for calculating constants ρ_i) and the coding algorithm used in a practical implementation. By virtue of this separation, better steganographic algorithms can be derived by using better coding or by using a better image model. One important consequence is that, in order to study the effect of the image model on steganographic security, no coding algorithm is needed at all! The optimal coding can be simulated by flipping each pixel with probability p_i as defined in (2).

We use this *separation principle* in Section 4 to find a good image model used to derive the costs ρ_i. The study of the loss introduced by a practical coding method is also included.

3 From Steganalysis to Steganography

Almost all state-of-the-art statistical steganalyzers (with the exception of steganalyzers for LSB replacement) are based on a combination of steganalytic features and pattern recognition algorithms. In steganalysis, steganalytic features are used to reduce the dimension of a space of all cover objects, so that the pattern recognition algorithms can learn (if possible) the difference between cover and stego objects in this reduced feature space. Using such a low-dimensional model for designing steganography usually leads to overtraining to a particular feature set (this issue of feature set completeness is discussed in [20,27]). Keeping this in mind, we believe that the features can serve as a good precursor of the image model to determine the embedding costs ρ_i. Although we show this transition from steganalytic features to a steganographic model on spatial domain steganography, we believe that the ideas and tools presented here can be used in other domains and with other steganalytic features as well.

We start by reviewing the recently proposed SPAM features [22] proposed to detect steganographic algorithms in spatial and transformed domains. Then, we discuss the problem of overfitting the steganographic model to steganalytic features as well as the remedy by expanding the model beyond the capabilities of contemporary pattern recognition algorithm. Finally, we propose a simple method to identify parts of the model that are more important for steganalysis.

3.1 SPAM Features

It is well known that values of neighboring pixels in natural images are not independent. This is not only caused by the inherent smoothness of natural images, but also by the image processing (de-mosaicking, sharpening, etc.) in the image acquisition device. This processing makes the noise, which is independent in the raw sensor output, dependent in the final image. The latter source of dependencies is very important for steganalysis because steganographic changes try to hide themselves within the image noise.

The SPAM [22] features model dependencies between neighboring pixels by means of higher-order Markov chains. They have been designed to provide a low-dimensional model of image noise that can be used for steganalytic purposes. The calculation of differences can be viewed as an application of high-pass filtering, which effectively suppresses the image content and exposes the noise. The success of SPAM features in detecting wide range of steganographic algorithms [21] suggests this model to be reasonable for steganalysis and steganography.

The SPAM features model transition probabilities between neighboring pixels along 8 directions $\{\leftarrow, \rightarrow, \downarrow, \uparrow, \nwarrow, \searrow, \swarrow, \nearrow\}$. Below, the calculation of the features is explained on horizontal left-to-right direction, because for the other directions the calculations differ only by different indexing. All direction-specific variables are denoted by a superscript showing the direction.

Let $\mathbf{I} \in \mathcal{X}$ be an image of size $n_1 \times n_2$. The calculation starts by computing the difference array $\mathbf{D}^{\bullet}$, which is for a horizontal left-to-right direction

$$\mathbf{D}^{\rightarrow}_{ij} = \mathbf{I}_{ij} - \mathbf{I}_{i,j+1},$$

for $i \in \{1, \dots, n_1\}$, $j \in \{1, \dots, n_2 - 1\}$. Depending on the desired order of the features, either the first-order Markov process is used,

$$\mathbf{M}^{\rightarrow}_{d_1 d_2} = Pr(\mathbf{D}^{\rightarrow}_{i,j+1} = d_1 | \mathbf{D}^{\rightarrow}_{ij} = d_2), \tag{4}$$

or the second-order Markov process is used,

$$\mathbf{M}^{\rightarrow}_{d_1 d_2 d_3} = Pr(\mathbf{D}^{\rightarrow}_{i,j+2} = d_1 | \mathbf{D}^{\rightarrow}_{i,j+1} = d_2, \mathbf{D}^{\rightarrow}_{ij} = d_3), \tag{5}$$

where $d_i \in \{-T, \ldots, T\}$. The calculation of the features is finished by separate averaging of the horizontal and vertical matrices and the diagonal matrices to form the final feature sets. With a slight abuse of notation, this averaging can be written as

$$\begin{aligned} \mathbf{F}^{\bullet}_{1,\ldots,k} &= \frac{1}{4}\left[\mathbf{M}^{\rightarrow}_{\bullet} + \mathbf{M}^{\leftarrow}_{\bullet} + \mathbf{M}^{\downarrow}_{\bullet} + \mathbf{M}^{\uparrow}_{\bullet}\right], \\ \mathbf{F}^{\bullet}_{k+1,\ldots,2k} &= \frac{1}{4}\left[\mathbf{M}^{\searrow}_{\bullet} + \mathbf{M}^{\nwarrow}_{\bullet} + \mathbf{M}^{\swarrow}_{\bullet} + \mathbf{M}^{\nearrow}_{\bullet}\right], \end{aligned} \tag{6}$$

where $k = (2T+1)^2$ for the first-order features and $k = (2T+1)^3$ for the second-order features. In [22], the authors used $T = 4$ for the first-order features (leading to 162 features) and $T = 3$ for the second-order features (leading to 686 features).

3.2 Decomposing SPAM Features

Although the second-order SPAM features use conditional probabilities to model pixel differences, their essential components are actually co-occurrence matrices

$$\mathbf{C}^{\rightarrow}_{d_1 d_2} = Pr(\mathbf{D}^{\rightarrow}_{ij} = d_1, \mathbf{D}^{\rightarrow}_{i,j+1} = d_2), \tag{7}$$

$$\mathbf{C}^{\rightarrow}_{d_1 d_2 d_3} = Pr(\mathbf{D}^{\rightarrow}_{ij} = d_1, \mathbf{D}^{\rightarrow}_{i,j+1} = d_2, \mathbf{D}^{\rightarrow}_{i,j+2} = d_3). \tag{8}$$

It is easy to show that the second order SPAM features with $T = 3$ can be directly obtained[3] from the set $\{\mathbf{C}^{k}_{d_1 d_2}, \mathbf{C}^{k}_{d_1 d_2 d_3} | k \in \{\rightarrow, \uparrow, \nwarrow, \nearrow\}, -3 \leq d_i \leq 3\}$. In fact, we observed that this set of $4 \times (343+49) = 1568$ co-occurrence features has only slightly inferior performance in detecting LSB matching, which we attribute to a smaller ratio of training samples per dimension (known as curse of dimensionality). From this point of view, the distortion measure used to derive embedding costs ρ_i should be designed to preserve the co-occurrence matrices (7) and (8), because their preservation implies the preservation of second-order SPAM features.

Although the idea of preservation of SPAM features is tempting, the distortion measure would not be general enough. The new scheme would be so tied to a particular steganalytic method that it can be expected to be detectable by a slight modification of the features. This problem of "overfitting" the distortion measure to a particular steganalytic method together with the need for a complete feature set has been already described [20,27] for the DCT domain. Here, we propose to resolve the issue of overfitting to a particular model by expanding it beyond practical limits of steganalysis (for this model). This can be easily done in the case of co-occurrence matrices by increasing the range of covered differences T.

At this point, it is important to clarify the difference between the effects of model dimensionality for steganography and for steganalysis. The high-dimensional models in steganalysis present a serious problem for subsequent machine learning due

[3] Observe that $\mathbf{C}^{\rightarrow}_{d_1 d_2 d_3} = \mathbf{C}^{\leftarrow}_{-d_3,-d_2,-d_1}$, and $\mathbf{M}^{\rightarrow}_{d_1 d_2 d_3} = \mathbf{C}^{\rightarrow}_{d_3 d_2 d_1} / \mathbf{C}^{\rightarrow}_{d_2 d_1}$.

to the curse of dimensionality and related overfitting. Although the actual ratio between the number of training samples and the model dimensionality depends on the used machine learning algorithm and the problem, the rule of thumb is to have ten times more samples than the model dimensionality (number of features). These drawbacks prevent the use of high-dimensional models in steganalysis. By contrast, high-dimensional models in steganography do not cause problems, because there is no statistical learning involved. The cover image provides the exact model to be preserved and, consequently, there is no curse of dimensionality, which justifies the use of high-dimensional models in steganography.

An additional important practical detail is that updating the co-occurrence matrices to reflect one pixel change is much easier than updating the conditional probabilities (the former involves only addition and subtraction of a few items of the matrices, while the latter involves division of the large part of the matrices). The efficient update of co-occurrence matrices enables modeling a wide range of differences between pixels (the use of large T) resulting in modeling most differences (and pixels) in the image (and better preservation of the SPAM features).

3.3 Identification of Detectable Parts of the Models

Unfortunately, the ideal case, when the image model is fully preserved during the embedding, is virtually impossible to realize in practice. It is therefore important to identify parts of the model important for steganalysis and set appropriate costs of pixel changes ρ_i.

The association of costs ρ_i to the modification of the model is in general very difficult because we do not know which parts of the model are important. Here, we suggest to evaluate the individual elements of the model independently of each other (any method for feature ranking can be used [15]) and set the costs ρ_i to reflect this ranking. The advantage of individual evaluation is that it can be done quickly even for a large number of features. On the other hand, the individual evaluation of the model elements is certainly not optimal, especially from the machine learning point of view. However, we believe (and our experiments confirm that) that the costs derived this way can be used as a good starting point. There is no doubt that other (and better) methods of deriving costs ρ_i exist.

Our approach works as follows. First, we create a set of images embedded with a simulated maximum payload by a given embedding operation (in our case of spatial domain steganography, this amounts to randomly increase or decrease the pixel value by one with probability 50%). Then, we use the criteria optimized in Fisher Linear Discriminant (FLD criteria) (9) to evaluate, how good are individual features for detecting given embedding changes. The values of FLD criteria (9) of individual elements may be either used directly to set the costs of embedding changes ρ_i, which might be dangerous due to the already discussed problem of overfitting. Alternatively, they can be used to obtain insight into the problem and set the costs heuristically, which is recommended. In the

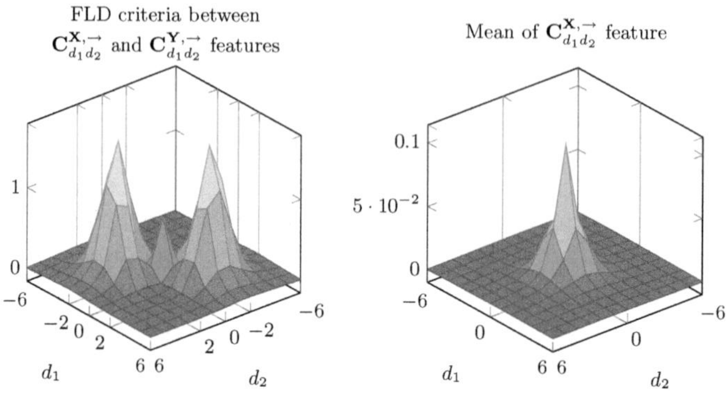

Fig. 1. Left: Values of FLD criteria (9) between the feature $\mathbf{C}^{\to}_{d_1d_2}$ calculated from cover images and stego images obtained by LSB matching with full payload. Right: mean of the feature $\mathbf{C}^{\to}_{d_1d_2}$ over the set of cover images from the BOWS2 database.

rest of this section, we use the analysis of the FLD criteria to identify parts of the co-occurrence model that can be used for embedding.

For co-occurrence matrices introduced in the previous subsection, the values of FLD criteria for a single feature $\mathbf{C}^{\to}_{d_1d_2}$ (for fixed d_1 and d_2) can be written as

$$\frac{\left(E[\mathbf{C}^{\mathbf{X},\to}_{d_1d_2}] - E[\mathbf{C}^{\mathbf{Y},\to}_{d_1d_2}]\right)^2}{E\left[\mathbf{C}^{\mathbf{X},\to}_{d_1d_2} - E[\mathbf{C}^{\mathbf{X},\to}_{d_1d_2}]\right]^2 + E\left[\mathbf{C}^{\mathbf{Y},\to}_{d_1d_2} - E[\mathbf{C}^{\mathbf{Y},\to}_{d_1d_2}]\right]^2}, \tag{9}$$

where $E[\cdot]$ stands for the empirical mean (obtained in our case over all images in the BOWS2[4] image database), and $\mathbf{C}^{X,\to}_{d_1d_2}$, $\mathbf{C}^{Y,\to}_{d_1d_2}$ stand for a single element of the co-occurrence matrix $\mathbf{C}^{\to}_{d_1d_2}$ calculated from the cover and stego image, respectively. The higher the value, the better the feature when used alone for detecting the LSB matching algorithm. Figure 1 shows the values estimated from cover and stego images obtained by embedding a full payload with LSB matching. We can see that the most influential features are $\mathbf{C}^{\to}_{-2,2}$ and $\mathbf{C}^{\to}_{2,-2}$ corresponding to regions containing noisy pixels in a smooth area. Also, it is interesting to see that regions having the same color (such as saturated pixels) represented by $\mathbf{C}^{\to}_{0,0}$, or pixels in smooth transitions represented by $\mathbf{C}^{\to}_{d,d}$, do not constitute a good *single* feature. This is most probably caused by their high variance, which makes features $\mathbf{C}^{\to}_{-2,2}$ and $\mathbf{C}^{\to}_{2,-2}$ more stable and more suitable for steganalysis. Although not easy to visualize, similar results and interpretation can be obtained from higher-order co-occurrence matrices $\mathbf{C}^{\bullet}_{d_1d_2d_3}$.

This analysis shows which parts of the image model should be preserved. We stress again that this analysis was performed from the evaluation of a single feature and its direct application may lead to overtraining. As was already mentioned above, we consider this analysis as a good guide to derive heuristics to build the embedding costs ρ_i.

[4] See `http://bows2.gipsa-lab.inpg.fr/BOWS2OrigEp3.tgz`

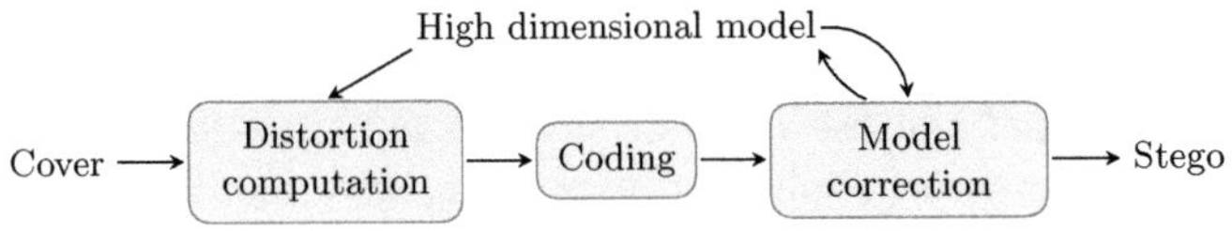

Fig. 2. High-level diagram of HUGO

4 From Theory to Practice

In this section, all pieces and ideas presented above are put together, in order to give life to a new steganographic algorithm called HUGO (Highly Undetectable steGO). The individual steps of this algorithm are depicted in Figure 2.

4.1 Evaluation Setting

The scheme was assessed using the BOWS2 image database, containing approximately 10800 images of fixed size 512×512. Thanks to the fixed size, all images have the same number of usable elements, which means that we do not have to take the Square Root Law [18,7] into the account. Prior to all experiments, the images were divided into two sets of equal size, one used exclusively for training, the other exclusively for evaluation of the accuracy. The chosen accuracy measure is the minimal average decision error under equal probability of cover and stego images, defined as

$$P_{\mathrm{E}} = \min \frac{1}{2}\left(P_{\mathrm{Fp}} + P_{\mathrm{Fn}}\right),$$

where P_{Fp} and P_{Fn} stand for the probability of false alarm or false positive (detecting cover as stego) and probability of missed detection (false negative). To observe the effect of over-fitting for a particular feature set, we create blind steganalyzers employing four different feature sets (first- and second-order SPAM features [22] with $T = 4$ and $T = 3$ respectively, WAM [14], and recently proposed Cross Domain Features[5] (CDF) [21]).

All steganalyzers were realized as soft-margin SVMs [28] with Gaussian kernel[6], $k(x, y) = exp(-\gamma \left\|x - y\right\|^2)$. The parameters γ and C were set to values corresponding to the least error estimated by five-fold cross-validation on the training set on the grid $(C, \gamma) \in \left\{(10^k, 2^j) | k \in \{-3, \ldots, 4\}, j \in \{-d-3, -d+3\}\right\}$, where d is the logarithm at the base 2 of the number of features.

Besides the SVM-based blind steganalyzers, we also use the Maximum Mean Discrepancy [23] (MMD) to quickly compare the security of different versions of the algorithm.

[5] CDF combines second-order SPAM features ($T = 3$) and cartesian calibrated features proposed originally for DCT domain. To extract the DCT domain features, we compressed the image with quality factor 100.

[6] We did some experiments with linear SVMs and never obtained better results. For a discussion related to linear SVMs, see [22].

4.2 Co-occurrence Model in Steganography

Section 3.2 motivated the use of co-occurrence matrices (SPAM features) as a reliable model for steganography and explained, why the distortion function D (not just constants ρ_i) is derived directly from them. In order to stress those parts of the co-occurrence matrices that are more important for steganalysis, the distortion function D is defined as a weighted sum of differences

$$D(\mathbf{X},\mathbf{Y}) = \sum_{d_1,d_2,d_3=-T}^{T} \left[w(d_1,d_2,d_3) \left| \sum_{k\in\{\rightarrow,\leftarrow,\uparrow,\downarrow\}} \mathbf{C}^{\mathbf{X},k}_{d_1 d_2 d_3} - \mathbf{C}^{\mathbf{Y},k}_{d_1 d_2 d_3} \right| + \right.$$

$$\left. + w(d_1,d_2,d_3) \left| \sum_{k\in\{\searrow,\nwarrow,\swarrow,\nearrow\}} \mathbf{C}^{\mathbf{X},k}_{d_1,d_2,d_3} - \mathbf{C}^{\mathbf{Y},k}_{d_1,d_2,d_3} \right| \right], \qquad (10)$$

where $w(d_1,d_2,d_3)$ is a weight function quantifying the detectability of the change in the co-occurrence matrix[7]. The weight function $w(d_1,d_2,d_3)$ has the following simple form

$$w(d_1,d_2,d_3) = \frac{1}{\left[\sqrt{d_1^2+d_2^2+d_3^2}+\sigma\right]^{\gamma}}, \qquad (11)$$

where $\sigma,\gamma > 0$ are parameters that can be tuned in order to minimize the detectability. This very conservative choice mimics the average number of samples available to Eve to estimate the individual features $\mathbf{C}^{\bullet}_{d_1 d_2 d_3}$ from a single image (see the right part of Figure 1). Motivated by the analysis performed in Section 3.3, the rationale of this choice is simple: the more samples Eve has, the better estimate of individual feature she can obtain and the more she can utilize it for steganalysis. By penalizing highly-populated features (in this case features extracted from pixels with low differences d_1, d_2, and d_3), we drive the algorithm to hide the message into parts of the image difficult for Eve to model. In practice, our choice of $w(d_1,d_2,d_3)$ correlates the distribution of the message bits with the local texture of the image.

Note that the distortion measure (10) is not additive in the sense of (1). This is a significant deviation from the assumptions of Theorem 1, because for this more general case near-optimal practical algorithms for minimizing such embedding impact do not exist yet. To make this measure additive, we approximate the costs of embedding change as

$$\rho_{i,j} = D(\mathbf{X},\mathbf{Y}^{i,j}), \qquad (12)$$

where $\mathbf{Y}^{i,j}$ is the stego image obtained by changing the (i,j)th pixel of cover image $\mathbf{X}$. As will be seen later, this approximation has a crucial impact on the detectability of the scheme.

[7] If the $w(d_1,d_2,d_3) = 1$ for all d_i and $T = 255$, then all ρ_i would be the same and the whole scheme would just minimize the number of embedding changes.

```
HUGO embedding algorithm
1  for (i,j) in PIXELS {            //function D is taken from (10)
2     Yp = X; Yp(i,j)++; rho_p(i,j) = D(X,Yp);  //calculate emb. impact
3     Ym = X; Ym(i,j)--; rho_m(i,j) = D(X,Ym);  //for each pixel
4  }
5  rho_min = min(rho_p, rho_m);     //elementwise; use minimum for embedding
6  PIXELS_TO_CHANGE = minimize_emb_impact(LSB(X), rho_min, message)
7  Y = X;                           //start making changes in cover image
8  for (i,j) in PIXELS_TO_CHANGE { //order given by the MC visit. strategy
9     if ( model_correction_step_enabled ) {
10       Yp = Y; Yp(i,j)++; dp = D(X,Yp); Ym = Y; Ym(i,j)--; dm = D(X,Ym);
11       if ( dp<dm ) { Y(i,j)++; } else { Y(i,j)--; }
12    } else {
13       if ( rho_p(i,j)<rho_m(i,j) ) { Y(i,j)++; } else { Y(i,j)--; }
14    }
15 }
```

Fig. 3. Pseudo-code of the HUGO embedding algorithm as described in Section 4.3

4.3 Implementation Details of HUGO

Figure 3 shows the pseudo-code of our implementation. On lines 1–5, the algorithm calculates distortions corresponding to modifying each pixel by ± 1 and sets the embedding cost of pixel change ($\rho_{i,j}$) to the minimum of these two numbers (for saturated pixels, there is only one choice).

Once the positions of pixel changes are determined (either by simulating the embedding by virtue of Theorem 1, or by using a practical algorithm, such as the syndrome-trellis codes [6], (function `minimize_emb_impact` on line 6 of the code)), there are two ways to ensure that the pixel's LSB communicates the message.

Without model correction: This version assumes that the assumption of the Theorem 1 holds, which means that we cannot do any better than change pixels to values determined in lines 1–5 (line 13 of the pseudo-code). The order in which the pixels are changed does not matter.

With model correction (MC): Since our distortion measure D (10) does not satisfy the assumptions of Theorem 1, we can further decrease the distortion by changing pixels to values (remember that there are two ways to match pixels' LSB to the desired bit) minimizing the overall distortion $D(\mathbf{X}, \mathbf{Y}^i)$, where $\mathbf{Y}^i$ denotes the cover image $\mathbf{X}$ after changing the ith pixel (see lines 10–11 in the pseudo-code). As will be seen in the experimental part below, the impact of model correction on the security is significant. In this case of model correction, the order in which the pixels requiring change of LSB are processed is important. In the next subsection, we experimentally evaluate the following strategies: (S1) top left to bottom right, (S2) from highest $\rho_{i,j}$ to lowest $\rho_{i,j}$, (S3) from lowest $\rho_{i,j}$ to highest $\rho_{i,j}$, (S4) random order.

Finally we note that our implementation of HUGO in C++ with $T = 90$, the model correction step, and practical Syndrome-Trellis Code (STC) embeds message with relative length 0.25bpp to image of size 512×512 in approximately 5s on Intel Core 2 Duo 2.8 GHz processor. We consider this time more than suitable for real applications. In practice, the algorithm may need to communicate a small number of parameters in order to be able to decode the message correctly. In HUGO, we need to communicate the size of the message in order to construct the same STC code at the receiver side. This is usually done by reserving a small portion of the image based on the stego key, where a known code is used for embedding.

4.4 HUGO's Maturing

The HUGO algorithm has several parameters: the range of modeled differences T, the parameters of the weight function γ and σ, and utilization of the model correction step. All these parameters need to be set before the actual use of the algorithm. Since we are not aware of any general guidance, we set them experimentally while comparing different versions of the algorithm by blind steganalysis. Although it can be argued that the parameters will be tied to the database, we prefer to see this step as tuning the algorithm to image source used by Alice and Bob.

The parameter setting proceeds as follows: (a) set the parameter T, (b) find suitable values of σ and γ in (11), (c) set the the strategy of pixel visits. In all experiments aimed to tune HUGO, the coding was simulated by virtue of Theorem 1.

The parameter T was set to $T = 90$ (the model has more then 10^7 features), causing more than 99% of the co-occurrences in the typical image to be covered by the model. By this choice of T, we strongly believe that the detectability of HUGO by SPAM features cannot be improved by increasing the range of modeled differences. In fact, our experiments showed that the increase of the range of modeled differences was not followed by a decrease of the classifier error (most probably due to the curse of dimensionality).

The search for suitable parameters of the weight function (11) was performed on a grid $(\sigma, \gamma) \in \{(10^k, 2^j) | k \in \{-3, \ldots, 1\}, j \in \{-1, 2\}\}$ for both versions of the algorithm (with and without MC). The embedding payload was fixed to 0.25bpp. In order to reduce the complexity of the search, the detectability was evaluated by means of the Maximum Mean Discrepancy [23]. Figure 4 shows the MMD values for HUGO with the MC step and S1 visiting strategy. Due to space constraints, we report graphs only for SPAM and WAM features with MC step S1. All other graphs even for the case of Hugo without MC step were of similar shape suggesting the choice parameters γ and σ to be reasonable. For all experiments presented in the rest of this section, we chose $\gamma = 4$ and $\sigma = 10$.

As we have already mentioned, the effect of the model correction on the security is substantial. For fixed classification error $P_{\mathrm{E}} = 40\%$ of an SVM-based steganalyzer utilizing second-order SPAM features, HUGO with model correction step increases the secure payload from 0.25bpp to 0.4bpp. This difference

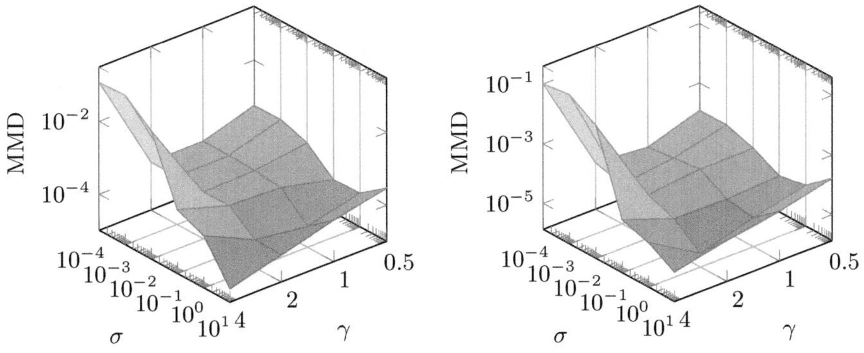

Fig. 4. Value of MMD (lower is better) plotted against parameters γ and σ for HUGO with model correction and S1 visiting strategy. Results for other features and even when MC step was not used were similar and are omitted due to space constraints.

is entirely due to the fact that our distortion measure is not additive. Since we do not know yet how to do optimal coding for non-additive measures, the model correction step is in this case a reasonably good remedy.

Finally, we have compared the strategies of pixel visits S1–S4 in the model correction step by training SVM-based steganalyzer utilizing second order SPAM features. From Figure 5 (a), strategy S2 seems to be the most secure wrt the SPAM features. Model correction strategies S3 and S4 were performing slightly worse than S2 and are not displayed. These results show that the model correction step should perform embedding changes from pixels causing the largest distortion to pixels causing the least distortion.

4.5 HUGO's Security

Figure 5 (a) compares the security of HUGO with simulated optimal coding utilizing different model correction strategies. For S2, which seems to be the best, we also report its practical implementation using syndrome-trellis code with constraint height $h = 10$ (STC) [6]. All algorithms are compared to ordinary LSB matching with optimal (simulated) ternary matrix embedding. The reported quantity P_{E} is the error of SVM-based steganalyzers. We did not compared HUGO to adaptive ternary LSB matching [14], or to MPSteg [3], because the reported improvement in the security of both schemes over standard LSB matching were not significant.

The impact of switching from the optimal (simulated) coding to the STC coder (STC) on the detectability of HUGO is also interesting and interpretable. Ideally, we would like to have code which would change each pixel with probability (2). To compare the effect of a practical coder for fixed distortion d, we evaluate the *coding loss* $l(d) = (\alpha_{\mathrm{OPT}} - \alpha_{\mathrm{ACT}})/\alpha_{\mathrm{OPT}}$, where α_{OPT} is the payload embedded by the optimal coder and α_{ACT} is the payload embedded by a practical algorithm

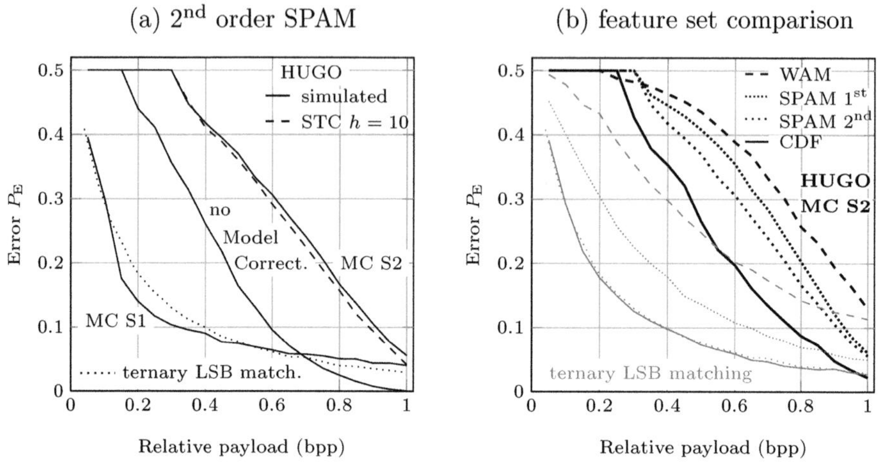

Fig. 5. (a) Comparison of security of different versions of HUGO by means of error P_E of steganalyzers utilizing second-order SPAM features with $T = 3$. (b) Comparison of different steganalytic features for detecting ordinary LSB matching with optimal ternary coding and HUGO with MC step S2. All steganalyzers are targeted to a given algorithm and message length.

while both of them achieve the same distortion d. Coding loss $0 \le l(d) \le 1$ tells us what portion of the ideal payload we are loosing due to practical embedding algorithm. For STCs, $l(d)$ was often around $3\% - 7\%$ depending on ρ and h. This finding is consistent with Figure 5 (a).

According to Figure 5 (b), HUGO offers very high security. Even for payloads as large as 0.30bpp, the error of all four steganalyzers targeted to detect HUGO with optimal coding and MC step is above 40%. It is expected that secure payload may be higher for cover sources without such strong pixel dependencies as present in BOWS2 database from scaling the original images.

Even though the improvement obtained from CDF features is significant when compared to second-order SPAM, the relative payload for which the scheme remains undetectable stays essentialy the same. This threshold may point to amount of pixels that are not modeled by either feature set (SPAM or DCT based). However, including such pixels in the steganalytic model may not be as beneficial as including them into steganographic model due to the statistical learning problem. Such pixels are expected to be part of very noisy end textured areas which will be challenging for steganalysis.

Last, but not least, if we compare HUGO with MC step S2 to the state-of-the-art LSB matching with optimal ternary coding, we can see that by using HUGO, Alice gains more than 700% of the capacity at $P_{\mathrm{E}} = 40\%$ on the BOWS2 database.

5 Conclusion

This paper presented a complete method for designing practical and secure steganographic schemes for real digital media. The main design principle is to minimize a suitably-defined distortion caused by the embedding. Since the distortion function is an essential input of the method, a large part of the paper was devoted to its design. We recommended to use weighted difference of *extended* state-of-the-art feature vectors already used in steganalysis. The extension of the feature sets, which can contain even 10^7 features, is important to avoid overfitting to a particular steganalyzer. The use of such large feature sets was justified by explaining the fundamental difference of their role in steganography and steganalysis.

The whole approach was demonstrated by designing a new steganographic algorithm for spatial domain (called HUGO), where the image model was derived from SPAM features. Parts of the model, i.e., the weights, responsible for detection of LSB matching were identified using criteria optimized in Fisher Linear Discriminant, which motivated the construction of an ad hoc distortion measure. The coding itself was performed using the syndrome-trellis codes which enable very fast implementation of the scheme in practice for arbitrary set of embedding costs ρ.

The security of HUGO was verified and compared to prior art (LSB matching) on a wide range of payloads for four different features sets. In contrast with LSB matching, HUGO allows the embedder to hide 7$\times$ longer message with the same level of security level. By concrete numbers, the payload of HUGO at detection error 40% is 0.3bpp, while for LSB matching it is 0.04bpp.

Acknowledgements

Tomáš Filler was supported by Air Force Office of Scientific Research under the research grant FA9550-08-1-0084. The U.S. Government is authorized to reproduce and distribute reprints for Governmental purposes notwithstanding any copyright notation there on. The views and conclusions contained herein are those of the authors and should not be interpreted as necessarily representing the official policies, either expressed or implied of AFOSR or the U.S. Government.

Tomáš Pevný and Patrick Bas are supported by the National French project Nebbiano ANR-06-SETIN-009. Tomáš Pevný is also supported by Czech Ministry of Education grant 6840770038.

Special thanks belong to Jessica Fridrich for many interesting discussions and for proofreading the paper.

References

1. Anderson, R.: Stretching the limits of steganography. In: Anderson, R. (ed.) IH 1996. LNCS, vol. 1174, pp. 39–48. Springer, Heidelberg (1996)
2. Cachin, C.: An information-theoretic model for steganography. In: Aucsmith, D. (ed.) IH 1998. LNCS, vol. 1525, pp. 306–318. Springer, Heidelberg (1998)
3. Cancelli, G., Barni, M., Menegaz, G.: Mpsteg: hiding a message in the matching pursuit domain. In: Proceedings SPIE, EI, Security, Steganography, and Watermarking of Multimedia Contents VIII, San Jose, CA, vol. 6072, p. 60720P (2006)

4. Crandall, R.: Some notes on steganography. Steganography Mailing List (1998), `http://os.inf.tu-dresden.de/~westfeld/crandall.pdf`
5. Filler, T., Fridrich, J.: Fisher information determines capacity of ϵ-secure steganography. In: Katzenbeisser, S., Sadeghi, A.-R. (eds.) IH 2009. LNCS, vol. 5806, pp. 31–47. Springer, Heidelberg (2009)
6. Filler, T., Fridrich, J., Judas, J.: Minimizing embedding impact in steganography using Trellis-Coded Quantization. In: Proceedings SPIE, EI, Media Forensics and Security XII, San Jose, CA, January 18-20, p. 05-1–05-14 (2010)
7. Filler, T., Ker, A.D., Fridrich, J.: The Square Root Law of steganographic capacity for Markov covers. In: Proceedings SPIE, EI, Security and Forensics of Multimedia XI, San Jose, CA, January 18-21, vol. 7254, p. 08-1–08-11 (2009)
8. Franz, E.: Steganography preserving statistical properties. In: Petitcolas, F.A.P. (ed.) IH 2002. LNCS, vol. 2578, pp. 278–294. Springer, Heidelberg (2002)
9. Franz, E., Rönisch, S., Bartel, R.: Improved embedding based on a set of cover images. In: Proceedings of the 11th ACM Multimedia & Security Workshop, Princeton, NJ, September 7-8, pp. 141–150 (2009)
10. Fridrich, J.: Steganography in Digital Media: Principles, Algorithms, and Applications. Cambridge University Press, Cambridge (2009)
11. Fridrich, J., Filler, T.: Practical methods for minimizing embedding impact in steganography. In: Proceedings SPIE, EI, Security, Steganography, and Watermarking of Multimedia Contents IX, San Jose, CA, January 29-February 1, vol. 6505, pp. 2–3 (2007)
12. Fridrich, J., Goljan, M., Soukal, D.: Perturbed quantization steganography. ACM Multimedia System Journal 11(2), 98–107 (2005)
13. Fridrich, J., Pevný, T., Kodovský, J.: Statistically undetectable JPEG steganography: Dead ends, challenges, and opportunities. In: Proceedings of the 9th ACM Multimedia & Security Workshop, Dallas, TX, September 20-21, pp. 3–14 (2007)
14. Goljan, M., Fridrich, J., Holotyak, T.: New blind steganalysis and its implications. In: Proceedings SPIE, EI, Security, Steganography, and Watermarking of Multimedia Contents VIII, San Jose, CA, vol. 6072, pp. 1–13 (2006)
15. Guyon, I., Gunn, S., Nikravesh, M., Zadeh, L.A.: Feature Extraction, Foundations and Applications. Springer, Heidelberg (2006)
16. Harmsen, J.J., Pearlman, W.A.: Steganalysis of additive noise modelable information hiding. In: Proceedings SPIE, EI, Security and Watermarking of Multimedia Contents V, Santa Clara, CA, January 21-24, vol. 5020, pp. 131–142 (2003)
17. Ker, A.D., Böhme, R.: Revisiting weighted stego-image steganalysis. In: Proceedings SPIE, EI, Security, Forensics, Steganography, and Watermarking of Multimedia Contents X, San Jose, CA, January 27-31, vol. 6819, p. 5-1–5-17 (2008)
18. Ker, A.D., Pevný, T., Kodovský, J., Fridrich, J.: The Square Root Law of steganographic capacity. In: Proceedings of the 10th ACM Multimedia & Security Workshop, Oxford, UK, September 22-23, pp. 107–116 (2008)
19. Kim, Y., Duric, Z., Richards, D.: Modified matrix encoding technique for minimal distortion steganography. In: Camenisch, J.L., Collberg, C.S., Johnson, N.F., Sallee, P. (eds.) IH 2006. LNCS, vol. 4437, pp. 314–327. Springer, Heidelberg (2006)
20. Kodovský, J., Fridrich, J.: On completeness of feature spaces in blind steganalysis. In: Proceedings of the 10th ACM Multimedia & Security Workshop, Oxford, UK, September 22-23, pp. 123–132 (2008)
21. Kodovský, J., Pevný, T., Fridrich, J.: Modern steganalysis can detect YASS. In: Proceedings SPIE, EI, Media Forensics and Security XII, San Jose, CA (2010)

22. Pevný, T., Bas, P., Fridrich, J.: Steganalysis by subtractive pixel adjacency matrix. In: Proceedings of the 11th ACM Multimedia & Security Workshop, Princeton, NJ, September 7-8, pp. 75–84 (2009)
23. Pevný, T., Fridrich, J.: Benchmarking for steganography. In: Solanki, K., Sullivan, K., Madhow, U. (eds.) IH 2008. LNCS, vol. 5284, pp. 251–267. Springer, Heidelberg (2008)
24. Ryabko, B., Ryabko, D.: Asymptotically optimal perfect steganographic systems. Problems of Information Transmission 45(2), 184–190 (2009)
25. Sachnev, V., Kim, H.J., Zhang, R.: Less detectable JPEG steganography method based on heuristic optimization and BCH syndrome coding. In: Proceedings of the 11th ACM Multimedia & Security Workshop, September 7-8, pp. 131–140 (2009)
26. Sallee, P.: Model-based steganography. In: Kalker, T., Cox, I., Ro, Y.M. (eds.) IWDW 2003. LNCS, vol. 2939, pp. 154–167. Springer, Heidelberg (2003)
27. Ullerich, C., Westfeld, A.: Weaknesses of MB2. In: Shi, Y.Q., Kim, H.-J., Katzenbeisser, S. (eds.) IWDW 2007. LNCS, vol. 5041, pp. 127–142. Springer, Heidelberg (2008)
28. Vapnik, V.N.: The Nature of Statistical Learning Theory. Springer, New York (1995)
29. Wang, Y., Moulin, P.: Perfectly secure steganography: Capacity, error exponents, and code constructions. IEEE Transactions on Information Theory, Special Issue on Security 55(6), 2706–2722 (2008)
30. Westfeld, A.: High capacity despite better steganalysis (F5 – a steganographic algorithm). In: Moskowitz, I.S. (ed.) IH 2001. LNCS, vol. 2137, pp. 289–302. Springer, Heidelberg (2001)
31. Zhang, X., Zhang, W., Wang, S.: Efficient double-layered steganographic embedding. Electronics Letters 43, 482–483 (2007)

Obtaining Higher Rates for Steganographic Schemes While Maintaining the Same Detectability

Anindya Sarkar[1], Kaushal Solanki[2], and B.S. Manjunath[1]

[1] Department of Electrical and Computer Engineering,
University of California,
Santa Barbara, CA 93106
[2] Mayachitra Inc.,
5266 Hollister Avenue,
Santa Barbara, CA 93111
anindya@ece.ucsb.edu, solanki@mayachitra.com, manj@ece.ucsb.edu

Abstract. This paper focuses on modifying the decoder module for an active steganographic scheme to increase the effective data-rate without affecting the embedding module. Three techniques are suggested to improve the error correction framework of an active steganographic scheme. The first involves puncturing where the code-length is increased by adding a suitable number of additional erasures. The second technique involves channel modeling and soft-decision decoding which is adaptive to the embeddable image coefficient. The third method adjusts the erasure threshold depending on the design hiding quantizer so as to achieve a higher data-rate. Combining these techniques, the effective data-rate is increased by 10%-50% for Yet Another Steganographic Scheme (YASS), a popular active steganographic scheme.

Keywords: data hiding, error correcting codes, puncturing, log-likelihood ratio, erasure rate, steganography.

1 Introduction

Steganography is the art of secure communication where the existence of the communication itself cannot be detected by an external agent. The art of detecting such secret communication is known as steganalysis. Covert communication is typically enabled by embedding the secret *message* into an innocuous looking *host* or *cover* signal to form a *composite* or *stego* signal. The task of an adversary, the steganalyst (the "warden"), is to discover the presence of covert communication, which requires use of statistical and/or perceptual analysis to distinguish between plain cover and stego signals. This is the scenario of *passive* steganalysis, wherein, the steganalyst can observe the communication but cannot modify the covers. In many cases, an adversary can simply thwart any covert communication by mildly modifying the signals being communicated without needing to know whether they are cover or stego, leading to what is commonly known

R. Böhme, P.W.L. Fong, and R. Safavi-Naini (Eds.): IH 2010, LNCS 6387, pp. 178–192, 2010.

as *active* steganography [3, 7, 12]. An active warden has a limited attack budget so as not to significantly affect innocent users, who typically are the majority.

The past decade has seen great strides being made in these competing fields of *steganography* and *steganalysis*. Images are, arguably, the most popular host media, which is evident from the vast amount of literature in image steganography and steganalysis. Blind steganalysis schemes, employing powerful machine learning algorithms and specifically designed image features that capture changes due to data hiding, are quite successful in detecting the presence of very low rate covert communication in image hosts [8, 16, 10].

In this paper we focus on practical aspects of active steganographic schemes, which has received relatively less attention in the literature so far. An active steganographic system can be modeled as a communication channel, wherein, both the data hider and the attacker have limited distortion budgets to modify the host signal. This is in addition to the statistical security that the underlying data hiding scheme must provide. Thus, an important component of a data hiding method that can survive attacks is the use of error correcting codes (ECC). This paper focuses entirely on the error correction aspects of a stego scheme to provide noticeable improvement in the operating point for the rate-detectability trade-off. We utilize a better modeling of the underlying data hiding channel (scalar quantization index modulation (QIM) [5] based hiding) to compute more accurate likelihood ratios, erasure thresholds, and code rates. In this paper, we consider a benign attack scenario (JPEG compression) for the channel attack.

We employ a popular active steganographic scheme, Yet Another Steganographic Scheme (YASS) [17], as a platform to demonstrate the improvements. YASS involves data embedding in the discrete cosine transform (DCT) domain of randomly chosen block locations. The error correction framework is provided by serial concatenated turbo codes (repeat accumulate codes [6]). The noise channel consists of the JPEG compression attack and our steganalysis framework comprises of a collection of calibrated and uncalibrated features, which was shown to be effective for detecting YASS-based hiding in [9].

Although much of the discussion in the paper is specific to YASS, we must note that if a proper model for the channel attack can be obtained, i.e., if it is possible to reliably estimate a suitable transition probability matrix for the given channel, the methods proposed in the paper can be applied. We do not change the embedder and thus the detectability against steganalysis remains unchanged. *However, by better channel modeling and appropriate modifications to the decoder, the effective data-rate can be increased, without compromising on the undetectability and robustness, which forms the crux of this paper.*

Paper Outline: The problem formulation and main contributions are presented in Sec. 2, followed by a brief overview of the YASS methodology in Sec. 3. A brief description of the composite data hiding channel is presented in Sec. 4. The puncturing scheme and its implications are explained in Sec. 5. Suitable channel modeling for the soft-decision decoding are discussed in Sec. 6. In Sec. 7, we explain how the erasure rate can be suitably varied to maximize the effective data-rate. Experimental results and overall performance improvements over the previous decoding methods are presented in Sec. 8.

2 Problem Setup

QIM-based Hiding Framework: The hiding is performed using quantization index modulation [5]. An embeddable coefficient is converted to the nearest even/odd integer depending on whether the bit to be embedded is 0/1, respectively. For perceptual transparency, we do not use quantized AC DCT coefficients in [-0.5,0.5] for hiding; a DCT coefficient in this range is converted to zero and the corresponding bit is assumed to be "erased". An erasure is denoted as e in the paper. A list of commonly used acronyms is presented in Table 1.

Table 1. Commonly used Acronyms

Acronym	Full form
QIM	Quantization Index Modulation (a commonly used embedding method)
RA code	Repeat Accumulate code (a coding scheme for providing error resilience)
QF	quality factor (determines extent of JPEG compression)
ECC	Error Correction Coding (adds redundant bits to survive channel attacks)
DCT	Discrete Cosine Transform
DWT	Discrete Wavelet Transform
LLR	log-likelihood ratio which denotes the soft confidence level in an embeddable coefficient while decoding
B	big-block size used in the YASS framework
λ	the number of top AC DCT coefficients, encountered during zigzag scan, used for hiding per 8×8 block
λ'	for a puncturing scheme, the hiding band, per 8×8 block, has λ' ($> \lambda$) coefficients
QF_h	the design quality factor used for hiding
QF_a	the output quality factor at which the stego image is JPEG compressed
q_{opt}	the minimum RA code redundancy factor which allows successful decoding
$[x]$	rounded off value of x
δ_{dec}	at the decoder side, coefficients in $[-\delta_{dec}, \delta_{dec}]$ are assumed to be erasures
N_c	fraction of the embeddable DCT coefficients in the range $[c - 0.5, c + 0.5]$
$N_{[a,b]}$	fraction of the embeddable DCT coefficients in the range $[a, b]$

Turbo codes have shown the most powerful performance for the additive white Gaussian noise (AWGN) and the erasures channels, among many specific channel models. Our approach is practical and the techniques are backed by experimental results and improvements. The following aspects are considered.

(i) **Puncturing:** After encoding the data-bits using a given ECC of known redundancy, the code-length can be further increased by placing erasures at certain locations. Puncturing has long been known to improve the performance of turbo codes [2,1]. It helps the encoding process in two ways: (i) it allows finer choice of embedding rates (rather than $1/q$ where q is a positive integer redundancy factor), and (ii) it allows the use of large codewords, a key factor contributing to the near-capacity performance of turbolike codes. We provide experimental evidence that puncturing increases the effective hiding rate for certain channels .

(ii) **Soft-decision decoding with coefficient-based LLR Allocation:** When the channel model is known a priori, the use of soft-decision decoding invariably improves the convergence probability and accuracy of an ECC decoder. We conduct experimental channel modeling to compute soft confidence values to be set for the log-likelihood ratio (LLR) to be used by the decoder (a sum-product algorithm).

(iii) **Varying erasure rate:** In the QIM-erasure data hiding channel considered in the paper, the decoder can vary a threshold to control the erasure rate. Such a control may not be available in conventional communication channels. We provide simple analysis to compute the optimal value of the threshold δ_{dec} (coefficients in the range $[-\delta_{dec}, \delta_{dec}]$ are assumed to be erased) that maximizes the data hiding rate, given the QIM quantizer (for JPEG case, the quality factor QF_h). These are verified by experiments with real image datasets.

Since the statistical security is demonstrated for the YASS [17] framework, we now provide a brief description of the YASS stego scheme.

3 Brief Overview of YASS

The security of YASS can be attributed to the randomized choice of hiding locations. The idea of YASS was conceived keeping in mind the fact that the steganalysis features (for JPEG images) mainly consist of block-based features, i.e. computed on the 8×8 block. If the hiding is performed using a block-based approach which is not aligned with the JPEG-based blocks, the steganalysis features will be out-of-sync with the features that are modified by the hiding. For hiding in randomly chosen block locations, the image needs to be converted from the compressed domain (if a JPEG image is the input) to the pixel domain.

YASS Framework: The input image is decompressed if it is in JPEG format and then divided into blocks of size $B \times B$ ($B > 8$), where B is called the big-block size. For each big-block, a 8×8 sub-block is pseudo-randomly chosen to hide data. *The encoder and decoder share the same key by which they can access the same set of 8×8 blocks.* For every sub-block, its 2D DCT is computed and then divided by a JPEG quantization matrix at a *design* quality factor (QF), denoted by QF_h. A band of λ AC DCT coefficients lying in the low and mid-frequency range is used for hiding. After data embedding, the resultant image is JPEG compressed at a QF of QF_a.

To emphasize the role of YASS, it decides the hiding locations (pseudo-randomly chosen block locations) and the hiding coefficients (AC DCT coefficients belonging to the hiding band for the randomly chosen blocks). Unless otherwise mentioned, the data-rate computation experiments use B=9, QF_a=75, for the YASS framework.

4 Brief Description of the Hiding Channel

The effective data-hiding channel and the ECC framework used are shown in Fig. 1. We use a repeat accumulate (RA) code [6] as the ECC framework due to the high erasure rate associated with quantized DCT-domain hiding channels.

Composite Hiding Channel: The RA coding framework determines how $\mathbf{u}$, a sequence of N' bits, is mapped to the encoded sequence $\mathbf{c}$, a sequence of $N'q$ bits, assuming the ECC to have a redundancy factor of q. The conversion from $\mathbf{u}$ to $\mathbf{c}$ is explained in Fig. 1. After RA-encoding, this sequence $\mathbf{c}$ acts as the input sequence for QIM-based embedding. For the coefficients in the erasure zone, the code-bits get mapped to e (erasures); for the remaining coefficients, the code-bits get properly embedded. The ternary sequence (with symbols $\{0, 1, e\}$) obtained after embedding is denoted by $\mathbf{z}$. Then, the stego image is JPEG compressed (other global noise attacks are also possible) and the same set of embeddable coefficients (as identified at the encoder) is identified at the decoder. The ternary sequence, derived at the decoder side, is denoted by $\hat{c}$. Thus, the effective hiding channel is represented by a 2×3 mapping, from $\mathbf{c}$ to $\hat{\mathbf{c}}$.

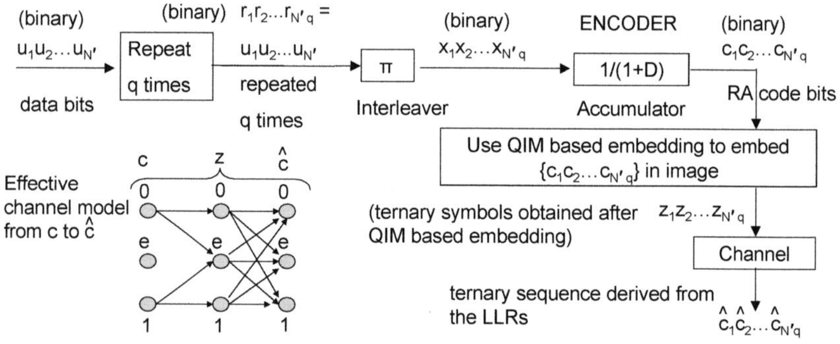

Fig. 1. The mapping between $\mathbf{c}$, the binary RA code-bit sequence to $\hat{\mathbf{c}}$, the ternary sequence obtained from the LLRs at the decoder output, is shown here for the QIM-RA framework - the "channel" between $\mathbf{z}$ and $\hat{\mathbf{c}}$ refers to the JPEG compression channel that introduces errors and erasures in the mapping from $\mathbf{z}$ to $\hat{\mathbf{c}}$

Numerical Examples: Some examples of the mapping between $\mathbf{c}$ and $\mathbf{z}$ ($\{0, 1\} \rightarrow \{0, 1, e\}$, denoted by a 2×3 matrix $P_{c,z}$), $\mathbf{z}$ and $\hat{\mathbf{c}}$ ($\{0, 1, e\} \rightarrow \{0, 1, e\}$, denoted by a 3×3 matrix $P_{z,\hat{c}}$), $\mathbf{c}$ and $\hat{\mathbf{c}}$ ($\{0, 1\} \rightarrow \{0, 1, e\}$, denoted by a 2×3 matrix $P_{c,\hat{c}}$), are presented below in Table 2. Since the modifications to the decoding framework require a thorough understanding of the channel transition probability matrices, we present these numerical examples.

The effective 2×3 mapping from $\mathbf{c}$ to $\hat{\mathbf{c}}$ is used to compute the channel capacity $\mathcal{C}$ by maximizing the mutual information $I(\mathbf{c}, \hat{\mathbf{c}})$ between the sequences $\mathbf{c}$ and $\hat{\mathbf{c}}$ (1) - a discrete memoryless channel is assumed.

$$\mathcal{C}_{c,\hat{c}} = \max_{p(c)} I(\mathbf{c}, \hat{\mathbf{c}}) = \max_{p(c)} \sum_{c \in \{0,1\}} \sum_{\hat{c} \in \{0,1,e\}} p(c, \hat{c}) \log \left(\frac{p(c|\hat{c})}{p(c)} \right) \quad (1)$$

Table 2. Using B=9, QF_a=75 for YASS, the transition probability matrices are computed for different hiding conditions. The results are averaged over 250 images.

Hiding Setup	$P_{c,z}$	$P_{z,\hat{c}}$	$P_{c,\hat{c}}$
QF_h=30, λ=6	$\begin{bmatrix} 0.23 & 0.00 & 0.77 \\ 0.00 & 0.37 & 0.63 \end{bmatrix}$	$\begin{bmatrix} 0.998 & 0.002 & 0.000 \\ 0.001 & 0.999 & 0.000 \\ 0.000 & 0.001 & 0.999 \end{bmatrix}$	$\begin{bmatrix} 0.23 & 0.00 & 0.77 \\ 0.00 & 0.37 & 0.63 \end{bmatrix}$
QF_h=75, λ=8	$\begin{bmatrix} 0.43 & 0.00 & 0.57 \\ 0.00 & 0.61 & 0.39 \end{bmatrix}$	$\begin{bmatrix} 0.833 & 0.167 & 0.000 \\ 0.096 & 0.888 & 0.016 \\ 0.000 & 0.079 & 0.921 \end{bmatrix}$	$\begin{bmatrix} 0.36 & 0.13 & 0.51 \\ 0.06 & 0.55 & 0.39 \end{bmatrix}$

Given an image, is it possible to obtain these transition probability matrices without actually simulating the entire channel? We now express these matrices in terms of $N_{[a,b]}$, the fraction of embeddable quantized DCT coefficients in $[a,b]$.

Table 3. Expression for a 2×3 transition probability matrix $P_{c,z}$

$p_{0,0} = 1 - p_{0,e}$	$p_{0,1} = 0$	$p_{0,e} = \frac{(N_{[-0.5,0.5]} + N_{(0.5,1)} + N_{(-1,-0.5)})}{2}$
$p_{1,0} = 0$	$p_{1,1} = 1 - p_{1,e}$	$p_{1,e} = \frac{(N_{[-0.5,0.5]})}{2}$

Explaining $P_{c,z}$: (as shown in Table 2) When a lower QF_h is used, the quantization applied to the DCT coefficients is coarser and hence, the fraction of embeddable DCT coefficients that lies in the erasure zone increases. Therefore, the $(1,3)^{th}$ ($0 \rightarrow e$ mapping) and $(2,3)^{th}$ ($1 \rightarrow e$ mapping) elements in $P_{c,z}$ are higher for QF_h of 30 (coarser JPEG quantization) than when $QF_h = 75$ (finer JPEG quantization), as seen from Table 3.

Why is $P_{c,z}(1,1) < P_{c,z}(2,2)$ (i.e. $p_{0,0} < p_{1,1}$) ? All quantized DCT coefficients in the range [-0.5,0.5] get mapped to erasures. For a coefficient in [-0.5,0.5], it is equally likely for the input bit to be 0 or 1. For DCT coefficients in the range (0.5,1) and (-0.5,-1), the coefficients get mapped to 1 and -1, respectively, when 1 is to be embedded; however, these coefficients are mapped to zero (erasures) when 0 is to embedded. Hence, $p_{1,1} > p_{0,0}$.

We now express the transition probability terms in $P_{z,\hat{c}}$ in terms of the noise distribution $Pr(n)$ (n is the noise signal that affects the mapping from $\mathbf{z}$ to $\hat{\mathbf{c}}$) and the decoder-side erasure cutoff δ.

$$p_{0,0} = \forall_{k \neq 0, k \in \mathbb{Z}} \sum_{x=2k}^{2k+0.5} N_x.Pr(n \leq ([x] + 0.5 - x)) +$$

$$\forall_{k \neq 0, k \in \mathbb{Z}} \sum_{x=2k-0.5}^{2k} N_x.Pr(n \geq ([x] - 0.5 - x)), \; p_{0,1} = 1 - p_{0,0}, \; p_{0,e} = 0$$

$$p_{1,1} = \forall_{k \neq 0, k \in \mathbb{Z}} \sum_{x=2k-1}^{2k-0.5} N_x.Pr(n \leq ([x] + 0.5 - x)) +$$
$$\forall_{k \neq 0, k \in \mathbb{Z}} \sum_{x=2k+0.5}^{2k+1} N_x.Pr(n \geq ([x] - 0.5 - x))$$
$$+ \sum_{0.5}^{1} N_x.Pr(n \geq (\delta - x)) + \sum_{-1}^{-0.5} N_x.Pr(n \leq (-\delta - x))$$
$$p_{1,e} = \sum_{0.5}^{1.5} N_x.Pr(n \leq (\delta - x)) + \sum_{-1.5}^{-0.5} N_x.Pr(n \geq (-\delta - x))$$
$$p_{1,0} = 1 - (p_{1,1} + p_{1,e})$$
$$p_{e,e} = \sum_{0}^{0.5} N_x.Pr(n \leq (\delta - x)) + \sum_{-0.5}^{0} N_x.Pr(n \geq (-\delta - x))$$
$$p_{e,1} = 1 - p_{e,e},\ p_{e,0} = 0$$

Explaining $P_{z,\hat{c}}$: (as shown in Table 2) Focussing now on $P_{z,\hat{c}}$, the question comes up as why this matrix is asymmetric? It is observed that $p_{e,1} > p_{e,0}$, and $p_{0,1} > p_{1,0}$. For coefficients in the range [-0.5, 0.5], channel noise can result in the coefficients being rounded off to ± 1. Therefore, $p_{e,1} > p_{e,0}$ (the noise value should be significantly high to shift a coefficient from the range [-0.5,0.5] to a range (1.5,2.5) or (-1.5,-2.5)).

For a coefficient to be mapped from $1 \rightarrow 0$, coefficients in the range [0.5,1.5] (or [-1.5,0.5]) can get mapped to (1.5,2.5) (or (-2.5,-1.5)). When a coefficient corresponds to a $0 \rightarrow 1$ mapping, a coefficient in the range [1.5,2.5] (or [-2.5,-1.5]) can get mapped to [0.5,1.5] or [2.5,3.5] (or [-1.5,0.5] or [-3.5,-2.5]). There is very low probability of a '0' getting mapped to an erasure, i.e. to the [-0.5,0.5] zone. On the other hand, it is more likely for a $1 \rightarrow e$ mapping to occur. This happens when a coefficient in [0.5,1.5] (or [-1.5,-0.5]) gets mapped to (-0.5,0.5).

5 Puncturing for Better Performance

Puncturing is a technique used to obtain a $\frac{m}{n}$ code from a "basic" rate $\frac{1}{2}$ code. Puncturing (code bit deletions) effectively decreases the code-length. E.g. when we have a RA codeword of 200 bits and the optimal redundancy factor q_{opt} is 4, we can embed $200/4 = 50$ data-bits. The effective codeword length is now increased to 300; the extra 300-200=100 bits are assumed to be erasures. *With respect to puncturing, the input is a 300-bit code-word and 100 bits are deleted from it, leaving a 200-bit code-word.* If the new value of the optimal redundancy factor ≤ 5, the new data rate will be increased as $\lfloor 300/5 \rfloor > 50$. When puncturing is done, the number of additional erasures needs to be decided; it is seen that the effective data-rate is increased for a certain range of additional erasures. Simply put, puncturing allows us to choose a finer embedding rate for a given setup.

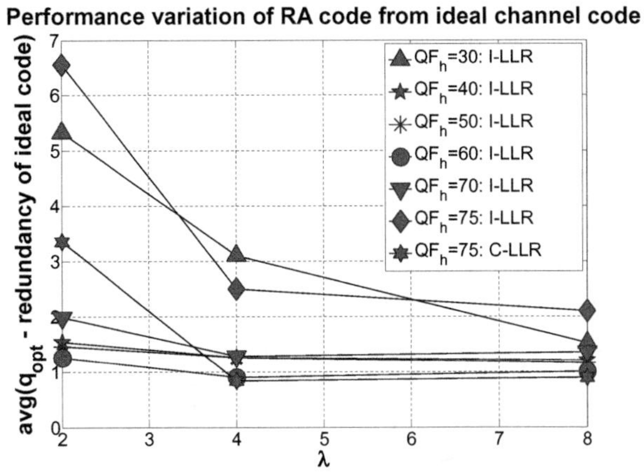

Fig. 2. The variation of the performance with λ and QF_h is computed over 250 images. Here, "I-LLR" and "C-LLR" refer to the "image-dependent" and "coefficient-dependent" LLR allocation methods, respectively.

The hiding band consists of λ AC DCT coefficients per 8×8 block. Let the number of $B \times B$ blocks used for hiding be N_B. Thus, the total number of coefficients available for hiding $N = \lambda.N_B$. While puncturing, it is assumed that the hiding band consists of λ' ($\lambda' > \lambda$) coefficients per 8×8 block.

We empirically observe that for more noisy channels, performance improvement is not obtained on using a higher number of erasures. The experimentally computed q_{opt} using RA codes is significantly higher than the redundancy factor for an ideal channel code, $\lceil \frac{1}{\mathcal{C}_{c,\hat{c}}} \rceil$ (1). The average value of ($q_{opt} - \lceil \frac{1}{\mathcal{C}_{c,\hat{c}}} \rceil$), computed for 250 images, is reported in Fig. 2. It is seen that the RA code performs closer to capacity (($q_{opt} - \lceil \frac{1}{\mathcal{C}_{c,\hat{c}}} \rceil$) becomes smaller) for larger λ, i.e. for longer code-lengths. It is seen that the RA code is most away from capacity for channels with very high error rates (when QF_h=75) or very high erasure rates (when QF_h=30). In Fig. 2, "I-LLR" and "C-LLR" refer to the "image-dependent" and "coefficient-dependent" LLR allocation methods, explained later in Sec. 6.1 and Sec. 6.2, respectively.

To vary the effective noise level in the channel (this affects the 3×3 transition probability matrix that denotes the mapping between $\mathbf{z}$ and $\hat{\mathbf{c}}$), the design quality factor used for hiding, QF_h, is varied. With a fixed value of QF_a, the output JPEG quality factor, the effective channel noise increases as QF_h is increased. Why does the effective noise increase as QF_h approaches QF_a? As QF_h increases, the DCT coefficients are divided element-wise by a finer quantization matrix (the elements in the JPEG quantization matrix become smaller). For a quantization matrix coefficient of Δ, the noise should exceed $\frac{\Delta}{2}$ to cause a decoding error. Therefore, as $QF_h \Uparrow \implies \Delta \Uparrow \implies$ noise robustness $\Downarrow$.

The bpnc improvements on using varying degrees of additional erasures for different hiding conditions are shown in Tables 4 and 5. For QF_h=70 and 75, the bpnc starts to decrease as $\lambda' > \lambda$, and hence, the corresponding bpnc results after erasure addition are not reported.

Table 4. The average bpnc values are computed over 250 images using the QIM-RA framework and puncturing. Here, $(2,6)$ denotes that λ=2 and λ'=6. The %-gain is expressed as $\frac{\max_{\lambda' \geq \lambda} \text{bpnc}(\lambda, \lambda') - \text{bpnc}(\lambda, \lambda)}{\text{bpnc}(\lambda, \lambda)}$. We use B=9, δ_{dec}=0.5 and QF_a=75.

$QF_h = 30$					$QF_h = 40$				
(2,2)	(2,3)	(2,4)	(2,5)	%-gain	(2,2)	(2,3)	(2,4)	(2,5)	%-gain
0.032	0.038	**0.043**	0.042	**30.50**	0.053	0.063	**0.064**	0.063	**22.43**
(4,4)	(4,5)	(4,6)	(4,7)	%-gain	(4,4)	(4,5)	(4,6)	(4,7)	%-gain
0.059	0.069	**0.073**	0.070	**24.91**	0.089	0.097	**0.100**	0.099	**12.88**
(8,8)	(8,9)	(8,10)	(8,12)	%-gain	(8,8)	(8,9)	(8,10)	(8,12)	%-gain
0.117	0.119	**0.121**	0.120	**3.76**	0.139	0.140	0.142	**0.145**	**4.24**
(12,12)	(12,13)	(12,14)	(12,15)	%-gain	(12,12)	(12,13)	(12,14)	(12,15)	%-gain
0.138	0.139	**0.140**	0.139	**1.01**	0.164	0.167	**0.169**	0.168	**2.80**

Table 5. The same experiments, as in Table 4, are now shown for QF_h of 50 and 60

$QF_h = 50$					$QF_h = 60$				
(2,2)	(2,3)	(2,4)	(2,5)	%-gain	(2,2)	(2,3)	(2,4)	(2,5)	%-gain
0.0508	0.0609	**0.0630**	0.0607	**24.02**	0.0427	**0.0446**	0.0421	0.0367	**4.45**
(4,4)	(4,5)	(4,6)	(4,7)	%-gain	(4,4)	(4,5)	(4,6)	(4,7)	%-gain
0.0932	0.1016	**0.1040**	0.1032	**11.59**	0.0865	0.0908	**0.0937**	0.0917	**8.32**
(8,8)	(8,9)	(8,10)	(8,12)	%-gain	(8,8)	(8,9)	(8,10)	(8,12)	%-gain
0.1485	0.1508	**0.1540**	0.1536	**3.70**	0.1427	0.1438	**0.1464**	0.1415	**2.59**
(12,12)	(12,13)	(12,14)	(12,15)	%-gain	(12,12)	(12,13)	(12,14)	(12,15)	%-gain
0.1748	0.1772	0.1792	**0.1811**	**3.60**	0.1738	0.1739	**0.1740**	0.1738	**0.12**

6 Suitable LLR Allocation for Soft Decision Decoding

RA codes are an example of serial concatenated turbo codes, where the component decoders are based on the BCJR algorithm [4]. The BCJR algorithm takes as input the a-posteriori probability of each code-bit, which is used to compute the log-likelihood ratio (LLR) at each code-bit location, as defined in (2). The forward and backward Viterbi-decoding algorithms running through the trellis depend on the initial estimates of the posterior probabilities which decide the LLR values. We have consolidated upon a recently proposed method to suitably initialize the LLR estimates at the decoder locations [15]. It has been experimentally observed that proper initialization of LLR values leads to faster convergence at the decoder, i.e. convergence using a lower redundancy factor. For a given image and an attack channel, the LLR values belonged to the 3-tuple, $(\alpha, -\alpha, 0)$, corresponding to 0, 1, and e, respectively, where α is a soft confidence value decided based on the composite channel parameters $(P_{c,\hat{c}})$. We repeat the discussion from [15] in Sec. 6.1 for ease of understanding. This LLR allocation scheme works well, except for very noisy channels. For such channels, we present a per-coefficient based, instead of a per-image based, LLR allocation method in Sec. 6.2.

6.1 Image-Based LLR Allocation

Let a certain image coefficient be equal to y and the corresponding embedded bit be b. The LLR value $LLR(y)$ denotes the logarithm of the ratio of the likelihood that a 0 was transmitted through that coefficient ($Pr(b=0|y)$) to the likelihood that a 1 was transmitted ($Pr(b=1|y)$).

$$LLR(y) = \log\left(\frac{Pr(b=0|y)}{Pr(b=1|y)}\right) \tag{2}$$

Let p_e denote the effective error probability in the channel (mapping from $\mathbf{c}$ to $\hat{\mathbf{c}}$) and N_c denotes the fraction of embeddable DCT coefficients whose value changes to c on rounding. In [15], the LLR value is estimated as follows:

$$LLR(y|[y]=c, c\neq 0) = \pm\log\left(\frac{(N_c + N_{c-1}/2 + N_{c+1}/2)(1-p_e)}{(N_c + N_{c-1} + N_{c+1} + N_{c-2}/2 + N_{c+2}/2)p_e/2}\right), \tag{3}$$

where the $\pm$ signs are for $c =$ even/odd, respectively, and $LLR(y)$ is kept at 0 when $[y]=0$.

The distribution of the AC DCT coefficients has been approximated as Laplacian [11, 14]. Always, $N_{c-1} > N_c > N_{c+1}$ holds, for $c \geq 1$, and $N_c \approx N_{-c}$, by symmetry. If we assume $N_c \approx (N_{c-1} + N_{c+1})/2$, then $LLR(y)$ reduces to:

$$LLR(y|[y]=c, c\neq 0) = \pm\log\left(\frac{1}{p_e} - 1\right) \tag{4}$$

It is experimentally observed that the LLR allocation methods using (3) and (4) result in similar embedding rates. Hence, in subsequent experiments, the image-dependent LLR is computed using the relatively simpler expression (4). The next issue is computing p_e for a given image and noise channel. In [15], it is seen that knowledge of the image histogram (that governs $P_{c,z}$) and the output JPEG QF (that governs $P_{z,\hat{c}}$) helps in accurate estimation of p_e.

6.2 Coefficient-Based LLR Allocation

This LLR allocation provides the same 3-tuple of LLR values $\left\{\pm\log\left(\frac{1}{p_e} - 1\right), 0\right\}$ for all the embeddable coefficients in a certain image. For very noisy channels, we observe that q_{opt} (the minimum RA-code redundancy factor for a given image and a hiding channel) is far-off from the $\lceil \frac{1}{C_{c,\hat{c}}} \rceil$. The solution is to perform a more in-depth analysis of the LLR allocation. We decide whether a 0/1 is embedded based on the fact that a certain image coefficient rounds off to an even/odd integer. E.g. if a coefficient in the received image is valued at 4, we are "more confident" that it corresponds to a 0-embedding than if the coefficient were valued at 3.6 or 4.4. Denoting the JPEG compression introduced noise signal by n, for the coefficient valued at 4 to correspond to a bit-error, the noise should exceed (4-3.5)=0.5 in magnitude. When the corresponding coefficients equal 3.6 (or 4.4), a bit-error can be caused when the noise exceeds 0.1 in magnitude. Due to the highly Laplacian-like pdf of n, $Pr(n > 0.5)$ is significantly less likely than $Pr(n > 0.1)$. *We have experimentally observed that using a per-coefficient LLR allocation scheme is more advantageous than using a per-image LLR allocation.*

Let y be the received DCT coefficient, $(x-1)$ be an even integer and $[y] = (x-1)$. For a decoding error to occur, the noise signal should exceed $((x-0.5)-y)$; then y would get mapped to x. It is assumed, due to the Laplacian pdf, that the value of the noise signal n is generally limited to [-1,1].

$$LLR(y) = \pm \log \left(\frac{Pr(b=0|y)}{Pr(b=1|y)} \right) = \pm \log \left(\frac{1 - Pr(n > (x - 0.5 - y))}{Pr(n > (x - 0.5 - y))} \right) \quad (5)$$

where $[y] = (x-1)$, the $\pm$ signs are for $(x-1)$ being an even/odd integer, respectively, and $LLR(y)$ is 0 when round$(y) = 0$.

When $[y] = x$ and x is an even integer, then $LLR(y)$ is expressed as:

$$LLR(y) = \pm \log \left(\frac{Pr(b=0|y)}{Pr(b=1|y)} \right) = \pm \log \left(\frac{1 - Pr(n < ((x - 0.5) - y))}{Pr(n < ((x - 0.5) - y))} \right) \quad (6)$$

where the $\pm$ signs are for x being an even/odd integer, respectively.

7 Variation of the Erasure Rate

Erasures are used to better account for symbols where the probability of a bit-error is quite high. By increasing the erasure threshold, the erasure rate is increased and the error rate is decreased - the flip-side is that the rate of correctly mapped symbols also decreases. It is experimentally observed that if the erasure rate is suitably adjusted, the decrease in the rate of correctly mapped symbols is offset by the decrease in the error rate, and the hiding rate is increased.

This method results in increased hiding rates for channels where the effective error rate is high, i.e. it dominates over the erasure probability term. It is experimentally observed that for channels with high error rates (e.g. channels with QF_h of 50-75), using an increased erasure rate results in a higher effective hiding rate. For channels with low error rates (e.g. channels with $QF_h \leq 30$), decreasing the erasure rate increases the hiding rate.

We now show an example of how the effective mapping $P_{c,\hat{c}}$ changes as the erasure cutoff (δ_{dec}) increases. As δ_{dec} is increased from 0.5 (assume that $\delta_{dec} \leq 1$), coefficients in the range $[0.5, \delta_{dec}]$ get mapped to erasures now, and these were mapped to 1 (bit '1' is embedded) when $\delta_{dec} = 0.5$. Thus, $p_{1,1}$ decreases and $p_{1,e}$ increases. Similarly, as the erasure range is increased (from (0.5,0.5) to $(-\delta_{dec}, \delta_{dec})$), $p_{e,e}$ increases while $p_{e,1}$ decreases. From (1), it is seen that $C_{c,\hat{c}}$ depends on the transition probability matrix $P_{c,\hat{c}}$, whose parameters are

Table 6. The average values for $P_{z,\hat{c}}$, $P_{c,\hat{c}}$ and $\mathcal{C}_{c,\hat{c}}$ are computed over 250 images, using B=9, QF_h=75, QF_a=75 and λ=8

δ_{dec}	$P_{z,\hat{c}}$	$P_{c,\hat{c}}$	$\mathcal{C}_{c,\hat{c}}$
0.5	$\begin{bmatrix} 0.8333 & 0.1673 & 0.0000 \\ 0.0963 & 0.8880 & 0.0157 \\ 0.0000 & 0.0792 & 0.9208 \end{bmatrix}$	$\begin{bmatrix} 0.3557 & 0.1317 & 0.5126 \\ 0.0602 & 0.5503 & 0.3895 \end{bmatrix}$	0.1910
0.6	$\begin{bmatrix} 0.8333 & 0.1673 & 0.0000 \\ 0.0963 & 0.8570 & 0.0467 \\ 0.0000 & 0.0378 & 0.9622 \end{bmatrix}$	$\begin{bmatrix} 0.3557 & 0.1009 & 0.5434 \\ 0.0602 & 0.5258 & 0.4140 \end{bmatrix}$	0.2050
0.7	$\begin{bmatrix} 0.8333 & 0.1673 & 0.0000 \\ 0.0963 & 0.8092 & 0.0944 \\ 0.0000 & 0.0163 & 0.9837 \end{bmatrix}$	$\begin{bmatrix} 0.3557 & 0.0834 & 0.5609 \\ 0.0602 & 0.4967 & 0.4431 \end{bmatrix}$	**0.2074**
0.8	$\begin{bmatrix} 0.8333 & 0.1673 & 0.0000 \\ 0.0963 & 0.7465 & 0.1572 \\ 0.0000 & 0.0061 & 0.9939 \end{bmatrix}$	$\begin{bmatrix} 0.3557 & 0.0745 & 0.5697 \\ 0.0602 & 0.4599 & 0.4799 \end{bmatrix}$	0.1986

expressed in terms of $N_{[a,b]}$, P_n and δ_{dec}. Thus, for a given distribution of the image DCT coefficients and an assumed noise distribution for fixed values of λ, QF_h and QF_a, the capacity can be expressed as a function of δ_{dec}.

We empirically show the best cutoffs to use for different values of QF_h in Fig. 3. The 2×3 mapping between c and z is image-dependent and is unaffected by δ_{dec}. Table 6 shows how the channel capacity varies with the erasure cutoff δ_{dec}, where the embedder side hiding parameters are left unchanged.

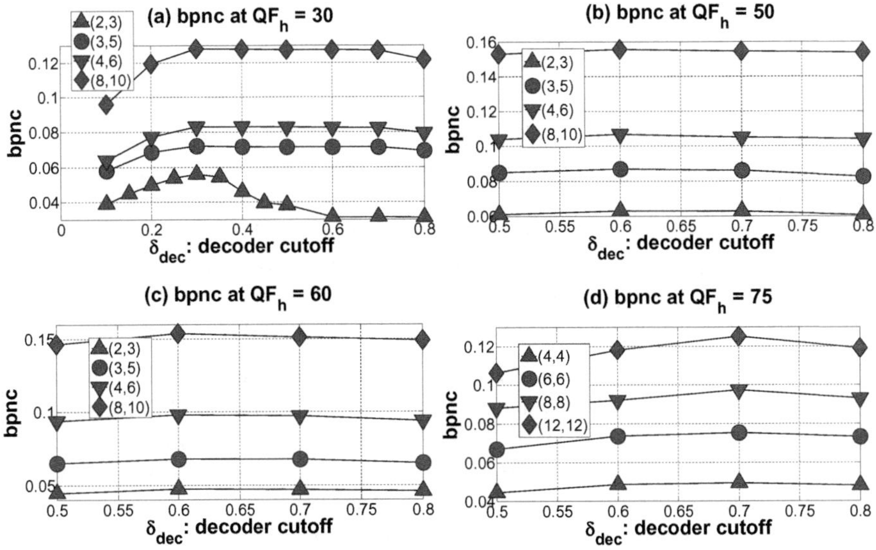

Fig. 3. Variation in bpnc with change in δ_{dec} for different choices of QF_h - the best choices for δ_{dec} are 0.3, 0.6, 0.6 and 0.7 for QF_h of 30, 50, 60 and 75, respectively

Table 7. The average bpnc results are presented for different hiding conditions. Here, I-LLR refers to the use of image-dependent LLR for decoding. "Puncture" refers to the use of the best combination (λ, λ') for a given λ that maximizes the bpnc. "Erasure" refers to the use of the best choice of δ_{dec}, the decoder cutoff, for a given QF_h, after puncturing. "C-LLR" refers to the use of the coefficient-dependent LLR, after puncturing and using best choice of δ_{dec}. Here, %-gain refers to the fractional gain obtained after using (puncture + erasure + C-LLR), as compared to using only I-LLR. **It is to be emphasized that P_d is unchanged as bpnc is increased for the same hiding parameters by varying the decoder module.**

QF_h	λ	I-LLR	puncture	Erasure	C-LLR	%-gain
30	2	**0.0318**	0.0425	0.0570	**0.0578**	81.76
30	4	**0.0586**	0.0732	0.0833	**0.0846**	44.37
30	8	**0.1169**	0.1213	0.1226	**0.1245**	6.50
30	12	**0.1387**	0.1401	0.1420	**0.1440**	3.82
50	2	**0.0508**	0.0630	0.0640	**0.0650**	27.95
50	4	**0.0932**	0.1040	0.1064	**0.1070**	14.81
50	8	**0.1485**	0.1540	0.1555	**0.1580**	6.40
50	12	**0.1748**	0.1811	0.1839	**0.1880**	7.55
60	2	**0.0427**	0.0446	0.0476	**0.0519**	21.55
60	4	**0.0865**	0.0937	0.0972	**0.1034**	18.38
60	8	**0.1427**	0.1464	0.1537	**0.1615**	13.17
60	12	**0.1738**	0.1745	0.1858	**0.1957**	12.60
70	2	**0.0230**	0.0230	0.0257	**0.0278**	20.87
70	4	**0.0644**	0.0644	0.0701	**0.0768**	19.25
70	8	**0.1132**	0.1132	0.1241	**0.1345**	18.82
70	12	**0.1379**	0.1379	0.1555	**0.1719**	24.66

8 Results

We show how the effective hiding rate is increased by a combination of the three factors. The %-age improvement in the bpnc is shown for different hiding parameters in Table 7.

For steganalysis, we use a set of 3400 high-quality JPEG images which were originally at a QF of 95 and they were JPEG compressed at a QF of 75 for the experiments. For the experiments, we crop out the central 512×512 region inside each image - the cropping is done for both the cover and stego images.

Steganalysis Feature Used: KF-548 - To improve upon the 274-dimensional calibrated feature [13], Kodovský and Fridrich [9] proposed the use of a 548-dimensional feature set **KF-548** which accounts for both calibrated and uncalibrated features. Here, the reference feature is used as an additional feature instead of being subtracted from the original feature.

Half of the images are used for training and the other half for testing. We use a support vector machine (SVM) based classifier for steganalysis, where the SVM is trained using the **KF-548** feature. The probability of classifying a test image correctly as cover or stego - the detection accuracy P_d (P_d=50% implies unde-

Table 8. The steganalysis results are reported using **KF-548**. P_d refers to the detection accuracy. "I-LLR" refers to the bpnc obtained using image-dependent LLRs while "final" refers to the bpnc obtained after using all the three proposed techniques - puncturing, suitably varied erasure cutoff, and coefficient-based LLRs.

Hiding Setup	P_d(%)	I-LLR	final	%-gain
QF_h=50, λ=2	65.00	0.0508	0.0650	27.95
QF_h=50, λ=3	69.00	0.0710	0.0860	21.13
QF_h=60, λ=3	65.80	0.0665	0.0795	19.55
QF_h=60, λ=4	70.06	0.0865	0.1034	19.54
QF_h=75, λ=4	55.00	0.0445	0.0564	26.74
QF_h=75, λ=6	60.40	0.0680	0.0849	24.85
QF_h=75, λ=8	68.00	0.0882	0.1099	24.60

tectable hiding, and as the detectability improves, P_d increases towards 100%) is obtained using **KF-548**. The steganalysis results are reported in Table 8. Based on the hiding parameters, it is seen that a higher QF_h gives better bpnc-vs-P_d trade-off, i.e. higher bpnc for similar P_d values. E.g. a bpnc of about 0.11 is obtained at a P_d of 0.68 at QF_h=75, while similar P_d values result in bpnc of 0.086 and 0.103 at QF_h = 50 and 60, respectively.

9 Conclusions

In this paper, we have demonstrated three simple methods to increase the effective data-rate at the decoder side without affecting the embedder process. These methods have been tested on the YASS framework where they have produced 10%-40% improvement in the hiding rate, without affecting the detectability. The methods have been based on the repeat accumulate code based framework, which has been shown to be close to capacity achieving for most hiding conditions. The only hiding condition where the RA code is still somewhat away from being capacity-achieving is for small code-lengths with channels having high error-rates or high erasure rates, and that leaves scope for further improvement. The decoding techniques proposed here are generic and can be incorporated in other steganographic schemes, which involve different methods for hiding coefficient selection, different transform domains for embedding, and different iterative decoding frameworks.

Acknowledgments. This research is supported in part by a grant from ONR # N00014-05-1-0816 and # N00014-10-1-0141.

References

1. Abbasfar, A.: Turbo-like Codes Design for High Speed Decoding. Springer, Heidelberg (2007)
2. Abbasfar, A., Divsalar, D., Yao, K., Inc, R., Los Altos, C.: Accumulate-repeat-accumulate codes. IEEE Transactions on Communications 55(4), 692–702 (2007)

3. Backes, M., Cachin, C.: Public-key steganography with active attacks. In: Kilian, J. (ed.) TCC 2005. LNCS, vol. 3378, pp. 210–226. Springer, Heidelberg (2005)
4. Bahl, L., Cocke, J., Jelinek, F., Raviv, J.: Optimal decoding of linear codes for minimizing symbol error rate. IEEE Trans. Inform. Theory 20(2), 284–287 (1974)
5. Chen, B., Wornell, G.W.: Quantization Index Modulation: A class of provably good methods for digital watermarking and information embedding. IEEE Trans. on Info. Theory 47(4), 1423–1443 (2001)
6. Divsalar, D., Jin, H., McEliece, R.J.: Coding theorems for turbo-like codes. In: 36th Allerton Conf. on Communications, Control, and Computing, pp. 201–210 (September 1998)
7. Kharrazi, M., Sencar, H., Memon, N.: Image steganography: Concepts and practice. Lecture Note Series, Institute for Mathematical Sciences, National University of Singapore (2004)
8. Kodovskỳ, J., Pevnỳ, T., Fridrich, J.: Modern steganalysis can detect YASS. In: Proceedings of SPIE, vol. 7541, p. 754102 (2010)
9. Kodovský, J., Fridrich, J.: Calibration revisited. In: MM & Sec 2009: Proceedings of the 11th ACM Workshop on Multimedia and Security, pp. 63–74. ACM, New York (2009)
10. Li, B., Shi, Y., Huang, J.: Steganalysis of YASS. In: Proceedings of the 10th ACM Workshop on Multimedia and Security, pp. 139–148. ACM, New York (2008)
11. Muller, F.: Distribution shape of two-dimensional DCT coefficients of natural images. Electronics Letters 29(22), 1935–1936 (1993)
12. Petitcolas, F.A.P., Anderson, R.J., Kuhn, M.G.: Information hiding — A survey. Proceedings of the IEEE, Special Issue on Identification and Protection of Multimedia Information 87(7), 1062–1078 (1999), `http://citeseer.nj.nec.com/petitcolas99information.html`
13. Pevný, T., Fridrich, J.: Merging Markov and DCT features for multi-class JPEG steganalysis. In: Proc. of SPIE, San Jose, CA, p. 3-1–3-14 (2007)
14. Reininger, R., Gibson, J.: Distributions of the two-dimensional DCT coefficients for images. IEEE Trans. on Comm. 31(6), 835–839 (1983)
15. Sarkar, A., Manjunath, B.S.: Image dependent log-likelihood ratio allocation for repeat accumulate code based decoding in data hiding channels. In: Proceedings of SPIE, 2010, Media Forensics and Security, vol. 7541, pp. 754113–754113–6 (January 2010), `http://vision.ece.ucsb.edu/publications/LLR_SPIE_2010_anindya.pdf`
16. Shi, Y.Q., Chen, C., Chen, W.: A Markov process based approach to effective attacking JPEG steganography. In: Camenisch, J.L., Collberg, C.S., Johnson, N.F., Sallee, P. (eds.) IH 2006. LNCS, vol. 4437, pp. 249–264. Springer, Heidelberg (2006)
17. Solanki, K., Sarkar, A., Manjunath, B.S.: YASS: Yet Another Steganographic Scheme that resists blind steganalysis. In: Furon, T., Cayre, F., Doërr, G., Bas, P. (eds.) IH 2007. LNCS, vol. 4567, pp. 16–31. Springer, Heidelberg (2007)

Robust and Undetectable Steganographic Timing Channels for i.i.d. Traffic

Yali Liu[1], Dipak Ghosal[2], Frederik Armknecht[3], Ahmad-Reza Sadeghi[3], Steffen Schulz[3], and Stefan Katzenbeisser[4]

[1] Department of Electrical and Computer Engineering, UC Davis, USA
[2] Department of Computer Science, University of California, Davis, USA
[3] Horst-Görtz Institute for IT-Security (HGI), Ruhr-University Bochum, Germany
[4] Department of Computer Science, Technische Universität Darmstadt, Germany

Abstract. Steganographic timing channels exploit inter-packet delays in network traffic to transmit secret messages. The two most important design goals are undetectability and robustness. In previous proposals undetectability has been validated only against a set of known statistical methods, leaving the resistance against possible future attacks unclear. Moreover, many existing schemes do not provide any robustness at all. In this paper, we introduce a steganographic timing channel that is both robust and provably undetectable for network traffic with independent and identically distributed (i.i.d.) inter-packet delays. I.i.d. traffic models are very useful because they are simple to analyze, and constitute essential elements of many advanced network traffic models. In contrast to previous work on i.i.d. traffic we do not rely on any strong assumptions, e.g., bounded jitter, but require only the existence of a cryptographically secure pseudorandom generator. We verify the effectiveness of our approach by conducting a series of experiments on Telnet traffic and discuss the trade off between various encoding and modulation parameters.

1 Introduction

Steganographic channels aim at establishing a communication channel hidden from any outsider. There is a large body of literature on steganographic channels in computer networks that exploit open overt communication (such as certain network protocols) as the carrier medium to transmit secret messages [1]. A specific class of steganographic channels concerns *timing channels* which exploit timing information and predominantly inter-packet delays to transmit messages [2, 3]. Most existing methods to establish steganographic timing channels have been successfully defeated using statistical tests [2, 4, 3] or entropy-based approaches [5].

Recent research have started to incorporate traffic models into the timing channel design process to evade detection [6, 7, 8]. However, these solutions have deficiencies regarding various aspects such as the required resources (e.g., to update and periodically transmit the system model parameters) [7], the underlying assumptions (e.g., of bounded jitter) [6] or the limited adversary model (e.g., to offer security only against specific statistical tests) [7, 8]. To the best of our knowledge, none of the existing solutions to steganographic timing channels provide a scheme that is both *provably undetectable* and *robust*. Here, undetectability refers to the incapability of the adversary to

R. Böhme, P.W.L. Fong, and R. Safavi-Naini (Eds.): IH 2010, LNCS 6387, pp. 193–207, 2010.

detect the steganographic channel by distinguishing between steganographic and overt communications. Robustness means that the steganographic message can be correctly decoded, even in the presence of (possibly maliciously added) additive noise. Note that advanced traffic obfuscation schemes such as traffic shaping can defeat any steganographic timing channel. However, since certain regular traffic patterns may be artificially generated by the obfuscation, the two communication parties may be aware that the communication has been compromised. In addition, for underlying traffic that is generated from applications that have real-time constraints or are interactive, the performance may be impacted as a result of changing the nature of the traffic pattern. The goal of robustness is to deal with inherent network noise (without any assumption of the underlying channel) and with additive noise that is introduced by a jamming device, which tries to defeat steganographic communication.

As recently proposed in [6], we consider legitimate traffic generated by applications for which the inter-packet delays are independent and identically distributed (i.i.d.). Among the set of possible traffic models, i.i.d. traffic models are extensively used in existing network analysis [9, 10]. Although there are only a limited number of real traffics that are strictly i.i.d., it is an essential element in many advanced traffic models and the methodology can be readily extended to other more general traffic models with reasonable analytical adjustments. For example, most multimedia applications, such as Voice over IP (VOIP) [11], can be modeled as Markov Renewal Processes. Typically, such models may consist of a multiple-state Markov chain with each state corresponding to a different i.i.d. traffic source. For more general scenarios, recent research [12] has shown that a Batch Renewal Process (BRP) with i.i.d. batch size and i.i.d. inter-batch time can be used to model traffic sources that have correlated inter-packet delays. Our proposed approach works for i.i.d inter-packet delays with any arbitrary distribution. When modeling network traffic, packet arrivals and connection requests are often assumed to follow a Poisson process because of its attractive theoretical properties [9]. However, as shown in [10], the Telnet traffic exhibits self-similar behavior. Consequently, the i.i.d. negative exponential distribution (i.e., a Poisson process) cannot be used to accurately model Telnet inter-packet delays while a Pareto-based i.i.d. can well capture the long range dependencies of the traffic. Therefore, in this paper, we evaluate the proposed steganographic timing channel using real Telnet trace samples as the legitimate traffic. We also examine the effectiveness of i.i.d. Pareto model to emulate the distribution of inter-packet delays of the real Telnet traffic.

Contribution and outline. In this paper, we propose a steganographic timing channel that is *both* provably undetectable and robust for any legitimate traffic whose inter-packet delays are i.i.d. following an arbitrary distribution. Toward this goal, in Section 4, we adopt the idea of spreading codes in the encoding process to mitigate the impact of transmission noise. Compared to the existing work [8] where the distribution of the steganographic traffic is only an approximation of the legitimate one and undetectability was only guaranteed with respect to two commonly used statistical tests, in Section 5, we design a modulation scheme for an i.i.d. traffic model to achieve undetectability against *any* (efficiently computable) statistical test. We will call this security property *polynomial undetectability* in the sequel. In Section 6, through an experimental study using different types of noise and noise power levels, we verify the effectiveness of our robust encoding scheme and compare our results to the most recently proposed steganographic timing channel [6]. The latter, to the best of our knowledge, is the only steganographic timing channel that is known to achieve polynomial undetectability for i.i.d. traffic. We show that our scheme improves the robustness of the steganographic communication and achieves true undetectability without relying on strong bounded jitters assumption required in [6]. Furthermore, we demonstrate the tradeoff between robustness and attainable transmission rate (by adjusting the encoding and modulation parameters) and validate the approach using Telnet traffic.

2 Related Work

Our work focuses on steganographic timing channels that involve modulating the packet transmission of legitimate traffic to embed secret information (the steganographic message) into packet arrival patterns. The simplest form of this class of steganographic timing channel is implemented by a binary on-off transmission scheme [13,2]: in a specific time interval, a packet arrival indicates the bit 1 and an absence indicates the bit 0. However, the inter-packet delays of such a steganographic traffic will be very regular and can be easily differentiated from legitimate traffic [2]. To avoid detection, [14] extended the idea by adding noise to the channel in an attempt to conceal the steganographic communication. Nonetheless, its security is only experimentally verified against regularity tests [4] that check the variance of inter-packet delays.

A more advanced type of steganographic timing channel encodes the steganographic message directly in the inter-packet delays. In the interval-time-replay scheme proposed in [15], the empirical range of the inter-packet delays of legitimate traffic is partitioned into two equal subsets, corresponding to "small-delays" and "large-delays". Then, a bit-1 of the steganographic message is sent by randomly replaying a large delay and a bit-0 by using a small delay. The keyboard JitterBug [3] exploits inter-packet delays of an existing interactive session to exfiltrate secret message. Specifically, small delays are added between key-presses to encode the steganographic message. Compared to the on-off schemes [13, 2], the design of these steganographic timing channels must take into account some of the characteristics of the legitimate traffic. Specifically, the shape of the distribution of the inter-packet delays of legitimate traffic is generally retained by the steganographic traffic. However, some statistical attribute of the inter-packet delays such as the distribution [2], the correlation [4], or the entropy [5], is altered by embedding the steganographic message, which can be exploited to detect the steganographic channel. Recently, a model-based steganographic timing channel was proposed in [7] to thwart these detection methods. The scheme first derives a filter model based on the statistics of the observed legitimate inter-packet delays and uses this model to generate the steganographic traffic. In order to adaptively fit a non-stationary traffic, however, this scheme requires frequent transmission of the model parameters to the decoder, which results in a considerable system overhead.

A powerful measure against steganographic timing channels is to actively disrupt the communication channel. For example, the study in [16] proposes to add random delays to the traffic using a jamming device and shows that the throughput of the steganographic information can be made vanishingly small in practice. Although jammers may also reduce the performance of the legitimate traffic at the same time, they introduce the requirement of robustness to steganographic timing channel design. Most existing steganographic timing channels only address the issue of detectability and the transmission efficiency in terms of channel robustness has only been considered under limited conditions. One of the earliest study to characterize the capacity of a steganographic timing channel was presented in [17]. Subsequently, a number of studies [18, 19, 20, 6] have investigated low complexity code design to achieve the desired capacity. Particularly, in [6], the authors proposed an encoding scheme that achieves near optimal data rate under normal network conditions. Based on this scheme, a timing channel was introduced to generate i.i.d. traffic undetectable for a polynomial time adversary. However, these schemes are limited by strong assumptions on traffic pattern modifications that are introduced during transmission. For example, the additive jitter must be bounded within a certain range [6]. Hence, security is only guaranteed under this condition and the robustness is not sufficient against noisy channels or a malicious jammer. Previous work [8] has shown that spreading codes can effectively increase the robustness against various intentional and/or unintentional channel distortions. In particular, the authors present a method to modulate the (encoded) secret message to mimic the distribution of the inter-packet delays of a non-stationary legitimate traffic. Nonetheless, only two commonly used classes of statistical tests were applied to validate the undetectability aspect. Moreover, the distribution of the steganographic traffic is only an approximation of the legitimate one and thus it may not provide security against future more sophisticated detection methods.

3 System Model and Design Criteria

3.1 Preliminaries

We will use the terms *steganographic communication* and *overt communication*, respectively, to refer to a communication with and without embedded steganographic channel and use *steganographic traffic* and *legitimate traffic* to refer to the packet stream associated with these two different channels.

We also introduce some basic definitions that we use later. A function $f : \mathbb{N} \rightarrow \mathbb{R}$ is called *negligible* if for any $j' \in \mathbb{N}$ there exists a polynomial $p(z)$ over $\mathbb{N}$ such that $|f(j)| < |1/p(j)|$ for all $j \geq j'$. For a probability distribution $\mathbb{D}$, the expression $\mathbf{z} \leftarrow \mathbb{D}$ denotes the event that an element $\mathbf{z}$ has been sampled according to $\mathbb{D}$. A *distinguisher* $\mathcal{D}$ is a (possibly probabilistic) algorithm that aims for distinguishing between two different distributions $\mathbb{D}$ and $\mathbb{D}'$ over the same space. More precisely, $\mathcal{D}$ receives a value $\mathbf{z}$ that is either sampled according to $\mathbb{D}$ or $\mathbb{D}'$ and outputs a bit $b \in \{0, 1\}$. The *advantage* of $\mathcal{D}$ is defined by

$$\mathrm{Adv}_{\mathcal{D}}(\mathbb{D}, \mathbb{D}') := |Pr[\mathcal{D}(\mathbf{z}) = 1|\mathbf{z} \leftarrow \mathbb{D}] - Pr[\mathcal{D}(\mathbf{z}) = 1|\mathbf{z} \leftarrow \mathbb{D}']| \quad (1)$$

Definition 1 (CSPRNG). *Let R denote a range of values and G a pseudorandom number generator that produces a sequence of values in R depending on a (secret) seed $s \in \{0, 1\}^{\sigma}$. For a positive integer $N \geq 1$, we consider two distributions on R^N: $\mathbb{U}_N$, the uniform distribution, and $\mathbb{G}_N$, the distribution of N outputs by G.*

G is called a cryptographically secure pseudorandom generator *(CSPRNG) with respect to the size σ of the seed if it holds for any integer $N \geq 1$ that is polynomial in σ and for any distinguisher $\mathcal{D}$ between $\mathbb{U}_N$ and $\mathbb{G}_N$ with a runtime polynomial in σ that $Adv_{\mathcal{D}}(\mathbb{U}_N, \mathbb{G}_N)$ is negligible in σ.*

3.2 System Model

We define the *sender* (S) and the *receiver* (R) entities as the two ends of a steganographic communication. S has access to some sensitive information (steganographic message) to be transmitted to R. To achieve this, S embeds the steganographic information into a legitimate packet stream. Here, we consider an active steganographic timing channel [5], where the sender generates the traffic and embeds steganographic information. He can therefore perfectly control inter-packet delays (and increase and decrease them at will). We consider a binary channel, where the steganographic message is coded as a binary sequence over the alphabet $\{-1, +1\}$. The steganographic message is $\{b_1, b_2, b_3, \ldots\}$, where b_i is the i-th *information bit*. The information bits are encoded to produce *code symbols* $\{s(1), s(2), s(3), \ldots\}$, which are finally modulated in the inter-packet delays $\{d(1), d(2), d(3), \ldots\}$ of a packet stream due to channel noise, the receiver obtains a slightly different stream $\{\hat{d}(1), \hat{d}(2), \hat{d}(3), \ldots\}$. To decode the steganographic message correctly, a secret key is shared between S and R prior to the steganographic transmission; this key may be used to derive a shared codebook (i.e., a set of common random codewords).

Adversary Model. The adversary (e.g., a timing channel jammer or an intrusion detection system) can monitor or manipulate the transmission between S and R. A passive adversary aims only at detecting and/or deciphering the steganographic timing channel without interfering with the transmission, while an active adversary aims at detecting and disrupting the communication by manipulating an ongoing steganographic communication. Our system is designed to resist active adversaries.

We assume that an adversary has access to (different) samples of both legitimate and steganographic traffic and can easily derive some characteristics (such as the distribution of the inter-packet delays). The adversary also has knowledge of the structure and the modulation algorithm of the steganographic timing channel. However, the adversary has no access to the shared key (and any information derived therefrom). Observe that in our system model, no backward channel exists from R to S. Thus, the embedding of the secret message is independent of any disruption in the communication.

3.3 Design Criteria

Undetectability. On a high level, undetectability means that no efficient algorithm can distinguish between inter-packet delays of the legitimate traffic and steganographic traffic. We define a sequence of N inter-packet delays which stems from the steganographic traffic as $\mathbf{d} = (d(1), d(2), \ldots, d(N))$ and from the legitimate traffic as $\tilde{\mathbf{d}} = (\tilde{d}(1), \tilde{d}(2), \ldots, \tilde{d}(N))$. We denote their corresponding distributions by $\mathbb{D}_N$ and $\tilde{\mathbb{D}}_N$, respectively. Next we define our notion of undetectability.

Definition 2 (Polynomial Undetectability). *A steganographic timing channel is called* polynomially undetectable *with respect to a security parameter σ if it holds for* any *distinguisher $\mathcal{D}$ with a runtime polynomial in σ and for any N that is polynomial in σ that $Adv_{\mathcal{D}}(\mathbb{D}_N, \widetilde{\mathbb{D}}_N)$ is negligible in σ.*

Robustness. The robustness can be measured as the capability to achieve a decoding bit error rate (BER) $P_e \leq \varepsilon$ under a given robustness requirement $\varepsilon \in \mathbb{R}^+$. P_e is inversely proportional to the Signal-to-Noise Ratio (SNR) E_s/E_x [21], where E_s is the signal power and E_x is the noise power. Considering that there is a one-to-one mapping between BER and SNR, a given robustness requirement measured by ε can be achieved by increasing the SNR. Hence we define the robustness as follows:

Definition 3 (Robustness). *A steganographic timing channel is called γ-gain robust if the SNR after performing the encoding and modulation process is γ times greater than the original SNR, where $\gamma \in \mathbb{R}^+$ (we call γ the* robustness gain*).*

4 Robust Encoding with Spreading Codes

It was shown in [8] that encoding the message using spreading codes can efficiently increase the robustness of steganographic timing channels. Any arbitrarily strong additive noise can be mitigated by selecting a sufficiently large spreading factor. In this section, we briefly review the general concept of spreading codes and then show how to utilize spreading codes for steganographic timing channel design.

In the simplest case each bit $b_k \in \{+1, -1\}$ of the steganographic message $\{b_1, b_2, \ldots\}$ is encoded into $\mathbf{s}_k = b_k \cdot \mathbf{c}$, where $\mathbf{c} = (c_1, c_2, \ldots c_N) \in \{\pm 1\}^N$ is an arbitrary code word. Hence a code word $\mathbf{c}$ of length N will be used to convey just one information bit b_k. The transmission rate R_t, which measures the transmission efficiency of each bit by the number of packets, after applying the spread encoding process decreases to $1/N$ bit per packet (bpp)[1]. To encode multiple bits at once, K code words $\mathbf{c}_1, \ldots, \mathbf{c}_K$ can be used to carry K information bits $b_1, \ldots, b_K$ over K parallel channels. The symbols from all K channels are combined to a single sequence

$$\mathbf{s} = (s(1), s(2), \ldots, s(N)) = \sum_{k=1}^{K} \mathbf{s}_k = \sum_{k=1}^{K} b_k \cdot \mathbf{c}_k, \tag{2}$$

with $s(i) \in [-K, K]$, for transmission. To differentiate the transmitted bits for each channel, the spreading codes must be orthogonal to each other, i.e., $\langle \mathbf{c}_i, \mathbf{c}_j \rangle$ equals N if $i = j$ and 0 otherwise.

Assuming additive noise, the received symbols can be expressed as $\hat{\mathbf{s}} = \mathbf{s} + \mathbf{x}$. Here $\mathbf{x}$ is the process noise after demodulation; $\mathbf{s}$ and $\hat{\mathbf{s}}$ are N-dimensional vectors. To decode the k-th information bit, we apply a threshold rule to the inner product of the received code symbols and the code word $\mathbf{c}_k$:

$$\hat{b}_k = \frac{1}{N}\langle \hat{\mathbf{s}}, \mathbf{c}_k \rangle = \sum_{i=1}^{K} \frac{b_i}{N}\langle \mathbf{c}_i, \mathbf{c}_k \rangle + \frac{1}{N}\langle \mathbf{x}, \mathbf{c}_k \rangle = b_k + \frac{1}{N}\langle \mathbf{x}, \mathbf{c}_k \rangle. \tag{3}$$

[1] Directly transmitting binary information bits per packet can achieve 1 bpp.

If $\mathbf{x}$ and $\mathbf{c}_k$ are uncorrelated, $\hat{b}_k$ is equal to b_k and can be recovered by choosing a proper threshold (e.g., 0 for a binary sequence). Since N is the length of the spreading code and the maximum number of orthogonal channels, K must be less or equal to N. Therefore, for a binary channel, the maximum transmission rate R_t we can achieve is 1 bpp in the case of $K = N$.

We assume that the input steganographic message is composed of random binary bits. This is always achievable, as the steganographic message can be encrypted under a semantically secure cipher or compressed. In addition, a good spreading code should have the properties of a pseudo random sequence, such as balance (i.e., on average the same number of 1's and -1's) and short run length (low average number of consecutive 1's or -1's). Therefore, the code symbol in each channel can be regarded as a random binary sequence with equal probability, that is, $Pr[s_k(n) = 1] = Pr[s_k(n) = -1] = 1/2$. For simplicity, we omit the code symbol index n in the following. Let k_1 be the number of channels with the code value $s_k = 1$ and k_2 be the one with the code value $s_k = -1$; k_1 and k_2 are random variables with $0 \leq k_1 \leq K, 0 \leq k_2 \leq K$ and $k_1, k_2 \in \mathbb{N}$. As there are K channel in total, we have $K = k_1 + k_2$. From Eq. (2), the code symbol s is the sum of encoded symbols at each channel. Then we have $s = k_1 - k_2$. Since each channel carries an independent bit of the input binary sequence, the probability mass distribution (PMF) of code symbol s is given by

$$P_s(l) = Pr[k_2 = \frac{K-l}{2}] = \begin{cases} \binom{K}{\frac{K-l}{2}}(\frac{1}{2})^K & K - l \text{ even} \qquad (4a) \\ 0 & \text{otherwise,} \qquad (4b) \end{cases}$$

where $-K \leq l \leq K$ and $l \in \mathbb{Z}$.

5 Construction

For legitimate traffic whose inter-packet delays $\{\tilde{d}(1), \tilde{d}(2), \ldots\}$ are i.i.d. random variables, the inter-packet delays are fully determined by their probability distribution, e.g., the cumulative distribution function (CDF) $F_{\tilde{d}}(\cdot)$. Therefore, the goal of undetectability is to provide an efficient generator of random inter-packet delays $\{d(1), d(2), \ldots\}$ that follow the specific known distribution, but encode a steganographic message.

5.1 Modulation

Since the transformation of a code symbol $s(n)$ to the inter-packet delay $d(n)$ must be invertible, we consider a one-to-one mapping:

$$d(n) := T(s(n)), \quad n = 1, 2, \ldots, \qquad (5)$$

where $T(\cdot)$ is an invertible function. The demodulation at the receiver is done by:

$$\hat{s}(n) := T^{-1}(\hat{d}(n)), \quad n = 1, 2, \ldots, \qquad (6)$$

where $\hat{d}(n)$ and $\hat{s}(n)$ are the received inter-packet delays and demodulated code symbols, respectively, at time n. In the following, we show how to obtain $T(\cdot)$ and $T^{-1}(\cdot)$.

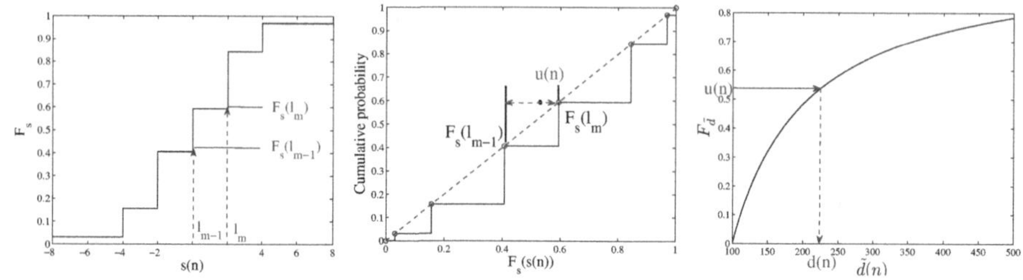

(a) (b) (c)

Fig. 1. An example of the mapping process (a) from $s(n) = l_m$ to $(F_s(l_{m-1}), F_s(l_m)]$, (b) from $(F_s(l_{m-1}), F_s(l_m)]$ to $u(n)$ and (c) from $u(n)$ to $d(n)$

After the encoding process, the amplitude of the code symbol $s(n)$ is a discrete random variable with a cumulative density function (CDF) denoted by $F_s(\cdot)$. Here $F_s(\cdot)$ can be calculated by accumulating the PMF of the code symbols (see Eq. (4)) as

$$F_s(l) = Pr[s \leq l] = \sum_{l_m \leq l} Pr[s = l_m] = \sum_{l_m \leq l} P_s(l_m). \tag{7}$$

Here, l_m are the unique values of code symbols sorted in ascending order.

To generate steganographic inter-packet delays following the distribution of the given i.i.d. traffic, we employ the commonly used inverse transform technique [22] (shown in Figure 1). In particular, a code symbol $s(n) = l_m$ $(1 \leq m \leq M)$ is first input into the function $F_s(\cdot)$ to obtain its cumulative density $F_s(l_m)$. Since the CDF is monotonic, this is a 1-to-1 mapping between $s(n)$ and $F_s(l_m)$. Second, with the help of a CSPRNG, we generate a uniform random number $v(n)$ in the range $(0, 1]$ to achieve the randomness of the inputs for the next step. Particularly, we construct $u(n)$ by linear interpolation:

$$u(n) = F_s(l_{m-1}) + [F_s(l_m) - F_s(l_{m-1})] \cdot v(n), \tag{8}$$

where $F_s(l_0) = 0$. It is easy to see that the overall sequence $\{u(1), u(2), \ldots\}$ consists of pseudo random numbers with a uniform distribution over $(0, 1]$. Now, by letting $d(n) = F_{\tilde{d}}^{-1}(u(n))$, we obtain a random variable $d(n)$ with the CDF $F_{\tilde{d}}(\cdot)$. The resulting values $\{d(1), d(2), \ldots\}$ are used as modulated inter-packet delays for the steganographic message $\{b_1, b_2, \ldots\}$.

At the receiver, the received inter-packet delay $\hat{d}(n)$ passes through the CDF of the legitimate traffic $F_{\tilde{d}}(\cdot)$ and generates $\hat{u}(n) = F_{\tilde{d}}(\hat{d}(n))$. To decode, we set $\hat{s}(n) = l_m$ if $\hat{u}(n) \in (F_s(l_{m-1}), F_s(l_m)]$. Hence the key used by CSPRNG to generate a uniform random number $v(n)$ does not need to be shared between the sender and receiver for demodulation. Finally, the information bits can be recovered by applying the decoding process of Eq. (3).

5.2 Proof of Undetectability

Theorem 1. *Consider the proposed steganographic timing channel mechanism and let G denote the deployed CSPRNG. If G is cryptographically secure with respect to a* *security parameter σ, then the generated steganographic timing channel is undetectable (cf. Def. 2) with respect to σ as well.*

Proof. Let $N \geq 1$ be an integer that is polynomial in σ and let $\mathbb{D}_N$ and $\widetilde{\mathbb{D}}_N$ denote the distribution of N inter-packet delays that stem from legitimate and steganographic traffic, respectively. According to Definition 2, we have to show that $Adv_{\mathcal{D}}(\mathbb{D}_N, \mathbb{D}_N)$ is negligible for any distinguisher $\mathcal{D}$ with a run-time polynomial in σ. By (8), if the values $v(n)$ are uniformly random in the range $[0, 1]$, then the values $u(n)$ are uniformly distributed in the range $[F_s(l_{m-1}), F_s(l_m)]$. Let $\mathbb{U}_N$ denote the uniform distribution on R^N (with R being the range of G) and $\mathbb{G}_N$ the distribution on R^N induced by G. We will prove that

$$\max_{\mathcal{D}} Adv_{\mathcal{D}}(\mathbb{D}_N, \widetilde{\mathbb{D}}_N,) \leq \max_{\mathcal{D}'} Adv_{\mathcal{D}'}(\mathbb{U}_N, \mathbb{G}_N) \tag{9}$$

where the maximum is taken over the set of all possible distinguishers $\mathcal{D}$ and $\mathcal{D}'$, respectively, and where Adv is defined as in Eq. (1). By assumption, G is cryptographically secure for any integer $N \geq 1$ that is polynomial in σ. Hence, the right-hand side of (9) is negligible by definition. Consequently, the left-hand side is negligible as well, being exactly the condition for undetectability.

Let $\mathcal{D}$ denote an arbitrary distinguisher between $\mathbb{D}_N$ and $\widetilde{\mathbb{D}}_N$ with a run-time polynomial in σ. We use $\mathcal{D}$ to construct a distinguisher $\mathcal{D}'$ for $\mathbb{U}_N$ and $\mathbb{G}_N$ that has the same efficiency and advantage as $\mathcal{D}$ and effectively the same run-time as $\mathcal{D}$. That is, it holds that

$$Adv_{\mathcal{D}}(\mathbb{D}_N, \widetilde{\mathbb{D}}_N) = Adv_{\mathcal{D}'}(\mathbb{U}_N, \mathbb{G}_N). \tag{10}$$

In particular, this proves Eq. (9) as $\mathcal{D}$ was an arbitrary (appropriate) distinguisher. $\mathcal{D}'$ is defined as follows. By definition, $\mathcal{D}'$ receives some values $\mathbf{u} = (u(1), u(2), \ldots, u(N))$ with either $\mathbf{u} \leftarrow \mathbb{U}_N$ or $\mathbf{u} \leftarrow \mathbb{G}_N$ and has to tell apart both cases. The distinguisher $\mathcal{D}'$ uses this input to create some steganographic traffic as described in Section 5. Observe that according to the system model (Sec. 3.2), the inter-packet delays of the legitimate traffic can be increased or decreased at wish. This results in some inter-packet delays $\mathbf{d} = (d(1), d(2), \ldots, d(N))$ that are handed to $\mathcal{D}$. Distinguisher $\mathcal{D}'$ receives a bit output $b \in \{0, 1\}$ from $\mathcal{D}$ and uses it as its own output.

Now observe that if $\mathbf{u} \leftarrow \mathbb{G}_N$, i.e., is a CSPRNG-sequence, then $\mathcal{D}'$ has correctly generated steganographic traffic and it holds that $\mathbf{d} \leftarrow \widetilde{\mathbb{D}}_N$. On the other hand, if $\mathbf{u} \leftarrow \mathbb{U}_N$, i.e., the sequence is truly random, then the generated traffic is truly i.i.d. as in the legitimate traffic, i.e., $\mathbf{d} \leftarrow \mathbb{D}_N$. Therefore, $\mathcal{D}'$ has the same advantage for distinguishing between $\mathbb{U}_N$ and $\mathbb{G}_N$ as $\mathcal{D}$ has for deciding between $\mathbb{D}_N$ and $\widetilde{\mathbb{D}}_N$ and almost the same run-time. This shows (10). □

5.3 Determining Encoding Parameters

According to the Definition 3, the system robustness is measured by the ratio of SNR before and after performing the encoding and modulation process. To generate undetectable steganographic traffic, the core idea is to generate random numbers uniformly distributed in the range $[0, 1]$. Beside the above encoding process to convert code symbol $s(n)$ into $u(n)$, the other method is to direct map the original information bit into the

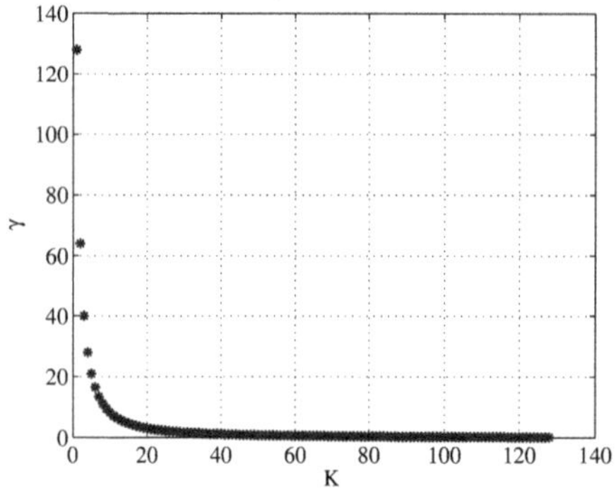

Fig. 2. The impact of K on the parameter γ ($N = 128$)

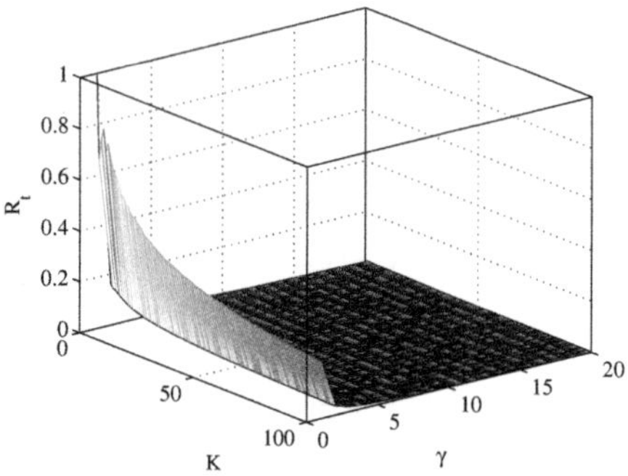

Fig. 3. Transmission rate R_t vs. the number of orthogonal channels K, and robustness gain γ

a number $u_1(n)$, which is uniformity distributed in the range $(0, 1/2]$ and $(1/2, 1]$, for the bit-0 and bit-1, respectively. The robustness gain is equivalent to the signal power ratio between $u(n)$ or $u_1(n)$ for the same information bits, provided the noise are the same for both the cases. Given Definition 3 and the spreading factor N from the encoding process, the robustness gain can be calculated by $\gamma = N \cdot E_K/E_1$, where E_K is the power of the random variable $u(n)$ when K channels are used and E_1 is the power of the $u_1(n)$. Since there are M different possible ranges for $u(n)$ with equal probability, the total robustness gain γ is given by

$$\gamma = \frac{N}{M} \sum_{m=1}^{M} \frac{[F_s(l_m) - F_s(l_{m-1})]^2}{(\frac{1}{2})^2} = \frac{4N}{M} \sum_{m=1}^{M} [P_s(l_m)]^2. \tag{11}$$

Given the distribution of $s(n)$ in Eq.(4) and $M = K + 1$, we rewrite Eq. (11) as:

$$\gamma = \frac{4N}{M} \sum_{m=1, -K \leq l_m \leq K}^{M} \binom{K}{\frac{K-l_m}{2}}^2 (\frac{1}{2})^{2K} = N \cdot \frac{1}{K+1} (\frac{1}{4})^{K-1} \sum_{k=0}^{K} \binom{K}{k}^2. \tag{12}$$

The effect of K on the gain γ is illustrated in Figure 2. With N fixed, it is easy to verify that γ is monotonically decreasing with K. At the same time, since γ is a linearly increasing function of N, one can achieve a higher robustness by decreasing K and increasing N. Therefore, any given gain requirement γ_0 can be achieved by choosing appropriate values for K and N. A reasonable choice of K and N is to maximize the data transmission rate $R_t = K/N$. Next we present a solution to achieve this goal.

Eq. (12) can be expressed as $\gamma = N \cdot q(K)$ with $q(K) = \frac{1}{K+1}(\frac{1}{4})^{K-1} \sum_{k=0}^{K} \binom{K}{k}^2$. To achieve a desired robustness gain requirement γ_0, we choose $N = \lceil \gamma_0 / q(K) \rceil$, the smallest value that satisfies the robustness requirement. Finally, we choose the best K to maximize $R_t = K / \lceil \gamma_0 / q(K) \rceil$. Figure 3 shows R_t as a function of K and γ_0. It shows that the best K for all values of γ occurs when $K = 1$. This also corresponds to the maximum robust gain γ, as mentioned above. From Eq. (12), $\gamma = N$ and the

Algorithm 5.1: INTERPACKETDELAYGENERATOR($\gamma_0, F_{\tilde{d}}$)

Input : robust gain γ_0, distribution of legitimate inter-packet delays $F_{\tilde{d}}(\cdot)$

Output : steganographic inter-packet delays $\mathbf{d}$

$N \leftarrow \lceil \gamma_0 \rceil$ // estimate spreading ratio with given robustness gain

for each information bit b

do
- generate a pseudorandom code word $\mathbf{c}$ using the shared key
- $(s(1), \ldots, s(N)) \leftarrow b \cdot \mathbf{c}$ //encoding
- generate $(v(1), \ldots, v(N)) \in_{\mathbb{R}}$ Uniform$(0, 1]$
- $u(n) \leftarrow F_s(s(n)) - \frac{1}{2} + \frac{1}{2}v(n)$ //modulation s to u
- $d(n) \leftarrow F_{\tilde{d}}^{-1}(u(n))$ //modulation u to d
- $\mathbf{d} := (d(1), d(2), \ldots, d(N))$

transmission rate R_t becomes $1/N$. Therefore, by setting $K = 1$ and $N = \lceil \gamma_0 \rceil$, our system can achieve a desired robustness requirement γ_0, corresponding to a specific BER P_e.

5.4 Algorithm Summary

The function *InterPacketDelayGenerator*$(\gamma_0, F_{\tilde{d}})$ in Algorithm 5.1 describes how to generate the steganographic inter-packet delays $\mathbf{d}$ under given robustness and security requirements. Since $K = 1$, that is, only one code word is used to transmit steganographic message at any time, the random code word $\mathbf{c}$ is dynamically generated based on the shared key between the sender and receiver. At the same time, the resulting code symbol $s(n)$ is also binary and its CDF must satisfy $F_s(-1) = \frac{1}{2}$ and $F_s(1) = 1$. Equivalently, we have $F_s(s(n)) = \frac{1}{2} + \frac{1}{4}(1 + s(n))$.

6 Experimental Results

6.1 Experimental Setup

Our testbed consists of a server and a client which act as the sender and the receiver of both the steganographic and the overt communication. We insert the inter-packet delays for steganographic channel modulation or additional noise using hooks in the Linux network stack at the sender. The receiver passively collects the inter-packet delays using tcpdump and decodes them with a shared key.

Test Scenarios. Two test scenarios are considered in our experimental evaluation. The first scenario is in a LAN environment in the campus network of UC Davis; client and server are hosted at two different departments. The second scenario is in the WAN environment to represent worldwide Round Trip Times (RTTs) between sender and receiver. The sender is hosted in the UC Davis and the receiver in Ruhr University. The network attributes of the two experimental scenarios are summarized in Table 1. Here the packet retransmission rate is measured using the ping command. We compute the jitter statistics based on the difference of delays between packets leaving the source and arriving at the destination.

Table 1. The network conditions for each test scenario

	LAN	WAN
Physical distance (miles)	1.5	5352
Packet retransmission rate (%)	0.60	1.03
Jitter(std) (ms)	12.6814	25.4023
Jitter(mean)(ms)	0.0550	0.10274

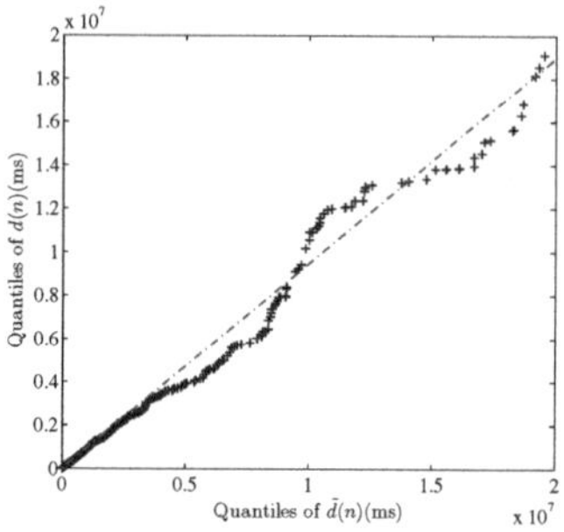

Fig. 4. Q-Q plot of $d(n)$ vs. $\tilde{d}(n)$

Dataset. To test our steganographic channel, we consider the Telnet traffic which uses TCP as the transport layer protocol. The legitimate samples that we use for our experiments are from the online MAWI working group traffic archive dataset [23]. They consist of traffic traces generated by tcpdump and in total of 32376 Telnet packets. Based on the trace samples, we derive an i.i.d. Pareto model to generate the steganographic traffic for our system. Specifically, the CDF of legitimate samples is modeled as, $F_{\tilde{d}}(s) = 1 - \left(\frac{\alpha}{s}\right)^{\beta}$, where $s > \alpha$. Here, $\alpha \in \mathbb{R}^+$ is a scale parameter and $\beta \in \mathbb{R}^+$ is a shift shift parameter and they are set $\alpha = 49$ ms and $\beta = 0.93$ respectively to emulate the statistics of the real traffic.

6.2 Performance Analysis

In our experiments, we generate a sequence of steganographic packets to carry $100,000$ random information bits. The time to transmit the steganographic message is determined by the transmission rate R_t (see Section 4) and the inter-packet delay of the traffic source. In our application, the average inter-packet delay of legitimate traffic is about 286 ms. The channel undetectability, robustness, and the transmission tradeoff are discussed below.

Undetectability. In addition to the rigorous proof that our i.i.d. based steganographic timing channel is undetectable against any polynomial distinguisher (see Section 5.2), we use the Quantile-Quantile (Q-Q) plot to visually examine the statistical similarity of inter-packet delays between the steganographic traffic based on our proposed method and the legitimate traffic from the trace database. Specifically, Figure 4 compares the real traffic $\tilde{d}(n)$ with artificially generated steganographic traffic $d(n)$. In a Q-Q plot, if the two sets have the same distribution, the points should fall approximately along a reference line. The results indicate that our steganographic traffic closely matches the legitimate one in terms of the distribution.

Robustness. To evaluate the robustness of the proposed algorithm, we consider three different types of noise during the transmission process. The first type represents the inherent network noise due to packet loss, delay, and jitter. The second and the third types of noise are jamming noises with different distributions which may be injected by an active adversary. Specifically, we use noise with a normal distribution with zero mean

Table 2. Summary of the bit error rate $P_e(\%)$ for the timing channel experiments

Test scenario	Encoding scheme		Channel noise	Gaussian σ(ms)				Uniform Δ(ms)			
				20	50	100	150	10	20	50	100
LAN	γ_0 spreading	1	0	5.37	11.90	16.63	20.37	1.87	5.30	13.63	28.33
		5	0	0	0.43	1.03	1.13	0	0.03	0.77	9.07
		10	0	0	0	0.01	0.06	0	0	0	2.13
	(L, n) encoding	(1, 1)	0	3.90	11.65	15.72	16.66	0	4.20	13.85	28.30
		(8, 2)	0	13.43	16.75	19.38	19.79	15.74	24.33	32.85	37.67
WAN	γ_0 spreading	1	0.02	6.22	15.43	18.04	21.24	2.83	6.71	18.61	30.62
		5	0.005	0.01	0.55	1.20	4.04	0.001	0.05	2.23	9.66
		10	0	0	0.002	0.03	0.10	0	0	0.002	3.08
	(L, n) encoding	(1, 1)	0.01	4.75	12.38	17.40	19.60	0	5.15	15.63	30.50
		(8, 2)	0.06	11.27	14.16	18.50	23.13	18.75	25.53	33.94	39.25

and variance σ^2 as the second type of noise. A uniformly distributed noise represents the worst case scenario in terms of the steganographic channel capacity [24]. This follows from the fact that a random variable with uniform distribution has maximum entropy among all random variables over a fixed range. Therefore, we choose the third type of noise to be uniformly distributed in the range $[0, \Delta]$. Note that, as long as the spreading codes used in the encoding process are orthogonal to $(1, 1, \ldots, 1)$, the mean of the noise does not impact the demodulation.

Table 2 summarizes the robustness results. In each experiment, steganographic inter-packet delays are generated with a given robustness requirement. This requirement is defined by the robustness gain γ_0. In our test, we set γ_0 as 1, 5, and 10 to model the effect of an absent, a moderate, and a strong spreading factor, respectively. In addition, we compare the robustness of our scheme to that of the (L, n) undetectable steganographic timing channel of [6]. Particularly, we choose the $(8, 2)$ scheme due to its high transmission rate. We also select the $(1, 1)$ scheme which is the most robust (L, n) scheme as a result of the largest range of each code symbol (the total range for all code symbols is fixed in the range $[0, 1]$), which ensures the largest signal power, so as the system SNR.

From these results we observe that when there is no jamming noise, there are no bit errors in the LAN scenario. When there is no spread encoding ($N = 1$), Gaussian noise with $\sigma = 20$ ms can result in more than 5.37% errors. When the noise is uniformly distributed in the range $[0, 20]$ ms, the bit error rate is 5.30%. These results are very similar to the performance of the $(1, 1)$ encoding scheme, which determines the upper bound of the robustness of the (L, n) encoding scheme. However, when the robustness gain γ_0 is increased to 5, the correct bit rate $(1-P_e)$ achieved by our proposed algorithm is more than 90.34% for both the LAN and WAN tests even with a jamming noise uniformly distributed in the range $[0, 100]$ ms. When γ_0 increases to 10, the correct bit rate is more than 99.90% for additive Gaussian noise with $\sigma = 150$ ms. Even when the upper limit of uniform noise is increased to 100 ms, we can still correctly transmit more than 96.92% of the total bits. These results show dramatic BER improvements compared to the $(8, 2)$ encoding scheme, which can achieve a good data rate.

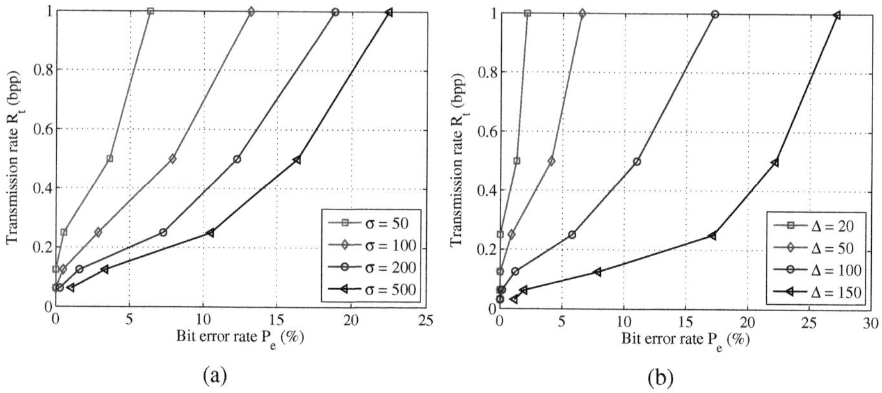

Fig. 5. Trade-off between the transmission rate R_t and the bit error rate P_e under (a) Gaussian and (b) uniform noise

Trade-off. By analyzing the results obtained in the LAN and WAN scenarios and comparing them to the (L, n) encoding scheme [6], we have shown that increasing the spreading ratio N can significantly reduce the BER P_e. However, this also requires more number of inter-packet delays to deliver one information bit which in turn reduces the transmission rate R_t. To investigate this relationship, Figure 5 plots the transmission rate R_t versus P_e under different noise scenarios in the LAN environment. It clearly shows that there is a trade-off between the transmission rate R_t and the robustness. In particular, the bit error rate increases monotonically with the transmission rate. This property can easily be verified by examining the definition of R_t, which is $1/N$, and the measure of robustness gain N. Therefore, our system provides a solution for balancing the system robustness and transmission rate. At the same time, we note that the achieved security level is independent of transmission rates and robustness.

7 Conclusion and Future Work

We propose a method to modulate a steganographic timing channel on network traffic with independent and identically distributed (i.i.d.) inter-packet delays. Our steganographic timing channel is both robust and provably undetectable. We discussed the choice for i.i.d. traffic and validated the theoretical results through experimental analysis using real Telnet traffic. Since i.i.d traffic models are building blocks of more complex traffic models, a natural open question is the extension of our approach for real applications such as video streaming or Voice over IP (VOIP).

References

1. Zander, S., Armitage, G., Branch, P.: A survey of covert channels and countermeasures in computer network protocols. IEEE Communications Surveys & Tutorials 9(3), 44–57 (2007)
2. Cabuk, S., Brodley, C.E., Shields, C.: IP covert timing channels: design and detection. In: CCS 2004: Proceedings of the 11th ACM Conference on Computer and Communications Security, New York, pp. 178–187 (2004)

3. Shah, G., Molina, A., Blaze, M.: Keyboards and covert channels. In: USENIX-SS 2006: Proceedings of the 15th Conference on USENIX Security Symposium, pp. 59–75 (2006)
4. Berk, V., Giant, A., Cybenko, G.: Detection of covert channel encoding in network packet delays. Technical Report. Darthmouth College (2005)
5. Gianvecchio, S., Wang, H.: Detecting covert timing channels: An entropy-based approach. In: CCS 2007: Proceedings of the 14th ACM Conference on Computer and Communications Security, pp. 307–316 (2007)
6. Sellke, S.H., Wang, C., Bagchi, S., Shroff, N.: TCP/IP timing channels: Theory to implementation. In: INFOCOM 2009: IEEE Conference on Computer Communications, pp. 2204–2212 (April 2009)
7. Gianvecchio, S., Wang, H., Wijesekera, D., Jajodia, S.: Model-based covert timing channels: Automated modeling and evasion. In: Lippmann, R., Kirda, E., Trachtenberg, A. (eds.) RAID 2008. LNCS, vol. 5230, pp. 211–230. Springer, Heidelberg (2008)
8. Liu, Y., Ghosal, D., Armknecht, F., Sadeghi, A., Schulz, S., Katzenbeisser, S.: Hide and seek in time - robust covert timing channels. In: Backes, M., Ning, P. (eds.) ESORICS 2009. LNCS, vol. 5789, pp. 120–135. Springer, Heidelberg (2009)
9. Kleinrock, L.: Queueing Systems. Wiley, New York (1976)
10. Paxson, V., Floyd, S.: Wide area traffic: the failure of poisson modeling. IEEE/ACM Transactions on Networking 3(3), 226–244 (1995)
11. Heffes, H., Lucantoni, D.: A markov modulated characterization of packetized voice and data traffic and related statistical multiplexer performance. IEEE Journal on Selected Areas in Communications 4, 856–868 (1986)
12. Li, W., Fretwell, R.J., Kouvatsos, D.D.: Analysis of correlated traffic by batch renewal process. In: EBISS 2009: International Conference on E-Business and Information System Security, pp. 1–5 (June 2009)
13. Padlipsky, M.A., Snow, D.W., Karger, P.A.: Limitations of end-to-end encryption in secure computer networks. Technical Report ESD TR-78-158, Mitre Corporation (1978)
14. Cabuk, S., Brodley, C.E., Shields, C.: IP covert channel detection. ACM Transaction of Information Systerm and Security 12(4), 1–29 (2009)
15. Cabuk, S.: Network covert channels: Design, analysis, detection, and elimination. PhD thesis, Purdue University (2006)
16. Giles, J., Hajek, B.: An information-theoretic and game-theoretic study of timing channels. IEEE Transactions on Information Theory 48(9), 2455–2477 (2002)
17. Anantharam, V., Verdu, S.: Bits through queues. IEEE Transactions on Information Theory 42, 4–18 (1996)
18. Coleman, T.P., Kiyavash, N.: Practical codes for queueing channels: An algebraic, state-space, message-passing approach. In: IEEE Information Theory Workshop on Networking and Information Theory, pp. 318–322 (May 2008)
19. Kiyavash, N., Coleman, T.: Covert timing channels codes for communication over interactive traffic. In: IEEE International Conference on Acoustics, Speech, and Signal Processing, pp. 1485–1488 (2009)
20. Dunn, B.P., Bloch, M., Laneman, J.N.: Secure bits through queues. In: IEEE Information Theory Workshop on Networking and Information Theory, pp. 37–41 (June 2009)
21. Proakis, J.: Digital Communications. McGraw Hill, Singapore (1995)
22. Devroye, L.: Non-Uniform Random Variate Generation. Springer, New York (1986)
23. Mawi working group traffic archive, `http://tracer.csl.sony.co.jp/mawi/`
24. Sellke, S.H., Wang, C., Shroff, N., Bagchi, S.: Capacity bounds on timing channels with bounded service times. In: IEEE International Symposium on Information Theory, pp. 981–985 (2007)

STBS: A Statistical Algorithm for Steganalysis of Translation-Based Steganography

Peng Meng[1], Liusheng Hang[1,2], Zhili Chen[1,2], Yuchong Hu[1], and Wei Yang[1,2]

[1] NHPCC, Depart. of CS. & Tech., USTC, Hefei 230027, China
[2] Suzhou Institute for Advanced Study, USTC, Suzhou 215123, China
mengpeng@mail.ustc.edu.cn

Abstract. Translation-Based Steganography is a secure text steganographic algorithm. In this paper, we present a novel statistical algorithm for steganalysis of Translation-Based Steganography (STBS). We first show that there are fewer high-frequency words in stegotexts than in normal texts. We then design a preprocessor to refine all the given texts to expand the frequency differences between normal texts and stegotexts. 12 dimensional feature vectors sensitive to frequency are derived from the refined texts. We finally use a SVM classifier to classify given texts to normal texts and stegotexts. A series of experiments is given to demonstrate the performance of STBS.

Keywords: steganalysis, natural language steganography, translation-based steganography, text, SVM, STBS.

1 Introduction

Text-based information, like web pages, academic papers, emails, e-books and so on, exchanged or distributed on Internet plays an important role in people's daily life. Because there are a huge number of texts available in which to hide information, a covert means of communication which is known as linguistic steganography [1] attracts more and more people's attention. Linguistic steganography makes use of writing natural language to conceal secret messages.

Traditional linguistic steganography has used syntactically-correct text generation and semantically-equivalent word substitutions within a cover text as a medium in which to hide messages [2]. TEXTO [3] is an early linguistic steganography program. It works just like a simple substitution cipher, with each of the 64 ASCII symbols or uuencode from secret data replaced by an English word. Wayner [4] introduced a method which uses precomputed context-free grammars to generate steganographic text without sacrificing syntactic and semantic correctness. Chapman and Davida [5] gave another steganographic method called NICETEXT. The texts generated by NICETEXT not only had syntactic and lexical variation, but whose consistent register and "style" could potentially pass a casual reading by a human observer.

For detecting the above linguistic steganography, some steganalytic algorithms have been proposed. Taskiran et al. [6] used a universal steganalytic method

R. Böhme, P.W.L. Fong, and R. Safavi-Naini (Eds.): IH 2010, LNCS 6387, pp. 208–220, 2010.

based on language models and support vector machines to differentiate sentences modified by a lexical steganography algorithm from unmodified sentences. Chen et al. [7] used the statistical characteristics of correlations between the general service words gathered in a dictionary to classify given text segments into stegotexts and normal texts. This method can accurately detect NICETEXT and TEXTO systems. The paper [8] also brought forward a detection method for NICETEXT, which took advantage of distribution of words. Another effective linguistic steganography detection method [9] uses an information entropy-like statistical variable of words together with its variance as two features to classify text segments.

Traditional linguistic steganography is relatively easy for a human to detect. Translation-Based Steganography (TBS) [2,10,11] which was introduced by Grothoff et al. is a novel method, which embeds information in the "noise" created by automatic translation of natural language texts. The key idea of TBS is: "When translating a non-trivial text between a pair of natural languages, there are typically many possible translations. Selecting one of these translations can be used to encode information" [2]. Because legitimate automatic translated texts have frequent errors, it is difficult for humans as well as for computers to distinguish the variations or even additional errors inserted by an information hiding mechanism from the normal noise associated with translation. The goal of this paper is to devise methods that can distinguish the inaccuracies caused by the use of steganography from the inaccuracies caused by deficiencies of the translation software.

We have previously introduced a steganalytic method [12] on TBS, but the method needs to know the Machine Translator (MT) set and the source text for the cover. Because the source text and the translator set may be part of the private secret of the sender [10], our previous method cannot be used in general to detect TBS. Also, the detection process of the method has to translate the given text two times by every translators of TBS, which may be too expensive for large-scale deployment.

Motivated by the challenge of detection TBS and based on our previous work, we present a novel statistical algorithm for steganalysis of TBS (STBS) in this paper. Our new method no longer needs to use the MT translation engines used by the TBS encoder during steganalysis and the cover source text is no longer used at all. We first show that there are fewer high-frequency words in stegotexts than in normal texts. For example, suppose the words "A1" and "A2" are the translation results of the word "A" by different machine translators, and "A" appears 5 times in the source text. For normal translation, there will be 5 "A1" or 5 "A2" occurrences in the translated text. But in TBS translated text, "A" may be translated twice to A1 and three times to A2 because sentences of the stegotext can come from different translators. As a result a high-frequency word in the cover text thus has a reasonable chance of becoming two or more low-frequency words in the stegotext.

We then design a preprocessor to refine all the test texts to expand the frequency differences between normal texts and stegotexts. 12 dimensional feature vectors sensitive to word frequency are derived from the refined texts. We finally use an SVM classifier to classify the test texts. A series of experiments is given to demonstrate the performance of STBS.

The organization of the paper is as follows: Section 2 briefly covers the basic operations of the TBS algorithm. Section 3 focuses on the statistical analysis of TBS, and gives a method to expand the word and 2-gram frequency difference between normal texts and stegotexts. Section 4 describes the process of steganalytic feature generation. In Section 5 we present the results of our steganography detection experiments. We give some discussions about STBS in Section 6. Finally, conclusions are presented in Section 7.

2 Translation-Based Steganography

In this section, we will briefly review the translation-based steganography protocol which our analysis tries to detect. Our work focuses on the "Lost in Just the Translation (LiJtT)" method [10] which extends the original Lost in Translation (Lit) [2] into one which allows the sender to only transmit the stegotext. In LiJtT, the source text and the MT systems are both private secrets of the sender.

Conceptually, TBS works as follows: The sender first needs to obtain a cover text in the source language. The cover text could be a secret of the sender or could have been obtained from public sources — for example, a news website. The sender then translates the sentences in the source text into the target language using many different translators. Because a sentence translated by a different translator may generate different translation results, the sender essentially creates multiple translations for each sentence and ultimately selects one of these to encode bits from the hidden message.

The protocol of LiJtT specifically works as follows: After generating multiple translations for a given cover text, the sender uses the secret key (which is shared

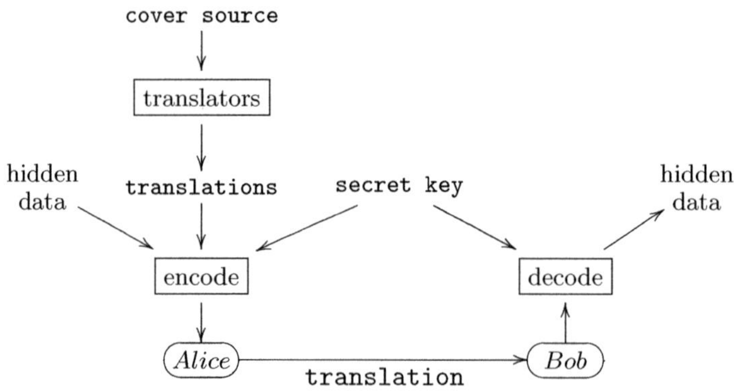

Fig. 1. Illustration of the basic protocol (from [10]). The adversary can observe the message between Alice and Bob containing the selected translation.

between the sender and receiver) to hash the individual translated sentences into bit strings. The lowest h bits of the hash strings, referred to as header bits, are interpreted as an integer $b \geq 0$. Then the sentence whose lowest $[h+1, h+b]$ bits corresponds to the bit-sequence that is to be encoded is selected.

When the receiver receives a translation which contains a hidden message, he first breaks this received text into sentences. Then he applies a keyed hash to each received sentence. The lowest $[h+1, h+b]$ bits in this hash contain the next b bits of the hidden message. Figure 1 illustrates the protocol.

3 Statistical Analysis of TBS

In this paper, natural language texts and ordinary translated texts are considered as normal texts, and stegotexts refer in particular to texts generated by TBS. Our work is to classify given texts to normal texts and stegotexts. In this section, we show the n-gram (n adjacent words, 1-gram equals to a word) frequency characteristic of different type texts. We mainly focus on word (1-gram) frequency differences between normal texts and stegotexts.

3.1 The Number of Words in Each Frequency

Normal texts have many inherent statistical characteristics which cannot be provided by stegotexts. In normal texts, there are a few high-frequency words (words that appear many times in a text), a middling number of medium-frequency words and many low-frequency words (words that appear only a few times in a text).

Because a word translated by different translators may generate different results and every sentence of the stegotext comes from different translators, a high-frequency word in the cover text may become two or more low-frequency words in the stegotext. For example, to translate from German to English using Google [13] and Systran [14] translators, the results of the German word "Unternehmen" are "Company" and "Enterprise" respectively. So in a German cover text which is talking about "Unternehmen", the German word " Unternehmen" will be a high-frequency word in the text, but in the stegotext, the word "Unternehmen" may be translated to two different words "Enterprise" and "Company". So compared with normal texts, stegotexts have fewer high-frequency words.

To present the word frequency difference between normal texts and stegotexts, for more than 1000 different types of texts, we counted the average number of words in each frequency, the rank of which is from 1 to 15. The texts contain natural language texts, stegotexts and translated texts by the Google, Systran and Promt [15] translators. Figure 2 shows the average number of words for each frequency when the text size is 20k bytes. We can find from Figure 2 that stegotexts have fewer high-frequency words than normal texts. The lines in Figure 2 are called word-frequency line in this paper.

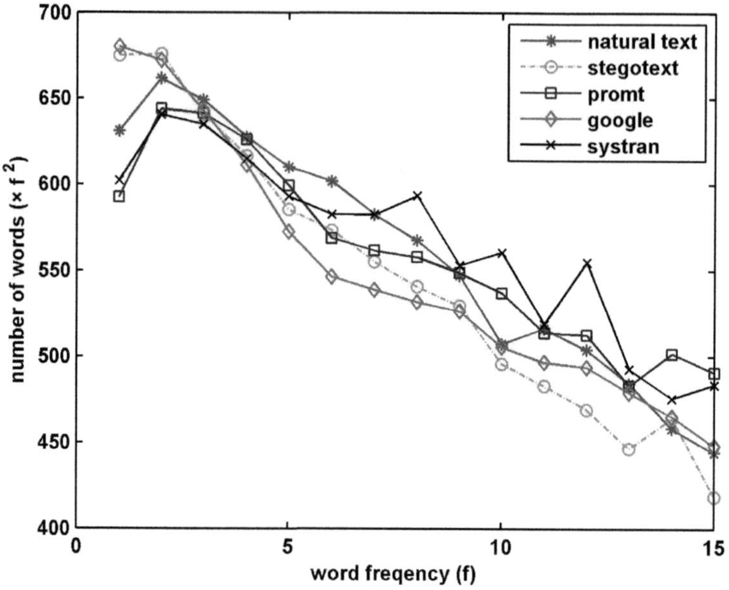

Fig. 2. The number of words for each frequency for 20kb texts

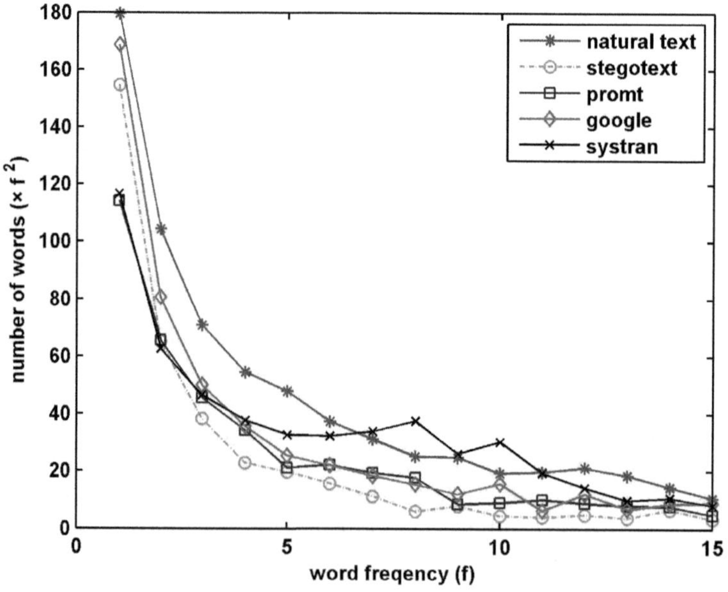

Fig. 3. The number of words for each frequency without one-to-one words

3.2 The Method to Expand Word Frequency Difference

Though stegotexts have fewer high-frequency words than normal texts, this statistical characteristic of word frequency is not adequate to accurately classify the given texts into stegotexts and normal texts. The word-frequency lines of small texts are volatile. Randomly selecting two normal texts, their word-frequency lines may be very different. To accurately classify short normal texts and stegotexts, we need to expand the word frequency difference between them.

We give two concepts before we describe our method to expand the word frequency difference:

One-to-one word: A word which is translated by different machine translators and always yields the same result is called a one-to-one word. If a word is a one-to-one word in the source language (cover text language), its translation in the destination language (stegotext language) is also called a one-to-one word.

One-to-many word: Words which are translated by different machine translators and generate different results are called one-to-many-words.

The word frequency difference between normal texts and stegotexts is caused by one-to-many words. Because some high-frequency one-to-many words in the cover text are translated to two or more low-frequency words in the stegotexts, stegotexts have fewer high-frequency words.

Our research shows most of words of the cover texts are one-to-one words, so the word-frequency lines of normal texts and stegotexts are close. If we delete all the one-to-one words in the cover text, then all the words in the refined cover text will be one-to-many words, and most of the high-frequency words in the refined cover text will be translated to two or more low-frequency words in the stegotexts. As a result, the word frequency difference between the refined cover text and stegotext will be expanded.

3.3 One-to-One Word Generation

For our experiments, we used a simple method to create a list of one-to-one words: The German texts from Europarl corpus were translated to English by Google, Systran and Promt translators sentence by sentence. Every sentence generated three translation results. The words which appeared in all the three translation sentences simultaneously were collected as one-to-one words. About 5M bytes German texts were processed in his method, yielding a total of 9784 one-to-one words.

For more than 1000 different type texts which contain natural language texts, stegotexts and translated texts by Google, Systran and Promt machine translators, we count the average number of words in each frequency after deleting one-to-one words. Figure 3 shows the word-frequency lines of refined normal texts and stegotexts. We find the word frequency difference between normal texts and stetotexts is greater than it is in Figure 2.

It should be noted that this method uses part of the secret key of the sender to collect one-to-one words: the MT translators, which technically may not be available to the adversary (especially if the sender uses human translators). We attempted other methods for generating one-to-one word lists, such as using a different language pair as TBS' or using certain high frequency words such as "a", "the", "an", "this" and "that". However, these methods have so far not been successful at achieving satisfactory detection results.

3.4 N-gram Frequency Difference between Normal Texts and Stegotexts

There exists not only word (1-gram) frequency differences between normal texts and stegotexts, but also 2-gram (two adjacent words) frequency differences between them, which is similar as word frequency characteristic. Deleting one-to-one 2-grams (one-to-one 2-grams can be similarly defined as one-to-one words) from both normal texts and stegotext also expands the 2-gram frequency difference between them. We can use the same method as one-to-one word to get the one-to-one 2-gram.

For n-grams where, when n is bigger than 2, there are too few high-frequency n-grams, especially when the text size is smaller than 40k bytes. As a result, there do not exist significant n-gram frequency differences between normal texts and stegotexts.

4 Features Generation

Our detection schema is an instance of two-class pattern recognition. A given text needs to be classified as either a stegotext (with hidden data) or as normal text (without hidden data). Therefore features generation is important in steganalysis.

Firstly, for a copy of the given text, we delete all the one-to-one words from it. Suppose the highest word frequency is m_1 in the refined text, we formalize the refined text as

$$T1 = \left\{n_1^1, n_2^1, ..., n_{m_1}^1\right\}, \tag{1}$$

where $n_i^1, 1 \leq i \leq m_1$ represents the number of different words whose frequency is i.

For example, given a text segment: "If a word is one to one word in the source language, its translation in the destination language is also called one to one word".

The text segment's word frequency is as follows:

8 words appear one time: If, a, source, its, translation, destination, also, called.

5 words appear two times: to, language, in, the, is.

1 word appears three times: word.

1 word appears four times: one.

So the text can be formalized to (refining process is omitted): $T1 = \{8, 5, 1, 1\}$

Secondly, for another copy of the given text, we delete all the one-to-one 2-grams from it. Because every word of a sentence except the first and the last one can constitute two 2-grams with its previous and following word, we cannot delete the one-to-one 2-grams directly from the given texts. We should first break the text down into a collection of 2-grams, then delete all the one-to-one 2-grams from the collection. We formalize the refined collection as:

$$T2 = \left\{n_1^2, n_2^2, ..., n_{m_2}^2\right\}, \tag{2}$$

where m_2 represents the highest 2-gram frequency, and $n_i^2, 1 \leq i \leq m_2$ represents the number of different 2-grams whose frequency is i.

For example, the above text segment's 2-gram frequency is as follows:

15 2-grams appear one time: If a, a word, word is, is one, word in, the source, source language, its translation, translation in, the destination, destination language, language is, is also, also called, called one.

4 2-grams appear two times: one to, to one, one word, in the.

So the text can be formalize to(refining process is omitted): $T2 = \{15, 4\}$

The features we extracted are defined as:

$$F_{b,c}^a = \sum_{i=c}^{m_b} (n_i^b * i^a) \tag{3}$$

$$a = \{0, 1, 2\}; b = \{1, 2\}; c = \{1, 5\}$$

We choose $[F_{1,1}^0, F_{1,1}^1, F_{1,1}^2, F_{1,5}^0, F_{1,5}^1, F_{1,5}^2, F_{2,1}^0, F_{2,1}^1, F_{2,1}^2, F_{2,5}^0, F_{2,5}^1, F_{2,5}^2]$ as the feature vector of the classifier. The first 6 features are generated from 1-gram (word) information of the given text and the last 6 features are generated from 2-gram information. We give the meanings of some features:

$F_{1,1}^0$: The number of different words, or in other words, the number of word types that appear in the refined text.

$F_{1,1}^1$: The number of word tokens in the refined text.

$F_{1,5}^0$: The number of different words whose frequency is bigger than 4 in the refined text.

$F_{1,5}^1$: The number of word tokens whose frequency is bigger than 4 in the refined text.

After computing the 12-dimensional features of a given text, we use a Support Vector Machine (SVM) Classifier to classify it as either normal text or stego-text. SVM is a popular technique for classification [16]. Our choice of the SVM classifier was motivated by the facts that they were previously used successfully for text classification [12]. In this paper, we use an existent SVM classifier [16] to classify our experiment texts.

5 Experiments and Results Analysis

In our experiment, the translator set of TBS contains Systran, Google and Promt translators. The texts were translated from German to English using the LiT prototype, with no semantic substitution, no article and preposition replacement enabled, and no "badness threshhold" [2].

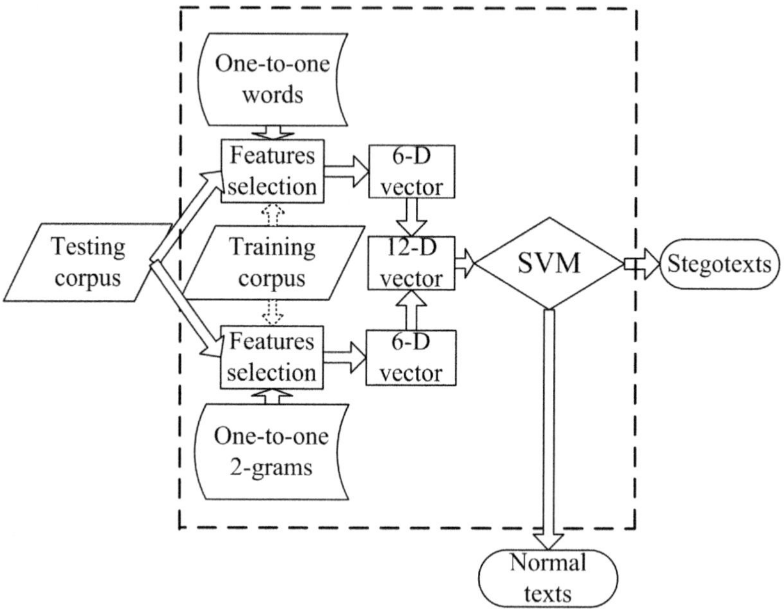

Fig. 4. Flow of STBS procedure

We built a training corpus and a testing corpus in our experiment. The text of both corpora contains: natural language texts, translated texts, and stegotexts. Translated texts were generated by Systran, Promt and Google translators. Both the natural language texts (English) and German language texts (used as cover text and the source language text of translated text) in our experiment come from the Europarl corpus [17].

Figure 4 shows the flow of the STBS procedure. The real line arrowhead represents the flow of data, while the dashed line arrowhead represents data transferred only when SVM classifier is training. The thick dashed rectangle indicates the whole detection system. Obviously, there are two key flows in the system: training and testing. The training process is always required before the testing process. But once the training process is completed, it does not not need to be repeated in each subsequent.

First, we used STBS to classify natural language texts, translated texts and stegotexts without deleting one-to-one words and 2-grams. Translated texts were generated from German to English by Google, Systran and Promt machine translators. Texts with size of 10k, 20k and 40k bytes were tested respectively. The detection results are listed in Table 1.

Second, we used STBS to classify natural language texts, stegotexts and translated texts with deleting one-to-one words and 2-grams. In the experiment, we collected 9784 one-to-one words and 39638 one-to-one 2-grams with the method described in Subsection 3.3. The detection results are listed in Table 2.

Table 1. Detection results without deleting one-to-one words and 2-grams

Text Size (byte)	Class	Train	Test	Non-stego (%)	Stego (%)
10k	Natural	0	985	91.1	8.9
	Promt	50	490	64.9	35.1
	Google	50	439	72.9	27.1
	Systran	50	489	57.5	42.5
	Stego	150	372	17.7	82.3
20k	Natural	0	493	98.2	1.2
	Promt	50	220	88.2	11.8
	Google	50	194	91.8	8.2
	Systran	50	219	65.3	34.7
	Stego	150	633	13.9	86.1
40k	Natural	0	246	100	0
	Promt	50	170	95.3	4.7
	Google	50	144	91.7	8.3
	Systran	50	168	58.3	41.7
	Stego	150	242	7.8	92.2

Finally, we used another language pair different from TBS' to get the one-to-one words and 2-grams. About 7M bytes French language texts from Europarl corpus were translated to English by Google, Systran and Promt translators sentence by sentence. In total, we collected 11897 one-to-one words and 57001 one-to-one 2-grams. The detection results are listed in Table 3.

6 Discussion

The data of Table 1 and Table 2 shows our method can effectively used to classify stegotexts and normal texts, and most of the time, the detection accuracy increases as the text size increases. Because machine translated texts are very noisy, even people cannot accurately distinguish stegotexts from machine translated texts, the detection accuracy on translated texts is not high when text size is smaller than 20k bytes.

In our experiment, we have used a simple and approximative method to get the one-to-one words and 2-grams. It needs to be emphasized that this method used the translator set and language pair of TBS. This information sometimes is not available to the adversary. Without this knowledge, we can use STBS without deleting one-to-one words and 2-grams.

We leave it to future work to determine other methods for finding good one-to-one words that give good detection accuracy without using the "secret" MT systems of the sender. Table 3 shows that this is not a trivial problem. When we use a language pair different from TBS' to get the one-to-one words and 2-grams, we get a worse detection result compared to not deleting any words.

To counter STBS, a clever steganographer might try to ensure that one word is always translated to the same word for all instances. We think STBS can still detect this method since it can not only consider word frequency differences

Table 2. Detection results of using the same language pair as TBS' to get the one-to-one words and 2-grams

Text Size (byte)	Class	Train	Test	Non-stego (%)	Stego (%)
10k	Natural	0	985	99.4	0.6
	Promt	50	490	68.0	32.0
	Google	50	439	57.7	42.3
	Systran	50	489	76.5	23.5
	Stego	150	372	12.3	87.7
20k	Natural	0	493	100	0
	Promt	50	220	80.9	19.1
	Google	50	194	71.1	28.9
	Systran	50	219	79.5	20.5
	Stego	150	633	10.7	89.3
40k	Natural	0	246	100	0
	Promt	50	170	98.8	1.2
	Google	50	144	84.7	15.3
	Systran	50	168	95.2	4.8
	Stego	150	242	5.4	94.6

Table 3. Detection results of using a language pair different from TBS' to get the one-to-one words and 2-grams

Text Size (byte)	Class	Train	Test	Non-stego (%)	Stego (%)
10k	Natural	0	985	58.1	41.9
	Promt	50	490	64.1	35.9
	Google	50	439	19.1	80.9
	Systran	50	489	70.6	29.4
	Stego	150	372	16.4	83.6
20k	Natural	0	493	71.8	28.2
	Promt	50	220	75.5	24.5
	Google	50	194	23.7	76.3
	Systran	50	219	74.9	25.1
	Stego	150	633	15.6	84.4
40k	Natural	0	246	71.5	28.5
	Promt	50	170	75.3	24.7
	Google	50	144	19.4	80.6
	Systran	50	168	85.7	14.3
	Stego	150	242	15.7	84.3

between normal texts and stegotexts but also consider n-gram frequency differences. Furthermore, any attempt to normalize word (or n-gram) distributions by the steganographer reduces the number of translations available for hiding information and hence results in the generation of longer cover texts — which in turn makes detection easier.

7 Conclusion

In this paper, we presented a statistical algorithm for steganalysis of Translation-Based Steganography (STBS). Our contributions are summarized as follows:

1) We have found a weakness in TBS: there are fewer high-frequency words in TBS generated stegotexts.

2) Based on the weakness we found, we designed a method to expand the word and 2-gram frequency difference between normal texts and stegotexts.

3) We have proposed to use a two-class SVM classifier to discriminate between normal texts and stegotexts. The detection accuracy is increases as the text size increases.

Stegotexts generated by TBS basically preserve the syntactic correctness and semantic coherence of the original translations, making it comparatively difficulty to detect this method. The accuracy of detection TBS also depends on many factors such as how many translators TBS used, which translators, the source language and the target language. However, we believe that our initial results show that STBS is a promising approach for the steganalysis of TBS.

Acknowledgment

The authors would like to thank Dr. Christian Grothoff and Dr. Yun-Qin Shi for helpful advice and the anonymous reviewers for their useful suggestions. This work was supported by the Major Research Plan of the National Natural Science Foundation of China (No. 90818005), the National Natural Science Foundation of China (Nos. 60903217 and 60773032), and the China Postdoctoral Science Foundation funded project (No. 20090450701).

References

1. Bennett, K.: Linguistic steganography: Survey, analysis, and robustness concerns for hiding information in text. Purdue University, CERIAS Tech. Report (2004)
2. Grothoff, C., Grothoff, K., Alkhutova, L., Stutsman, R., Atallah, M.: Translation-based steganography. In: Barni, M., Herrera-Joancomartí, J., Katzenbeisser, S., Pérez-González, F. (eds.) IH 2005. LNCS, vol. 3727, pp. 219–233. Springer, Heidelberg (2005)
3. Maker, K.: TEXTO, `ftp://ftp.funet.fi/pub/crypt/steganography/texto.tar.gz`
4. Wayner, P.: Disappearing cryptography: information hiding: steganography and watermarking. Morgan Kaufmann Pub., San Francisco (2008)
5. Chapman, M., Davida, D.: Hiding the hidden: A software system for concealing ciphertext as innocuous text. LNCS, pp. 335–345. Springer, Heidelberg (1997)
6. Taskiran, C., Topkara, U., Topkara, M., Delp, E.: Attacks on lexical natural language steganography systems. In: Proceedings of SPIE, vol. 6072, pp. 97–105 (2006)
7. Zhili, C., Liusheng, H., Zhenshan, Y., Wei, Y., Lingjun, L., Xueling, Z., Xinxin, Z.: Linguistic steganography detection using statistical characteristics of correlations between words. In: Solanki, K., Sullivan, K., Madhow, U. (eds.) IH 2008. LNCS, vol. 5284, pp. 224–234. Springer, Heidelberg (2008)

8. Zhili, C., Liusheng, H., Zhenshan, Y., Lingjun, L., Wei, Y.: A statistical algorithm for linguistic steganography detection based on distribution of words. In: Third International Conference on Availability, Reliability and Security, ARES 2008, pp. 558–563 (2008)
9. Zhili, C., Liusheng, H., Zhenshan, Y., Xinxin, Z.: Effective linguistic steganography detection. In: IEEE 8th International Conference on Computer and Information Technology Workshops, CIT Workshops 2008, pp. 224–229 (2008)
10. Stutsman, R., Atallah, M., Grothoff, K.: Lost in just the translation. In: Proceedings of the 2006 ACM Symposium on Applied Computing, pp. 338–345. ACM, New York (2006)
11. Grothoff, C., Grothoff, K., Stutsman, R., Alkhutova, L., Atallah, M.: Translation-based steganography. Journal of Computer Security 17(3), 269–303 (2009)
12. Peng, M., Liusheng, H., Wei, Y., Zhili, C.: Attacks on translation based steganography. In: Proceedings of the 2009 IEEE Youth Conference on Information, Computing and Telecommunication, pp. 227–230 (2009)
13. Google: Google translator (2009), http://translate.google.cn
14. Systran: Systran translator (2009), https://www.systransoft.com
15. PROMT: Promt translation software (2009), http://www.promt.com
16. Chang, C.C., Lin, C.J.: LIBSVM: a library for support vector machines (2001), http://www.csie.ntu.edu.tw/~cjlin/libsvm
17. Koehn, P.: Europarl: A parallel corpus for statistical machine translation. In: MT Summit, vol. 5 (2005)

The Reverse Statistical Disclosure Attack

Nayantara Mallesh and Matthew Wright

Department of Computer Science and Engineering,
The University of Texas at Arlington, Arlington, TX, USA
nayantara.mallesh@mavs.uta.edu, mwright@uta.edu
http://isec.uta.edu/

Abstract. Statistical disclosure is a well-studied technique that an attacker can use to uncover relations between users in mix-based anonymity systems. Prior work has focused on finding the receivers to whom a given targeted user sends. In this paper, we investigate the effectiveness of statistical disclosure in finding all of a users' contacts, including those from whom she receives messages. To this end, we propose a new attack called the Reverse Statistical Disclosure Attack (RSDA). RSDA uses observations of all users sending patterns to estimate both the targeted user's sending pattern and her receiving pattern. The estimated patterns are combined to find a set of the targeted user's most likely contacts. We study the performance of RSDA in simulation using different mix network configurations and also study the effectiveness of cover traffic as a countermeasure. Our results show that that RSDA outperforms the traditional SDA in finding the user's contacts, particularly as the amounts of user traffic and cover traffic rise.

1 Introduction

Mix-based anonymity systems [1] provide privacy by keeping eavesdroppers from linking communicating parties. Long term intersection attacks are particularly effective in reducing user anonymity in such systems. The most well known practical *traffic-confirmation* attack on mix systems is the Statistical Dislosure Attack [2] in which the attacker targets a single user with the aim of exposing the user's communication partners.

In the traditional form of this attack, the attacker eavesdrops on messages from senders to the mix and messages from the mix to receivers. The attacker uses the frequency of communication between parties to expose links between participating users. The aim of the attacker is to expose the contacts of a target user. Replies and other traffic sent to the targeted user is not considered. In reality, much of the communication in the Internet is two-way. The attacker, often assumed to be a *global eavesdropper* that can see all messages, would likely attempt to extract information from the patterns of traffic sent to the targeted user to learn more about her behavior.

1.1 Contributions

In this paper, we explore how the attacker could extract information from other users' sending patterns to learn more about the target user and her contacts. In

R. Böhme, P.W.L. Fong, and R. Safavi-Naini (Eds.): IH 2010, LNCS 6387, pp. 221–234, 2010.

particular, we introduce a new attack called the Reverse Statistical Disclosure Attack (RSDA) (§4). In the RSDA, the attacker simply applies the SDA to each user who sends messages. Some of the contacts of the targeted user — henceforth, we will refer to her as Alice — can be guessed based on the SDA applied to Alice. Additionally contacts of Alice can be guessed by examining the SDA results of other users. Let us consider Alice's friend Bob, who have replies regularly to Alice's messages or may simply send new messages to Alice. The attacker applies the SDA to Bob, and may be able to guess that Alice is one of his receivers. The RSDA leverages this information to note that Bob is a likely contact of Alice, even if the SDA did not allow the attacker to identify Bob as a receiver of Alice.

Note that the RSDA has a different model of what the attacker is interested in (§2), compared with the SDA. In the SDA, the attacker is only interested in the receivers to whom Alice sends. In the RSDA, the attacker wants to know all of the contacts with whom Alice communicates, whether sending or receiving. We believe that this is more realistic; traffic analysis is not generally confined to finding relationships in one direction.

The way RSDA uses information gained about other senders to learn about Alice is unique. In particular, we know of two other approaches that use similar information: the Two-Sided SDA (TS-SDA) [3] and the Perfect Matching Disclosure Attack (PMDA) [7]. We discuss these in more detail in §3, but briefly point out the key differences here. The TS-SDA assumes that the attacker is only interested in receivers to whom Alice *initiates* a message and attempts to filter out the statistical influence of Alice's replies on her SDA values. This is the opposite assumption from the RSDA model, in which the attacker is interested in any contacts of Alice, whether Alice initiates messages to them or not. The PMDA compares Alice's sending behavior to other senders' behavior with the intention of matching the senders to their most likely receivers in each *batch* of messages. PMDA is not looking for senders to Alice; RSDA is.

We use detailed simulation (§5) to study RSDA using different mix network configurations. Cover traffic has been recognized as an effective way to counter Statistical Disclosure Attacks [6,5]. Hence, we also study the effectiveness of cover traffic, including *background cover* and *receiver-bound cover*, as a countermeasure. Our results (§6) show that RSDA outperforms SDA particularly as the amounts of user traffic and cover traffic increase. Cover traffic from Alice affects SDA adversely and increases the time to 900 rounds; an increase of over three times compared with no cover. RSDA is extremely resilient to user cover and succeeds in only 250 rounds with cover and 100 rounds when no cover is present. We also found that as the total number of messages mixed in each round increases, both SDA and RSDA need more time to succeed. However, RSDA takes close to half the number of rounds compared to SDA as the mix batch size increases from 100 to 500 messages. When a binomial mix, having a more complex mixing strategy than the threshold mix is used, RSDA still proves to be a much faster attack compared to SDA. Furthermore, in the presence of increasing Alice cover, RSDA increases from 1000 rounds to only 1800 rounds while the increase in time for SDA is almost four times more going from 3000 to 6000 rounds with increasing Alice cover. RSDA is also affected very little in the presence of receiver-bound cover traffic. We conclude that RSDA is a much speedier attack than the traditional SDA. It shows a sizeable improvement over SDA and achieves high performance even in the presence of counter-measures like user and receiver-bound cover traffic.

2 Model

We now describe a model for our study of RSDA. We start by describing how we model mixes and users' communication patterns, and then we discuss our attacker model.

2.1 Mixes

We investigate statistical disclosure attacks against a simplified model of mixes. We use the term *mix* to refer to the entire mix network or mix cascade and abstract away details such as the number of mixes and their configuration. All users send their messages and cover traffic to the mix, and the mix sends messages on to all the receiving users.

We investigate RSDA's effectiveness against two types of mixes:

Threshold Mix. The threshold mix [1] collects a fixed number B (the batch size) of input messages before relaying the messages in a random order en route to their destinations. Each cycle of input and output together is called a *round*.

Binomial Mix. In a binomial mix [4], each incoming message is subject to a biased coin toss to decide whether the message leaves the mix in the current round or is delayed until a later round. The mix uses P_{delay} as the delay probability to bias the decision.

2.2 Communication Patterns

As we study the effectiveness of statistical attacks based on profiling users, the communication patterns of the users are critical to our evaluation. The three main features of the model are contacts (who sends to whom), sending behavior (how often does each user send to each of her contacts), and cover traffic.

We assume that there are N users, and we use a uniform model for establishing contacts between them. Specifically, each user, including Alice, has a fixed number of receivers m. The receivers are chosen uniformly at random from the set of other users. Unlike prior work in statistical disclosure attacks [6,3,7], we do not have separate sets of senders and receivers. Rather, each user will be a receiver for some of the other users. All of the users that communicate with a given user are included in that user's *contacts*. The total number of contacts per node will vary, but will be $2m$ on average.

Since the attacker focuses on a targeted user, Alice, we distinguish between Alice's behavior and other users' behavior. Alice sends n_A messages in a given round. n_A is a random variable selected from a Poission distribution with average rate λ_A. Alice chooses the recipients of her messages uniformly from her set of contacts. Users other than Alice are called *background senders*. When the mix uses a fixed batch size as in the case of a threshold mix, background senders together send $n_B = B - n_A$ messages. If the batch size is variable, as in the case of a binomial mix, background senders together send n_B messages, where n_B is chosen from a normal distribution with mean μ.

Cover traffic consists of fake messages called *dummy messages* that are inserted into the network along with real messages. Dummy messages are meant to look like real messages and cannot easily be distinguished from real messages. Usually, this means that the content of real messages that would be encrypted is replaced with random bits. The receiver of the dummy messages can recognize that they are fake, as they do not decrypt properly, and drops such messages on arrival. In our model, we use two types of cover traffic for the simulations. *Alice cover* consists of dummy messages that Alice sends to the mix. These messages are dropped at the mix. In each round in which Alice participates, she inserts zero or more dummy messages along with real messages. Alice may send dummy messages with no real messages in some rounds. *Receiver-bound cover* (RBC) consists of dummy messages from the mix to receivers. See [5] for details on how RBC is used to counter SDA.

2.3 Attacker Model

We model the attacker as a global eavesdropper who can observe all links from senders to the mix and all links from the mix to recipients. The target of the adversary is Alice and the adversary's aim is to determine with whom Alice communicates, i.e. to identify her contacts. The attacker observes all communications into and out of the mix during a number of rounds, including rounds with and without Alice's participation . The attacker observes only the incoming and outgoing links from the mix and does not observe activity inside the mix. This assumption is for the simplicity of the model, as there are many configurations for a mix network, but also because SDA and RSDA are effective without observations of activity inside the mix network.

3 Statistical Disclosure Attacks

In this section, we describe the Statistical Disclosure Attack, which is central to the function of RSDA and which we use as a basis for comparison. We also describe the Two-Sided Statistical Disclosure Attack and the Perfect Matching Disclosure Attack, both of which use observations about other senders to inform their method. RSDA uses these observations in a very different way from these existing techniques.

3.1 The (Original) Statistical Disclosure Attack

The Statistical Disclosure Attack (SDA) is a statistical technique for finding the receivers of a single targeted user Alice based on observed inputs to and output from the mix network. The attacker makes observations in a number of *rounds*, i.e. periods during which Alice participates. In each round of observation, the attacker records three pieces of information: n_A, the number of messages sent by Alice; n_B the number of messages sent by senders other than Alice; and $\overrightarrow{o}$ the distribution of messages received by receivers in that round. The attacker records the behavior of senders other than Alice, known as the *background*, by recording their activity when Alice does not participate. Vector $\overrightarrow{u}$ captures the distribution of messages from background senders to receivers in each round in which Alice is not present. The attacker sums $\overrightarrow{u}$ values over a large number of observations to obtain $\overline{U}$ which represents the sending behavior of background senders. The attacker sums $\overrightarrow{o}$ values over a large number of observations to obtain $\overline{O}$. Since $\overrightarrow{o}$ is recorded when both Alice and background senders participate, it represents their combined sending behavior. Thus, $\overline{O}$ represents the combined sending behavior of both Alice and the background during the observed rounds, and this can be written as:

$$\overline{O} = \overline{n_A}.D_A + \overline{n_B}.D_N \tag{1}$$

Here $\overline{n_A}$ and $\overline{n_B}$ are the total number of messages sent by Alice and the background, respectively, during the attacker's observation period. D_A is a vector that represents Alice sending behavior. For receiver i, who is Alice's contact, $0 < D_A[i] < 1$, and for receiver j who is not Alice's contact, $D_A[j] = 0$. D_N represents the sending behavior of background senders and is obtained by observing rounds in which Alice does not participate i.e. $D_N = \overline{U}/\overline{n_B}$. If the attacker is unable to collect background statistics before Alice begins communicating, D_N can be approximated as $D_N[i] = \frac{1}{N}\ \forall\ i$, meaning that the background sends in a uniform manner to all receivers. Alice's most likely set of contacts are determined by solving for D_A in equation (1) and picking m receivers with the highest $D_A[i]$ values.

Mathewson and Dingledine developed a simulation, including the use of a binomial mix (called a pool mix in their paper), to investigate the effect of a number of parameters on the performance of SDA [6]. They found that as the number of Alice's contacts grew, the rounds of observation to expose her full contact list correspondingly increased. They also found that cover traffic from Alice was effective in slowing, but not preventing, SDA. Cover traffic from Alice was found to be more effective when the delay probability of the binomial mix was increased. Increasing the mix delay spreads out the incoming traffic over a number of outgoing rounds, making it more difficult for the attacker to estimate which set of receivers might have gotten the messages from Alice.

3.2 Two-Sided Statistical Disclosure Attack

When Alice sends a message, she may be *initiating* the message or she may be replying to a message initiated by another user. If the attacker is only interested in knowing to whom Alice initiates messages, the SDA may have problems, as it is not designed to distinguish replies from initiated messages. The Two-sided

Statistical Disclosure Attack (TS-SDA) [3] extends the original SDA with observations of messages sent to Alice. TS-SDA uses these additional observations to estimate the likelihood that a given message from Alice is a reply to a previously received message and discounts possible replies accordingly.

As we note in Section 1, TS-SDA is based on very different assumptions from the RSDA. In particular, the assumption that the attacker is only interested in receivers of Alice's initiated messages leads TS-SDA to filter out the statistical influence of possible replies. In the current work, we assume that the attacker is interested in all of Alice's contacts, whether Alice initates the communication or not. TS-SDA would thus be worse than SDA in our model.

3.3 Perfect Matching Disclosure Attack

In the Perfect Matching Disclosure Attack (PMDA) [7], the attacker attempts to improve on SDA by using the insight that only one sender could have sent a particular message. This is best explained by a simple example in the threshold mix setting. Suppose that Alice and Bob are senders and Carol and Dave are receivers. In a given round, suppose that Alice and Bob each send one message and Carol and Dave each receive one message. Based on prior observations (profiling using SDA), both Alice and Bob are more likely to have sent to Carol than Dave. Since only one of them sent to Carol, however, PMDA finds the most likely matching of senders to receviers with, say, Alice sending to Carol and Bob sending to Dave. This matching is used to inform the profile of each sender and improve the attacker's chances of finding Alice's contacts.

This use of other senders' profiles is used in an entirely different way from RSDA. In particular, Alice is never a receiver and messages received by senders are never used in the profiling. We believe that the traffic analysis improvement in PMDA is therefore largely orthogonal to RSDA. Since both techniques require profiling of the users, however, combining the insights of PMDA with those of RSDA is challenging and we leave this for future work.

4 Reverse Statistical Disclosure Attack

In the Reverse Statistical Disclosure Attack (RSDA), the attacker first applies the SDA (as described in Section 3.1) to all N users. The attacker learns two pieces of information from this step. First, the attacker applies the SDA to Alice to learn about to whom Alice sends messages. Second, by applying the SDA to other users, he can determine which of them send to Alice. The attacker then combines this information to find the most likely contacts of Alice.

We break up the attack into three parts: (1) *forward observation*, or observations of Alice's sending behavior; (2) *reverse observation*, observations of other users' sending behavior; and (3), combining forward and reverse observations.

Forward Observation. In each round of observation the attacker records information in the forward direction as described in Section 3.1. This allows the attacker to calculate D_A, a set of scores representing Alice's estimated sending behavior.

Reverse Observation. In each round of observation the attacker also records information in the reverse direction. For a user X, the attacker records n_X, the number of messages sent by X, n_B, the number of messages sent by other users, and $\overrightarrow{o}$, the distribution of messages received by users in rounds that X sends. The eavesdropper also records D_N^X, the distribution of of messages received by users in rounds that X does not send. Using these observations, the eavesdropper does the SDA on X by using the following equation:

$$\overline{O} = \overline{n_X}.D_X + \overline{n_B}.D_N^X \tag{2}$$

With these observations, the attacker can apply Eqn. 2 to estimate D_X, the scores representing X's sending behavior.

Now let $D_X[A]$ represent the attacker's estimate of user X's sending behavior to Alice. We create a new vector D_R, such that $D_R[X] = D_X[A]$. In other words, D_R represents the estimated sending behavior of all other users with respect to Alice.

Combining Observations. The RSDA estimate of Alice's most likely contacts, $\widehat{D}_A$, can be determined by combining D_A and D_R calculated from the forward and reverse observations, respectively. D_A and D_R are combined by first normalizing and then obtaining a weighted mean of the two distributions. If v_f is the volume of traffic observed in the forward direction and v_r is the volume of traffic in the reverse direction, then we obtain:

$$\widehat{D}_A = \frac{v_f.D_A + v_r.D_R}{v_f + v_r} \tag{3}$$

Note that we could keep the information separate and simply determine Alice's receivers and those who send to Alice in isolation of each other. However, Alice's receivers will reply to her and vice versa. Since we assume that the attacker is interested in all of Alice's contacts, combining the information helps him learn more.

To see this, let us consider two users, Bob and Carol. Bob is a contact of Alice who occasionally sends to Alice and receives replies, while Carol is not Alice's contact. Over a very large number of rounds, the SDA alone will distinguish between Bob and Carol with respect their contact with Alice. In fewer rounds, however, Bob and Carol may have very similar statistical links to Alice. Since Alice replies to Bob, combining their SDA observations should provide better evidence that they are contacts. On the other hand, since Alice never sends to Carol, combining their SDA observations will likely weaken the evidence for them being contacts. Thus, combining scores should improve the relative evidence for real contacts.

5 Simulation Setup

We simulated the process of sending and receiving messages via a mix network according to the model described in Section 2. The parameters used in our

Table 1. Simulation parameter values

Parameter	Value	Description
N	100	Number of users in the system
m	10	Number of Alice's contacts
B	100 to 500	Batchsize of threshold mix
P_{delay}	0.1 to 0.9	Probability of delay of binomial mix
P_{reply}	0.5	User's reply probability
λ_A	5.0	Alice message initiation rate i.e. messages/round
λ_U	1.0 to 10.0	User message initiation rate i.e. messages/round per user
λ_{A_d}	1.0 to 10.0	Alice dummy initiation rate per round
$RBCVOL$	$10\% to 100\%$	RBC volume as per cent of real messages/round
$CUTOFF$	10^5 10^6	Simulation cutoff, Threshold Mix Simulation cutoff, Binomial Mix

simulations are discussed in this section and summarized in Table 1. The number of users in the system N is set to 100. The number of contacts for Alice is $m = 20$. The simulations were carried out for the two attacks that we are comparing: SDA and RSDA.

5.1 Mix Behavior

- Threshold Mix: For the threshold mix we set the batch size $B = 200$ messages a round.
- Binomial Mix: For the binomial mix, the probability that an incoming message is delayed is set to $P_{delay} = 0.2$.

5.2 Message Generation

- Alice Initiation: The number of messages Alice initiates is based on a poisson distribution with an average rate of $\lambda_A = 5.0$ messages per round.
- Other Users Initiation: The number of messages sent by users apart from Alice is based on a poisson distribution with an average rate of $\lambda_U = 5.0$ messages per round.
- User Reply Behavior: Users, including Alice, reply to messages they receive from other users with a probability of $P_{reply} = 0.5$. If users decide to reply, they do so in the very next round.

5.3 Cover Traffic

- Alice cover: The number of dummy messages per round is determined using a Poisson distribution with rate λ_{A_d}, which is varied from 1.0 to 10.0 messages per round for our simulations.

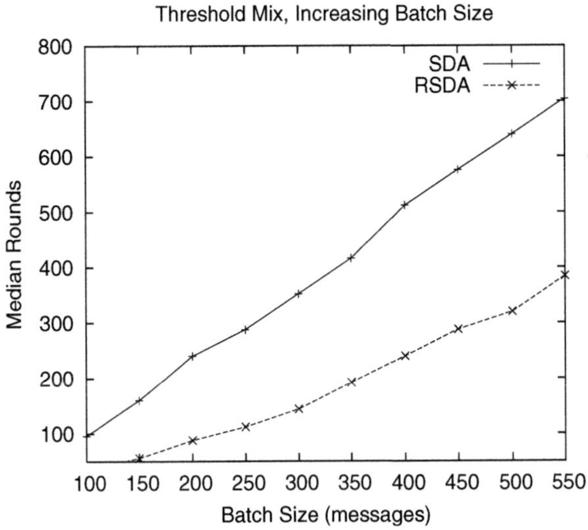

Fig. 1. Median rounds to identify a 50% of Alice's recipients. Threshold mix with no cover traffic.

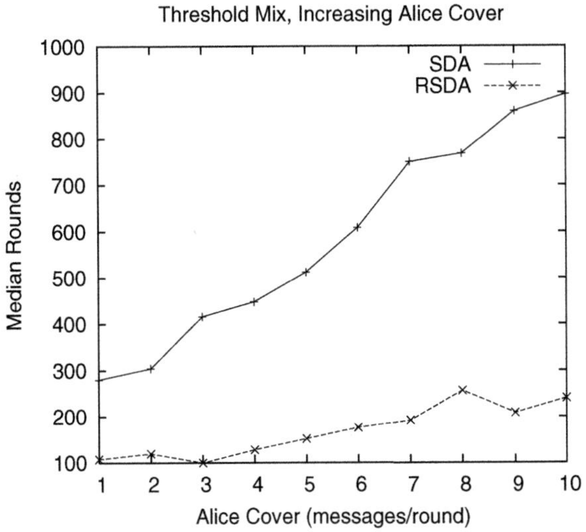

Fig. 2. Median rounds to identify a 50% of Alice's recipients with Alice Cover. Threshold mix with $B = 200$.

- Receiver-bound Cover: For the threshold mix simulations, the volume of receiver-bound cover is set to $RBCVOL = 100\%$. This means the number of dummy messages sent from the mix to users per round is 100% of the number of real outgoing messages from the mix to users in that round. For the binomial mix simulations, the volume of receiver-bound cover is set varied from $RBCVOL = 10\%$ to $RBCVOL = 90\%$.

5.4 Measuring Attacker Success

The attacker eavesdrops on communications between users over a period of time that is divided into rounds. We use the median number of rounds for the attacker to find 50% of Alice's recipients as a measure of the attacker's success. In [5] we discuss why exposing a fraction and not all of Alice's contacts sufficiently degrades her anonymity. The number of rounds of attacker observation is bounded by a $CUTOFF$ value, so that the simulation can end in finite time when the attack does not converge. The observation CUTOFF is set to 10^5 when the median rounds to identify Alice's contacts is lower than 50000 rounds. The CUTOFF is set to 10^6 rounds when the median rounds is higher. Generally, we observed lower median rounds for the threshold mix and higher median rounds for the binomial mix.

6 Results

In this section we discuss the results of our simulations. Please note the use of a logarithmic y-axis in some graphs.

6.1 Simple Threshold Mix

No Cover Traffic. We ran multiple simulations to compare the performances of SDA and RSDA. We studied the effectiveness of the attacks for different batch sizes ranging from $= 100$ to 500. Alice sends at a rate of $\lambda_A = 5.0$ messages per round. The results are shown in Figure 1. We see that when a threshold mix is used, RSDA outperforms SDA especially for higher batch sizes.

Alice Cover. For the next simulation we fixed Alice's message initiation rate and other user's message intiation rate at, $\lambda_A = \lambda_U = 5.0$ messages/round. Alice sends cover traffic increasing from 1.0 to 10.0 messages/round. Figure 2 compares the performance of SDA and RSDA in the presence of increasing cover traffic from Alice. For 10.0 messages/round of Alice cover, the number of rounds for SDA increases by a factor of three to 900 rounds. RSDA on the other hand is able to perform well even when Alice cover twice her real message rate and remains below well 250 rounds.

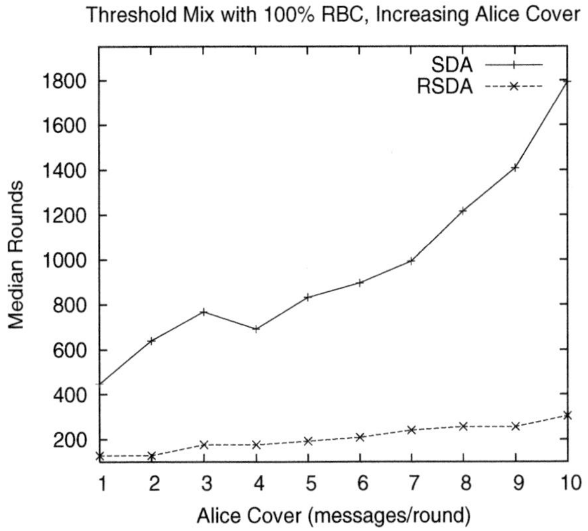

Fig. 3. Median rounds to identify 50% Alice's recipients. Threshold mix with $B = 200$, $RBCVOL = 100\%$.

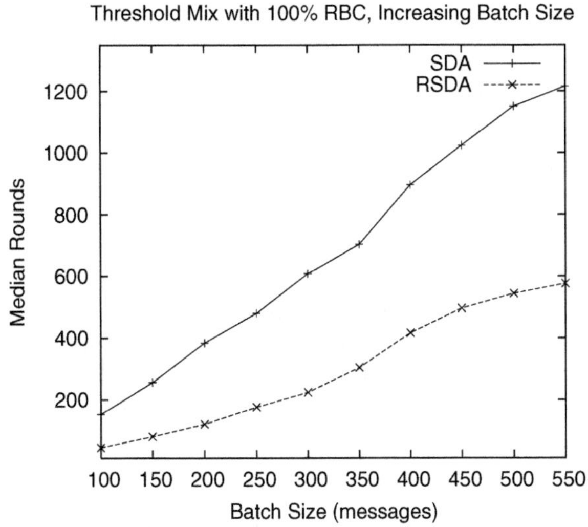

Fig. 4. Median rounds to identify 50% Alice's recipients. Threshold mix, $RBCVOL = 100\%$.

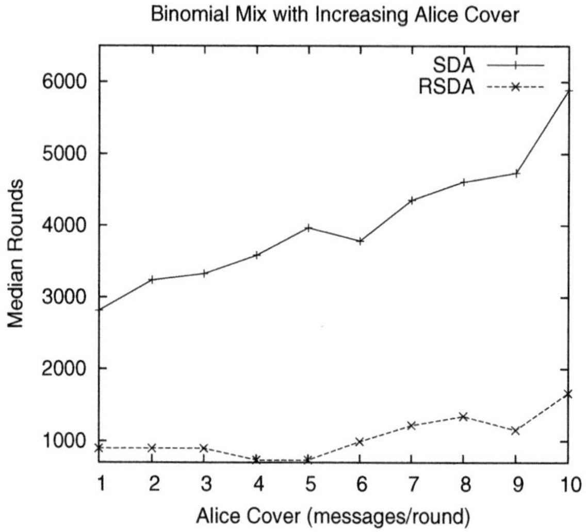

Fig. 5. Median rounds to identify 50% Alice's recipients. Binomial mix with increasing Alice cover.

Receiver-bound Cover. In addition to Alice cover we added receiver-bound cover with $RBCVOL = 100\%$ and compared the performance of SDA and RSDA. The results are shown in Figure 3. The median rounds for attacker success with SDA goes to about 1800 rounds when Alice cover is $\lambda_{A_d} = 10.0$ messages/round. In the presence of RBC and high volume of Alice cover, RSDA is still able to expose 50% of Alice's contacts within about 300 rounds which is 6 times lesser than SDA.

We also studied the performance of RSDA when Alice cover is not used and only the mix sends RBC to users. The results of this scenario is shown in Figure 4. SDA and RSDA are compared for increasing values of mix batch size.

6.2 Binomial Mix

Only Alice Cover Traffic. In our next simulation we compared the performance of SDA and RSDA using a binomial mix. The results are shown in Figure 5. We see that, like in the threshold case, RSDA outperforms SDA. When Alice sends one dummy message/round, RSDA succeeds in a third of the time taken for SDA to succeed. At higher volumes of Alice cover, RSDA continues to succeed faster than SDA.

Alice and Receiver-bound Cover. In this simulation we show the impact of introducing RBCVOL=20% along with increasing Alice cover. We compare the performance of both the attacks when the mix does and does not send receiver-bound cover. The results are shown in Figure 6. We see that the time needed for SDA more than doubles in the presence of 20% RBC. RSDA on the other hand is not affected by RBC to the same degree as SDA.

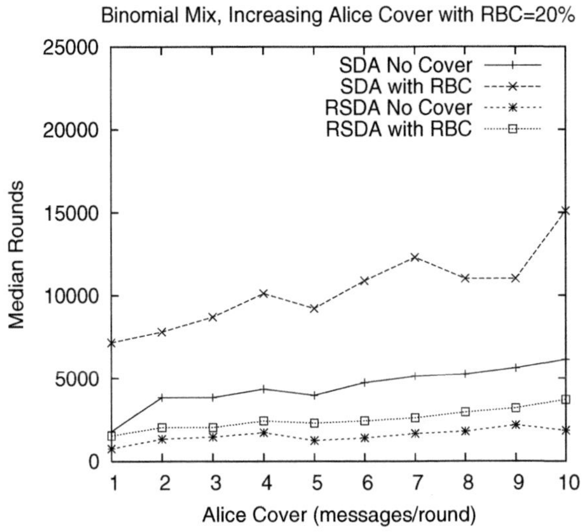

Fig. 6. Median rounds to identify 50% Alice's recipients. Binomial mix with RBC=20%.

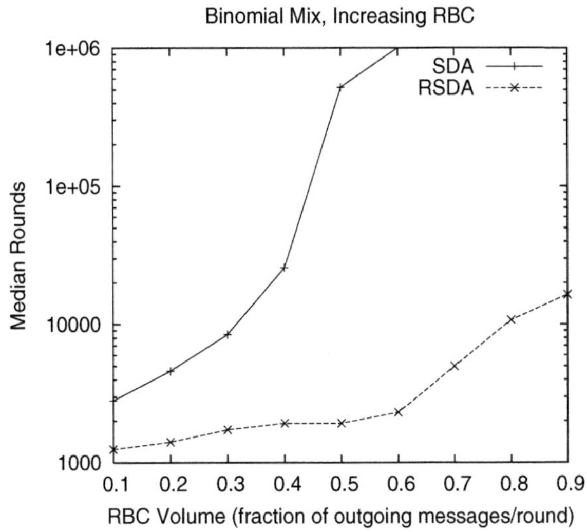

Fig. 7. Median rounds to identify 50% Alice's recipients. Binomial mix with increasing RBC.

Only Receiver-bound Cover. We study whether higher amounts of RBC affect the performance of RSDA and compare the results with the performance of SDA. In order to understand how RBC affects RSDA, we set Alice cover to zero and increased the volume of RBC generated by the mix from $RBCVOL = 10\%$ to $RBCVOL = 90\%$. The results are shown in Figure 7. We see that as the amount of RBC increases the taken for SDA dramatically increases from 2816 rounds to over a million rounds. RSDA shows a ten-fold increase from about 1000 rounds to 10000 rounds of observation when RBC is increased from 10% to 90%. However, compared to SDA, RSDA is significantly more tolerant to receiver-bound cover traffic.

7 Conclusions

In this paper, we have described and evaluated RSDA. RSDA is based on the assumption that the attacker is interested in all of Alice's contacts, not just the people to whom she initiates messages. We believe that this is a more realistic representation of the attacker's goals. In this model, we showed that taking other users' sending behavior into account, the attacker could better identify Alice's contacts than when just using SDA.

References

1. Chaum, D.: Untraceable Electronic Mail, Return Addresses, and Digital Pseudonyms. Communications of the ACM 24(2), 84–88 (1981)
2. Danezis, G.: Statistical disclosure attacks: Traffic confirmation in open environments. In: Proc. Security and Privacy in the Age of Uncertainty (SEC) (May 2003)
3. Danezis, G., Diaz, C., Troncoso, C.: Two-sided statistical disclosure attack. In: Borisov, N., Golle, P. (eds.) PET 2007. LNCS, vol. 4776, pp. 30–44. Springer, Heidelberg (2007)
4. Diaz, C., Serjantov, A.: Generalising mixes. In: Dingledine, R. (ed.) PET 2003. LNCS, vol. 2760, pp. 18–31. Springer, Heidelberg (2003)
5. Mallesh, N., Wright, M.: Countering statistical disclosure with receiver-bound cover traffic. In: Biskup, J., López, J. (eds.) ESORICS 2007. LNCS, vol. 4734, pp. 547–562. Springer, Heidelberg (2007)
6. Mathewson, N., Dingledine, R.: Practical traffic analysis: Extending and resisting statistical disclosure. In: Martin, D., Serjantov, A. (eds.) PET 2004. LNCS, vol. 3424, pp. 17–34. Springer, Heidelberg (2004)
7. Troncoso, C., Gierlichs, B., Preneel, B., Verbauwhede, I.: Perfect matching statistical disclosure attacks. In: Borisov, N., Goldberg, I. (eds.) PETS 2008. LNCS, vol. 5134, pp. 2–23. Springer, Heidelberg (2008)

Security Analysis of ISS Watermarking Using Natural Scene Statistics

Dong Zhang[1,*], Jiangqun Ni[1], Qiping Zeng[1], Dah-Jye Lee[2], and Jiwu Huang[1]

[1] School of Information Science and Technology, Sun Yat-sen University, Guangzhou, China, 510275

[2] Department of Electrical and Computer Engineering, Brigham Young University, Provo, Utah U.S.A. 84602

zhangd@mail.sysu.edu.cn

Abstract. Watermarking security has captured great attention from researchers in recent years. The security of watermarking is determined by the difficulty of estimating the secret key used in embedding/detecting schemes. As a widely used scheme, Improved Spread-Spectrum (ISS) watermarking performs better than Additive Spread-Spectrum (Add-SS) and is able to include Add-SS watermarking as a specialized case. Because of its popularity, the investigation on the security of ISS watermarking has been reported. Previous works on evaluating the security of ISS watermarking mainly focus on the assumption of Gaussian host and ignore the effects from the non-Gaussian characteristics of natural images. This paper analyzes the security of ISS watermarking from the viewpoint of Shannon information theory by using Gaussian Scale Mixture (GSM) model to characterize the natural scene statistics and reveals the relationship between the security and its related factors. Theoretical analysis and simulation results show that the security of ISS watermarking with the Gaussian host assumption is over-stated in previous work.

Keywords: Watermarking security, Improved Spread-Spectrum Watermarking, Gaussian Scale Mixture Model.

1 Introduction

In recent years, great attention has been paid on the research of watermarking security, which is defined as the difficulty to estimate the secret key used in embedding/ detecting schemes [1],[2]. The research on watermarking security tries to reveal the relationship between the security level and relative factors [3]. And relevant work in this field is expected to contribute to the development of a new generation of robust and secure watermarking schemes.

Security analysis on Spread-Spectrum based watermarking has been reported in the literatures [1],[3]-[5]. Cayre *et al.* firstly differentiated security from robustness and investigated the security of Additive Spread-Spectrum (Add-SS)

* The authors appreciate the support received from the National Natural Science Foundation of China (60970145, 60633030) and 973 Program (2006CB303104).

R. Böhme, P.W.L. Fong, and R. Safavi-Naini (Eds.): IH 2010, LNCS 6387, pp. 235–248, 2010.

watermarking with the help of Fisher Information [3]. Comesaña *et al.* analyzed the security issue of Add-SS watermarking from the viewpoint of Shannon information theory under the assumption of Gaussian host [4]. Ni and Zhang *et al.* presented the security analysis on Add-SS watermarking by incorporating a Natural Scene Statistics model - GSM (Gaussian Scale Mixture)[5],[6]. Pérez-Freire *et al.* discussed the security of Improved Spread-Spectrum (ISS) watermarking under the assumption of Gaussian host for the convenience of mathematical manipulations [7].

Compared with Add-SS watermarking, ISS watermarking adapts embedding with host and attenuates embedding strength to reduce the influence of host [7]. ISS watermarking is widely accepted because it has better performance in terms of robustness with constrained distortion than Add-SS and is able to include Add-SS as a specialized case. The security of ISS watermarking is affected not only by the spread-spectrum sequence, which is used as a secret carrier, but also by the distribution of its host as well. Consequently, it is critical to characterize the host distribution accurately in security analysis. In the development of practical watermarking system, the wavelet coefficients of natural images are widely employed as hosts. Since natural images exhibit strong non-Gaussian behaviors in wavelet domain, e.g. sharp peak and heavy tails, Gaussian model is not capable of describing the statistics of the natural image host.

To our best knowledge, security analysis on ISS watermarking was only reported in [7], where the Gaussian host assumption was employed. This paper uses Gaussian Scale Mixture (GSM) [8],[9] to characterize the statistics of natural images, and investigates the security of ISS watermarking from Shannon information theoretic point of view [10]. By using the properties of projection matrix, this paper obtains information theoretical security measures in closed form under the assumption of GSM host. Compared with the results based on the assumption of Gaussian host, this paper obtains more accurate security bound with the help of GSM.

The remainder of the paper is organized as follows. Section 2 presents the model of ISS watermarking incorporating GSM. After the Shannon information theoretic approach for watermarking security is introduced in Section 3, Section 4 analyses the security of ISS watermarking with the assumption of GSM host. Simulation results and discussion are included in Section 5. Section 6 gives concluding remarks at the end of this paper.

2 Model of ISS Watermarking Incorporating GSM

The embedding function of ISS watermarking can be expressed as

$$\mathbf{Y}_j = \mathbf{X}_j + \mu(\mathbf{X}_j, M_j)\mathbf{Z}, \tag{1}$$

where, $\mathbf{X}_j$ and $\mathbf{Y}_j$ are denoted as the host and the watermarked signal in the jth observation, respectively [7]. $\mathbf{Z}$ denotes the secret carrier. The host and the secret carrier are assumed to have the same length, N_v. For the convenience of description and without loss of generality, the situation with one embedded bit is considered. M_j is the embedded message in the jth observation. The embedding is determined by host and secret carrier, and expressed as $\mu(\mathbf{X}_j, M_j)$. Without introducing other noise, $\mathbf{Y}_j$ is the observed signal in the detection end. Compared with Add-SS watermarking, the embedding function of ISS watermarking is dependent on the host and is regarded as an informed embedding. Practically, most schemes of ISS watermarking restrict $\mu(\cdot)$ to be a linear function such that Eq.(1) is simplified into

$$\mathbf{Y}_j = \mathbf{X}_j + (-1)^{M_j}\nu\mathbf{Z} - \lambda\frac{\mathbf{X}_j^T\mathbf{Z}}{\|\mathbf{Z}\|^2}\mathbf{Z}, \tag{2}$$

where ν is a parameter to control the embedding distortion and $\lambda(0 \leq \lambda \leq 1)$ is a host-rejection parameter [2],[7]. It can be seen that ISS watermarking is fundamentally different from Add-SS watermarking in that ISS attenuates the host only in the direction of secret carrier, thus minimizes embedding distortion and improves the performance of ISS in terms of robustness.

In this paper, the wavelet coefficients of natural image host in one sub-band are modeled as a GSM random field, $\mathbf{X}$. It is formulated as the product of a Gaussian random field and a random scaling variable S, i.e. $\mathbf{X} = S \cdot \mathbf{U} = \{s_i \cdot \mathbf{u}_i, i \in I\}$, where I is the index of location, and $\mathbf{U} \sim N(0, \mathbf{Q})$ is a Gaussian random field with zero mean and covariance $\mathbf{Q}$. S is a positive random scalar and independent of $\mathbf{U}$. For natural image, $S = s$ can be estimated by Maximum Likelihood method [8],[9]. Conditioned on s, $\mathbf{X}$ is Gaussian with the probability density expressed as

$$p_{\mathbf{X}|S}(\mathbf{x}|s) = \frac{1}{(2\pi)^{\frac{N}{2}}|s^2\mathbf{Q}|^{\frac{1}{2}}}\exp\left(-\frac{\mathbf{x}^T\mathbf{Q}^{-1}\mathbf{x}}{2s^2}\right). \tag{3}$$

A simplified relationship can be obtained for a scalar model of GSM. With a scalar GSM model, when s_i is given, X_i distributes as Gaussian $p_{X_i|S_i}(x_i|s_i) \sim N(0, s_i^2\sigma_u^2)$ [4]. Without loss of generality, σ_u^2 can be assumed to be unity and X_i is independent of X_j for different i and j.

A scalar GSM model is used in this paper to model the statistics of natural image host. $\mathbf{X}^{N_o} = (\mathbf{X}_1, \mathbf{X}_2, \cdots, \mathbf{X}_{N_o})$ and $\mathbf{Y}^{N_o} = (\mathbf{Y}_1, \mathbf{Y}_2, \cdots, \mathbf{Y}_{N_o})$ denote the N_o hosts and N_o watermarked signals, respectively. $\mathbf{M}^{N_o} = (M_1, M_2, \cdots, M_{N_o})$ are the corresponding messages in the N_o observations. It is assumed that M_i takes 0 and 1 with the same probability. The secret carrier, $\mathbf{Z}$, is an i.i.d. Gaussain random vector, that is $\mathbf{Z} \sim N(0, \sigma_z^2\mathbf{I}_{N_v})$.

It can be seen from Formula (2) that the embedding distortion in the jth observation for ISS watermarking is

$$D_w = \frac{1}{N_v}E\left[\|\mathbf{Y}_j - \mathbf{X}_j\|^2|\mathbf{S} = \mathbf{s}\right] = \nu^2\sigma_z^2 + \frac{\lambda^2}{N_v^2}\sum_{i=1}^{N_v} s_{j,i}^2\sigma_u^2. \tag{4}$$

where $s_{j,i}$ is the ith scale parameter of the jth observation. So the Document to Watermark Ratio is defined as $DWR = 10\log_{10}\left(\frac{\sigma_x^2}{D_w}\right)$, in which $\sigma_x^2 = \frac{1}{N_v}\sum_{i=1}^{N_v} s_{j,i}^2\sigma_u^2$ is the average power of the host.

3 Method for Measuring Watermarking Security

According to Kerckhoff's principle [11], the security of a watermarking system depends only on the secret key [1]. The security of watermarking can be evaluated by the information leakage of secret key in the communication or the residual uncertainty of the secret key after the communication [3]. The lower information leakage or higher residual uncertainty implies that less information about secret key could be acquired by the attacker and hence provides higher security level for the watermarking system.

In the scheme of ISS watermarking, the secret key is the secret carrier [6][7]. $h(\mathbf{Z})$ denotes the differential entropy of secrete carrier. $h(\mathbf{Z}|\mathbf{Y}^{N_o})$ is the residual entropy of secret carrier given N_o watermarked images, and $I(\mathbf{Z};\mathbf{Y}^{N_o})$ is the mutual information between secret carrier and watermarked image. Taking into account the definition of mutual information [10], we have

$$h(\mathbf{Z}|\mathbf{Y}^{N_o}) = h(\mathbf{Z}) - I(\mathbf{Z};\mathbf{Y}^{N_o}), \tag{5}$$

$$I(\mathbf{Z};\mathbf{Y}^{N_o}) = h(\mathbf{Y}^{N_o}) - h(\mathbf{Y}^{N_o}|\mathbf{Z}), \tag{6}$$

in which $h(\mathbf{Y}^{N_o})$ is the differential entropy of N_o watermarked image and $h(\mathbf{Y}^{N_o}|\mathbf{Z})$ is the entropy of watermarked image conditioned the secret carrier is known. It is clear to see that $h(\mathbf{Z}|\mathbf{Y}^{N_o})$ represents the uncertainty of the secret key given the N_o watermarked images and $I(\mathbf{Z};\mathbf{Y}^{N_o})$ measures the information leakage of secret key during the communication. The larger the $h(\mathbf{Z}|\mathbf{Y}^{N_o})$ or the smaller the $I(\mathbf{Z};\mathbf{Y}^{N_o})$ is, the higher security level the watermarking scheme will have.

4 Security Analysis on ISS Watermarking

Based on the knowledge available to the attacker, three different attacks are defined, i.e. Known Message Attack (KMA), Known Original Attack (KOA), and Watermarked Only Attack (WOA) [1],[2]. KMA is the situation the attacker has the access to watermarked images and corresponding secret messages. KOA implies the attacker can access both the watermarked images and the original host images. Under WOA situation, the attacker has only watermarked images.

In this section, by modeling the natural image host with GSM, the security analysis is performed for ISS watermarking based on Shannon information theory. Only results for the cases of KMA and WOA are investigated in this paper. Although GSM model is considered, the result for KOA is similar to the work presented in [7].

4.1 Case under KMA

Since the scale parameter $\mathbf{S}$ can be estimated by the observation when the host is modeled by GSM, taking into account the embedded message is independent of secret carrier and host, the residual entropy of secret carrier and the mutual information between the observations and secret carrier under KMA situation are expressed as Eq.(7) and Eq.(8), respectively.

$$h(\mathbf{Z}|\mathbf{Y}^{N_o},\mathbf{S}^{N_o},M^{N_o}) = h(\mathbf{Z}) - I(\mathbf{Z};\mathbf{Y}^{N_o}|\mathbf{S}^{N_o},M^{N_o}), \tag{7}$$

$$I(\mathbf{Z};\mathbf{Y}^{N_o}|\mathbf{S}^{N_o},M^{N_o}) = h(\mathbf{Y}^{N_o}|\mathbf{S}^{N_o},M^{N_o}) - h(\mathbf{Y}^{N_o}|\mathbf{Z},\mathbf{S}^{N_o},M^{N_o}). \tag{8}$$

With the Gaussian assumption, the differential entropy of the secret carrier is calculated as

$$h(\mathbf{Z}) = \frac{N_v}{2}\log(2\pi e\sigma_z^2), \tag{9}$$

in which the logarithm is based on e [10].

In order to derive the value of $h(\mathbf{Y}^{N_o}|\mathbf{Z},\mathbf{S}^{N_o},M^{N_o})$, incorporating GSM model, a projection matrix is introduced to simplify the calculation [12]. The detailed derivation is attached in Appendix A by which we obtain $h(\mathbf{Y}^{N_o}|\mathbf{Z},\mathbf{S}^{N_o},M^{N_o})$ as Eq. (10).

$$h(\mathbf{Y}^{N_o}|\mathbf{Z},\mathbf{S}^{N_o},M^{N_o}) = \frac{1}{2}\sum_{j=1}^{N_o}\log\left[(2\pi e)^{N_v}(1-\lambda)^2\sigma_u^{2N_v}\prod_{i=1}^{N_v}s_{j,i}^2\right]. \tag{10}$$

In the calculation of $h(\mathbf{Y}^{N_o}|\mathbf{S}^{N_o},M^{N_o})$, two situations are considered. When only one observation is available, that is $N_o = 1$, we have

$$\begin{aligned} h(\mathbf{Y}|\mathbf{S},M) =& \frac{1}{2}\sum_{l=1}^{N_v}\log 2\pi e\{s_i^2\sigma_u^2 + \nu^2\sigma_Z^2 \\ &+ \frac{\sigma_u^2}{N_v(N_v+2)}\left[-2\lambda s_i^2(N_v+2) + 2\lambda^2 s_i^2 + \lambda^2\sum_{l=1}^{N_v}s_l^2\right]\}. \end{aligned} \tag{11}$$

For the case of $N_o > 1$, considering the inequality

$$\begin{aligned} &h(\mathbf{Y}^{N_o}|\mathbf{S}^{N_o},M^{N_o}) \\ &\le \sum_{i=1}^{N_v} E\left[h(Y_{1,i},\cdots,Y_{N_o,i}|S_{1,i}=s_{1,i},\cdots,S_{N_o,i}=s_{N_o,i},M_1=m_1,\cdots,M_{N_o}=m_{N_o})\right] \\ &\le \frac{1}{2}\sum_{i=1}^{N_v} E\left[\log\left((2\pi e)^{N_o}|\mathbf{\Sigma}_{\mathbf{Y}_i}|\right)\right], \end{aligned} \tag{12}$$

where $Y_{1,i}, Y_{2,i},\cdots,Y_{N_o,i}$ are the ith component of each watermarked image, $\mathbf{\Sigma}_{\mathbf{Y}_i}$ is the covariance matrix of $(Y_{1,i},Y_{2,i},\cdots,Y_{N_o,i})$, and $|\mathbf{\Sigma}_{\mathbf{Y}_i}|$ denotes the determinant of $\mathbf{\Sigma}_{\mathbf{Y}_i}$. We have

$$h(\mathbf{Y}^{N_o}|\mathbf{S}^{N_o},M^{N_o}) \le \frac{1}{2}\sum_{i=1}^{N_v}\log\left[(2\pi e)^{N_o}\left(1+\sum_{j=1}^{N_o}\frac{\nu^2\sigma_Z^2}{s_{j,i}^2\sigma_u^2+T_{j,i}}\right)\prod_{j=1}^{N_o}(s_{j,i}^2\sigma_u^2+T_{j,i})\right]. \tag{13}$$

where

$$T_{j,i} = \frac{\sigma_u^2}{N_v(N_v+2)}[-2\lambda s_{j,i}^2(N_v+2) + 2\lambda^2 s_{j,i}^2 + \lambda^2 \sum_{l=1}^{N_v} s_{j,l}^2].$$

The derivation of $h(\mathbf{Y}^{N_o}|\mathbf{S}^{N_o}, M^{N_o})$ is included in Appendix B.

By taking into account the results of $h(\mathbf{Y}^{N_o}|\mathbf{S}^{N_o}, M^{N_o})$ and $h(\mathbf{Y}^{N_o}|\mathbf{Z}, \mathbf{S}^{N_o}, M^{N_o})$, the information leakage about the secret carrier under KMA situation can be obtained.

For the case $N_o = 1$, with the results of Formula (10) and (11), we have

$$\begin{aligned} I(\mathbf{Y};\mathbf{Z}|\mathbf{S}, M) &= \frac{1}{2}\sum_{i=1}^{N_v} \log\{s_i^2\sigma_u^2 + \nu^2\sigma_Z^2 \\ &+ \frac{\sigma_u^2}{N_v(N_v+2)}\left[-2\lambda s_i^2(N_v+2) + 2\lambda^2 s_i^2 + \lambda^2\sum_{l=1}^{N_v} s_l^2\right]\} \\ &- \frac{1}{2}\log\left((1-\lambda)^2\sigma_u^{2N_v}\prod_{i=1}^{Nv} s_i^2\right). \end{aligned} \tag{14}$$

For the case $N_o > 1$, according to the results of Formula (12) and (13), it can be obtained that

$$\begin{aligned} &I(\mathbf{Z};\mathbf{Y}^{N_o}|\mathbf{S}^{N_o}, M^{N_o}) \\ &\le \frac{1}{2}\sum_{i=1}^{N_v}\log[(2\pi e)^{N_o}\left(1+\sum_{j=1}^{N_o}\frac{\nu^2\sigma_Z^2}{s_{j,i}^2\sigma_u^2 + T_{j,i}}\right)\cdot\prod_{j=1}^{N_o}(s_{j,i}^2\sigma_u^2 + T_{j,i})] \\ &- \frac{1}{2}\sum_{j=1}^{N_o}\log\left[(2\pi e)^{N_v}(1-\lambda)^2\sigma_u^{2N_v}\prod_{i=1}^{Nv} s_{j,i}^2\right]. \end{aligned} \tag{15}$$

4.2 Case under WOA

Under WOA situation, the attacker has the access to N_o observations of the watermarked images. By modeling host image with GSM, the residual entropy of secret carrier under WOA is expressed as (16).

$$h(\mathbf{Z}|\mathbf{S}^{N_o}, \mathbf{Y}^{N_o}) = h(\mathbf{Z}|\mathbf{S}^{N_o}) - I(\mathbf{Z};\mathbf{Y}^{N_o}|\mathbf{S}^{N_o}) = h(\mathbf{Z}) - I(\mathbf{Z};\mathbf{Y}^{N_o}|\mathbf{S}^{N_o}), \tag{16}$$

where

$$I(\mathbf{Z};\mathbf{Y}^{N_o}|\mathbf{S}^{N_o}) = h(\mathbf{Y}^{N_o}|\mathbf{S}^{N_o}) - h(\mathbf{Y}^{N_o}|\mathbf{S}^{N_o}, \mathbf{Z}). \tag{17}$$

When only one observation is available, the mutual information between the secret carrier and watermarked image is

$$\begin{aligned} I(\mathbf{Z};\mathbf{Y}|\mathbf{S}) &= h(\mathbf{Y}|\mathbf{S}) - h(\mathbf{Y}|\mathbf{S},\mathbf{Z}) \\ &= h(\mathbf{Y}|\mathbf{S}, M=0) - h(\mathbf{Y}|\mathbf{S},\mathbf{Z}, M) - I(\mathbf{Y}; M|\mathbf{S},\mathbf{Z}) \\ &\ge h(\mathbf{Y}|\mathbf{S}, M=0) - h(\mathbf{Y}|\mathbf{S},\mathbf{Z}, M) - \log 2. \end{aligned} \tag{18}$$

We achieve the result because $h(\mathbf{Y}|\mathbf{S}) = h(\mathbf{Y}|\mathbf{S}, M = 0)$ and M is assumed taking 0 or 1 with equal probability. Recall that each component of $\mathbf{X}$ distributes as $X_i \sim N(0, s_i^2\sigma_u^2)$ with GSM and each component of $\mathbf{Z}$ follows $Z_i \sim N(0, \sigma_Z^2)$. No matter what value M takes, the distribution of $\mathbf{Y}$ has the same expression, i.e. $p(\mathbf{Y}|\mathbf{S}, M = 1) = p(\mathbf{Y}|\mathbf{S}, M = 0)$. So, we have

$$\begin{aligned} h(\mathbf{Y}|\mathbf{S}) &= -E[\log p(\mathbf{Y}|\mathbf{S}, M = 0)] \\ &= h(\mathbf{Y}|\mathbf{S}, M = 0), \end{aligned} \tag{19}$$

and

$$\begin{aligned} I(\mathbf{Z};\mathbf{Y}|\mathbf{S}, M) &\geq I(\mathbf{Z};\mathbf{Y}|\mathbf{S}) \\ &= I(\mathbf{Z};\mathbf{Y}|\mathbf{S}, M) - I(\mathbf{Y}; M|\mathbf{S}, \mathbf{Z}) \\ &\geq I(\mathbf{Z};\mathbf{Y}|\mathbf{S}, M) - \log 2. \end{aligned} \tag{20}$$

From Formula (20), the information leakage of secret carrier under WOA situation is not larger than that under KMA situation. The maximum gap is $log2$ nat, which is due to the uncertainty of the embedded message.

4.3 Connection with the Result Based on Gaussian Host Assumption

The security of ISS watermarking given in [7] is based on the assumption of Gaussian host and can be regarded as a special case of the results derived in this paper. If each wavelet coefficient of host has the same variance, σ_x^2, the GSM model degrades to the Gaussian model. This simplification leads to

$$s_{j,i}^2\sigma_u^2 = \sigma_x^2 \tag{21}$$

By substituting Formula (21) into Eq. (10), Eq. (11), and Eq. (13), we have

$$h(\mathbf{Y}^{N_o}|\mathbf{Z}, \mathbf{S}^{N_o}, M^{N_o}) = \frac{N_o}{2}\log[(2\pi e)^{N_v}(1-\lambda)^2\sigma_x^{2N_v}] \tag{22}$$

$$\begin{aligned} h(\mathbf{Y}|\mathbf{S}, M) &= \frac{N_v}{2}\log(2\pi e) + \frac{N_v}{2}\log[\sigma_x^2 + \nu^2\sigma_Z^2 + \sigma_x^2\frac{\lambda(\lambda-2)}{N_v}] \\ &(\text{ for } N_o = 1) \end{aligned} \tag{23}$$

and

$$\begin{aligned} &h(\mathbf{Y}^{N_o}|\mathbf{S}^{N_o}, M^{N_o}) \\ &\leq \frac{N_v}{2}\log[(2\pi e)^{N_o}(\sigma_x^2)^{N_o}(1+\frac{\lambda(\lambda-2)}{N_v})^{N_o}(1+\frac{N_o\nu^2\sigma_Z^2}{\sigma_x^2(1+\frac{\lambda(\lambda-2)}{N_v})})] \\ &(\text{ for } N_o > 1) \end{aligned} \tag{24}$$

which are the same as $h(\mathbf{Y}^{N_o}|\mathbf{Z}, M^{N_o})$ and $h(\mathbf{Y}^{N_o}|M^{N_o})$ given in [7].

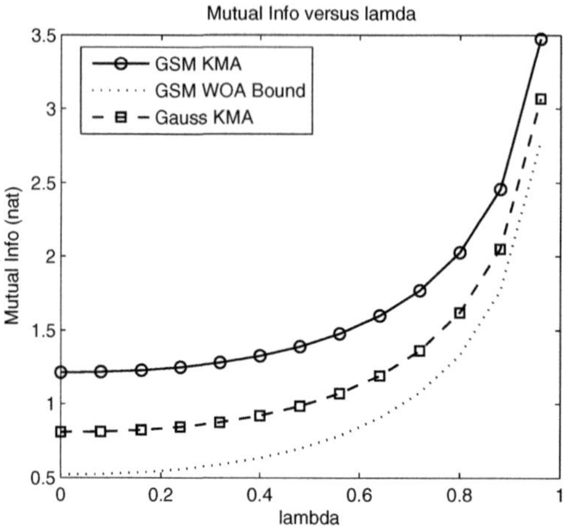

Fig. 1. Relationship between the information leakage of secret carrier and the host-rejection parameter

5 Simulation Results and Discussion

Based on the theoretical analysis above, the security of ISS watermarking was simulated with the assumption of GSM hosts. An i.i.d. Gaussian vector was used as secret carrier, i.e. $\mathbf{Z} \sim N(0, \sigma_z^2 \mathbf{I}_{N_v})$, according to the assumption in Section 4. By using eight natural images, with different texture characteristics, including aerial, baboon, barb, boat, f16, lena, peppers and sailboat, as hosts, the information leakage of secret carrier were calculated. Bi-orthogonal 9/7 wavelet was used to decompose these images into 2 layers. Coefficients from HL2, LH2 and HH2 were randomly selected as hosts.

Fig. 1 shows the relationship between the information leakage of the secret carrier and the host-rejection parameter λ with $N_o = 1$, $N_v = 512$ and DWR $=$ $25dB$. The value of λ ranges from 0 to 1. The solid line with circles and the dashed line with squares correspond to the information leakages with GSM model and Gaussian model under KMA situation, respectively. The dotted line shows the lower bound of information leakage under WOA with GSM. With the increase of the value of λ, information leakage increases for both GSM and Gaussian models. Note that $\lambda = 0$ means not rejection exerted on host such that ISS watermarking degrades to Add-SS watermarking. Although ISS watermarking performs better in robustness than Add-SS watermarking, the security is degraded as shown in Fig. 1. Compared with the results based on Gaussian, GSM based results show more information leakage of the secret carrier in ISS watermarking with the reason that GSM characterizes natural image more accurately and enables the attacker to improve his estimations on secret carrier.

The information leakage against the number of observations is shown in Fig. 2 when the length of secret carrier was fixed at 512, λ at 0.5 and DWR at $25dB$. It is observed that the information leakage of the secret carrier increases with

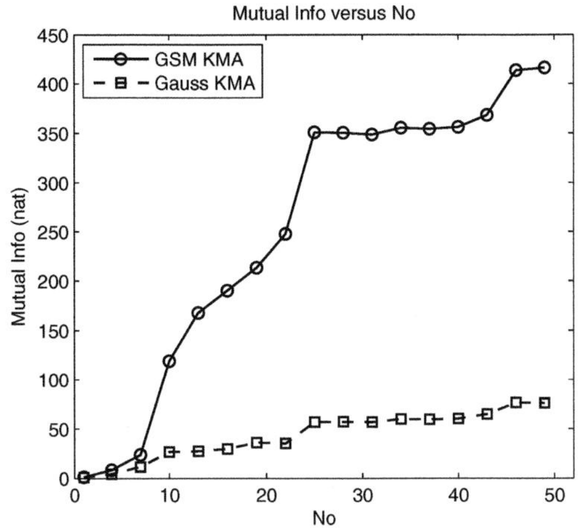

Fig. 2. Relationship between the upper bound of the information leakage of secret carrier and the number of observations

the accumulation of observations. The attacker could improve his knowledge about the secret carrier by increasing the number of observations. Due to the correlation among observations, the information leakage of secret carrier grows non-linearly against the number of observations for both GSM based method and Gaussian based method.

Fig. 3 shows the information leakage of secret carrier per-dimension against the length of carrier ranging from 128 to 512 when DWR and λ were respectively fixed at $25dB$ and 0.5 under KMA situation. Note the information leakage per-dimension decreases with the increase of the length of carrier. Results in this figure indicate that using longer secret carrier will provide higher security to ISS watermarking.

The relationship between the information leakage of secret carrier and DWR is shown in Fig. 4 when just one observation is available. The length of carrier and the value of λ were fixed at 512 and 0.5, respectively. Both GSM based and Gaussian based results were shown under KMA situation. The lower bound of information leakage was also drawn by dotted line. It can be seen from this figure that the information leakage of secret carrier decreases against the increasing of DWR. Since DWR is the power ratio between host and watermark, higher DWR implies lower strength has been embedded and thus corresponds to less information leakage of the secret carrier. So, the security level will be improved by increasing the value of DWR.

In all the aforementioned situations, the information leakage of the secret carries with GSM host is consistently greater than that with Gaussian host, which shows the over-estimated security level of ISS watermarking when the host is characterized with Gaussian.

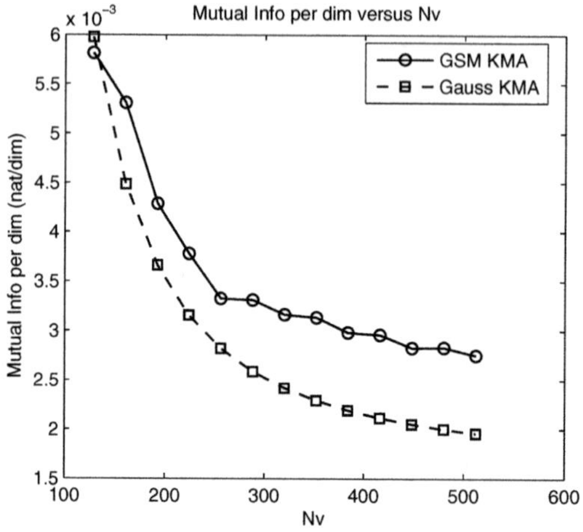

Fig. 3. Relationship between the information leakage of secret carrier and the length of carrier

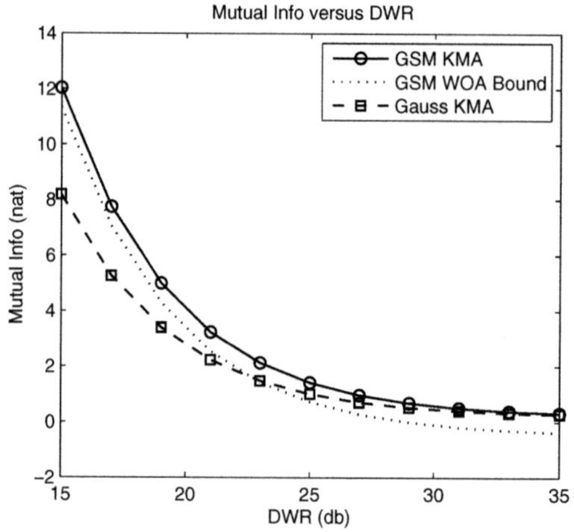

Fig. 4. Relationship between the information leakage of secret carrier and DWR

6 Conclusions

This paper presents a theoretical analysis on the security of ISS watermarking by taking advantage of the GSM model. Compared with the Gaussian model, the GSM model characterizes the statistics of natural images more accurately and thus reduces the uncertainty of host description. The results of the paper show the security level of ISS watermarking was over-estimated with the Gaussian model. The work of this paper is expected to contribute to the development of a new generation of robust watermarking schemes with improved security.

References

1. Pérez-Freire, L., Comesaña, P., Trocoso-Pastoriza, J.R., Pérez-González, F.: Watermarking security: a survey. In: Shi, Y.Q. (ed.) Transactions on Data Hiding and Multimedia Security I. LNCS, vol. 4300, pp. 41–72. Springer, Heidelberg (2006)
2. Malvar, H.S., Florencio, D.A.F.: Improved spread spectrum: a new modulation technique for robust watermarking. IEEE Transactions on Signal Processing 51(4), 898–905 (2003)
3. Cayre, F., Fontaine, C., Furon, T.: Watermarking security: theory and practice. IEEE Transactions on Signal Processing 53(10), 3976–3987 (2005)
4. Comesaña, P., Pérez-Freire, L., Pérez-González, F.: Fundamentals of data-hiding security and their application to spread spectrum analysis. In: Zhao, W., Gong, S., Tang, X. (eds.) AMFG 2005. LNCS, vol. 3723, pp. 146–160. Springer, Heidelberg (2005)
5. Ni, J.Q., Zhang, R.Y., Fang, C., Huang, J.W., et al.: Watermarking security incorporating natural scene statistics. In: Solanki, K., Sullivan, K., Madhow, U. (eds.) IH 2008. LNCS, vol. 5284, pp. 132–146. Springer, Heidelberg (2008)
6. Zhang, D., Ni, J.Q., Lee, D.J.: Security analysis on Add-SS watermarking with GSM. Acta Automatica Sinica 35(7), 841–850 (2009) (in Chinese)
7. Pérez-Freire, L., Pérez-González, F.: Spread spectrum watermarking security. IEEE Transactions on Inoformation Forensics and Security 4(1), 2–24 (2009)
8. Portilla, J., Strela, V., Wainwright, M.J., Simocelli, E.P.: Image denoising using scale mixtures of gaussians in the wavelet domain. IEEE Transactions on Image Processing 12(11), 1338–1351 (2003)
9. Sheikh, H.R., Bovik, A.C., de Veciana, G.: An information fidelity criterion for image quality assessment using natural scene statistics. IEEE Transactions on Image Processing 14(12), 2117–2128 (2005)
10. Cover, T.M., Thomas, J.A.: Elements of Information Theory, 2nd edn. Wiley, New York (2006)
11. Kerckhoffs, A.: La cruptgraphie militaire. Journal des Sciences Militaires IX, 161–191 (1883)
12. Zhang, X.D.: Matrix Analysis and Applications. Publishing House of Tsinghua University, Beijing (2005) (in Chinese)
13. Fang, K.T., Kotz, S., Ng, K.W.: Symmetric Multivariate and Related Distributions. Chapman & Hall, London (1990)

Appendix

A. The Calculation of $h(\mathbf{Y}^{N_o}|\mathbf{Z}, \mathbf{S}^{N_o}, M^{N_o})$

Taking into account the host is independent of each other conditioned on the scale parameter $\mathbf{S}$, each observation is conditionally independent given the secret carrier and secret message. Thus, $h(\mathbf{Y}^{N_o}|\mathbf{Z}, \mathbf{S}^{N_o}, M^{N_o})$ is the sum of the conditional entropies of each observation.

$$h(\mathbf{Y}^{N_o}|\mathbf{Z}, \mathbf{S}^{N_o}, M^{N_o}) = \sum_{j=1}^{N_o} h(\mathbf{Y}_j|\mathbf{Z}, \mathbf{S}_j, M_j) \tag{25}$$

It can be seen from Formula (2) that a watermarked image is composed of the host, the secret carrier modulated by secret message, and the projection of host in the direction of the secret carrier. For the convenience of analysis, $\mathbf{P}_v \triangleq \frac{\mathbf{Z}\mathbf{Z}^T}{\|\mathbf{Z}\|^2}$ is denoted as a matrix which projects a vector onto the direction of secret carrier, $\mathbf{Z}$. So we have

$$\lambda \frac{\mathbf{X}^T\mathbf{Z}}{\| \mathbf{Z} \|^2}\mathbf{Z} = \lambda \mathbf{P}_v \mathbf{X} \tag{26}$$

It can be proved that $\mathbf{P}_v$ has the following features [12]. (i) $\mathbf{P}_v$ is symmetrical, (ii) $\mathbf{P}_v$ is an idempotent matrix, i.e. $\mathbf{P}_v^2 = \mathbf{P}_v$, (iii) the eigenvalue of $\mathbf{P}_v$ is either 1 or 0, and (iv) the rank of $\mathbf{P}_v$ equals to the trace of $\mathbf{P}_v$. According to the assumption of $\mathbf{Z}$, the trace of $\mathbf{P}_v$ can be calculated as Formula (27).

$$tr(\mathbf{P}_v) = tr(\frac{\mathbf{Z}\mathbf{Z}^T}{\| \mathbf{Z} \|^2}) = \frac{1}{\| \mathbf{Z} \|^2} tr(\mathbf{Z}\mathbf{Z}^T) = 1 \tag{27}$$

So $\mathbf{P}_v$ has only one eigenvalue that equals to 1. The determinant of $\mathbf{P}_v$ is $|\mathbf{P}_v| = 0$ for $N_v > 1$ or $|\mathbf{P}_v| = 1$ for $N_v = 1$.

Recalling that the host distributes as Gaussian given the scale parameter in GSM model, and $\frac{\mathbf{X}^T\mathbf{Z}}{\|\mathbf{Z}\|^2}\mathbf{Z}$ projects the host onto the secret carrier, we have the jth observation is Gaussian with mean $(-1)^{M_j}\nu\mathbf{Z}$ and covariance as (28), given corresponding embedded message and secret carrier.

$$\begin{aligned} \mathbf{\Sigma}_j|_{\mathbf{Z}=z,\mathbf{S}=s_j,M=m_j} &\triangleq E[(\mathbf{X}_j - \lambda \frac{\mathbf{X}_j^T\mathbf{z}}{\| \mathbf{z} \|^2}\mathbf{z})(\mathbf{X}_j - \lambda \frac{\mathbf{X}_j^T\mathbf{z}}{\| \mathbf{z} \|^2}\mathbf{z})^T] \\ &= E[(\mathbf{X}_j - \lambda \mathbf{P}_v \mathbf{X}_j)(\mathbf{X}_j - \lambda \mathbf{P}_v \mathbf{X}_j)^T] \end{aligned} \tag{28}$$

Considering that $\mathbf{P}_v$ can be factorized as $\mathbf{U}\mathbf{\Gamma}_\nu\mathbf{U}^T$ with an orthogonal $\mathbf{U}$ and an eigenvalue matrix $\mathbf{\Gamma}_\nu$, the determinant of the above covariance matrix can be calculated by

$$\begin{aligned} |\mathbf{\Sigma}_j|_{\mathbf{Z}=\mathbf{z},\mathbf{S}=\mathbf{s}_j,M=m_j} &= |[(\mathbf{I} - \lambda \mathbf{P}_\nu)E(\mathbf{X}_j\mathbf{X}_j^T)(\mathbf{I} - \lambda \mathbf{P}_\nu)|\mathbf{z}, \mathbf{s}_j, m_j]| \\ &= [|\mathbf{C}_{\mathbf{X}_j}||(\mathbf{I} - \lambda \mathbf{\Gamma}_\nu)|^2|\mathbf{z}, \mathbf{s}_j, m_j] \\ &= (1-\lambda)^2 \sigma_u^{2N_v} \prod_{i=1}^{N_v} s_{j,i}^2 \end{aligned} \tag{29}$$

in which $\mathbf{\Gamma}_\nu = \begin{bmatrix} 1\, 0 \dots 0 \\ 0\, 0 \dots 0 \\ \vdots\ \vdots\ \ddots\ \vdots \\ 0\, 0 \dots 0 \end{bmatrix}$. Note that the result of (29) is independent of m_j, we have

$$h(\mathbf{Y}_j|\mathbf{Z}, \mathbf{S}_j, M_j) = \frac{1}{2}\log[(2\pi e)^{N_v}(1-\lambda)^2\sigma_u^{2N_v}\prod_{i=1}^{N_v} s_{j,i}^2] \tag{30}$$

and

$$h(\mathbf{Y}^{N_o}|\mathbf{Z}, \mathbf{S}^{N_o}, M^{N_o}) = \frac{1}{2}\sum_{j=1}^{N_o} \log[(2\pi e)^{N_v}(1-\lambda)^2\sigma_u^{2N_v} \prod_{i=1}^{N_v} s_{j,i}^2)] \tag{31}$$

B. The Calculation of $h(\mathbf{Y}^{N_o}|\mathbf{S}^{N_o}, M^{N_o})$

According to the embedding function in Formula (2), given the scale parameter $\mathbf{S}$ under KMA situation, the jth observation is the sum of an i.i.d. Gaussian $\mathbf{Z}$ and another Gaussian whose covariance matrix depends on $\mathbf{Z}$. The watermarked signal is still Gaussian [7].

1. Case $N_o = 1$

When only one observation is available, each component of the observation is independent, conditioned on the embedded message and the scale parameter, Given the embedded message, the entropy of the observation is

$$h(\mathbf{Y}^{N_o}|\mathbf{S}^{N_o}, M^{N_o}) = h(\mathbf{Y}|\mathbf{S}, M) = \sum_{i=1}^{N_v} E[h(Y_i|\mathbf{S}=\mathbf{s}, M=m)] \tag{32}$$

where Y_i is the ith component of the observation. Since the component of X_i is zero-mean, it is easy to see that $E[Y_i|\mathbf{S}=\mathbf{s}, M=m] = 0$. The variance of Y_i conditioned $\mathbf{S}$ and M is given by

$$E[Y_i^2|\mathbf{S}=\mathbf{s}, M=m] = s_i^2\sigma_u^2 - 2\lambda s_i^2\sigma_u^2 E[\frac{Z_i^2}{\| \mathbf{Z} \|^2}]$$
$$+ \lambda^2 s_i^2\sigma_u^2 E[\frac{Z_i^4}{\| \mathbf{Z} \|^4}] + \lambda^2\sigma_u^2 E[\sum_{l=1,l\neq i}^{N_v} \frac{s_l^2 Z_l^2 Z_i^2}{\| \mathbf{Z} \|^4}] + \nu^2\sigma_Z^2 \tag{33}$$

As $\mathbf{Z}$ is an N_v- dimensional Gaussian vector with i.i.d. components, $\frac{\mathbf{Z}}{\|\mathbf{Z}\|}$ is uniformly distributed on the surface of the N_v dimensional sphere of unit radius. $\frac{Z_i}{\|\mathbf{Z}\|}$ is the marginal distribution of $\frac{\mathbf{Z}}{\|\mathbf{Z}\|}$. $E[\frac{Z_i^2}{\|\mathbf{Z}\|^2}]$, $E[\frac{Z_i^4}{\|\mathbf{Z}\|^4}]$ and $E[\frac{Z_l^2 Z_i^2}{\|\mathbf{Z}\|^4}]$ are respectively the 2nd-order moment about origin, 4th-order moment about origin and 4th-order mixed moment. Taking into account the results in statistics for a uniformly distributed vector on the surface of an N_v dimensional sphere with unit radius [13], Formula (33) can be written as (34).

$$E[Y_i^2|\mathbf{S}=\mathbf{s}, M=m]$$
$$= s_i^2\sigma_u^2 + \frac{\sigma_u^2}{N_v(N_v+2)}[2\lambda^2 s_i^2 - 2\lambda s_i^2(N_v+2) + \lambda^2\sum_{l=1}^{N_v} s_l^2] + \nu^2\sigma_Z^2 \tag{34}$$

Note that the embedded message takes value by 1 and 0 with equal probability and the result of (34) is independent of M, so we have

$$h(\mathbf{Y}|\mathbf{S},M) = \frac{1}{2}\sum_{i=1}^{N_v}\log 2\pi e[s_i^2\sigma_u^2 + \frac{\sigma_u^2}{N_v(N_v+2)}(2\lambda^2 s_i^2 - 2\lambda s_i^2(N_v+2) + \lambda^2\sum_{l=1}^{N_v} s_l^2) + \nu^2\sigma_Z^2] \tag{35}$$

2. Case $N_o > 1$

In Formula (12), each component of $\mathbf{\Sigma}_{\mathbf{Y}_i}$ can be calculated by (36).

$$\mathbf{\Sigma}_{\mathbf{Y}_i}(j,k) = E[Y_{j,i}Y_{k,i}|S_{1,i}=s_{1,i},\ldots,S_{N_o,i}=s_{N_o,i},M_1=m_1,\ldots,M_{N_o}=m_{N_o}] = \begin{cases} s_{j,i}^2\sigma_u^2 + T_{j,i} + \nu^2\sigma_z^2 & \text{if } j=k \\ (-1)^{m_j+m_k}\nu^2\sigma_Z^2 & \text{if } j\neq k \end{cases} \tag{36}$$

where $T_{j,i} = \frac{\sigma_u^2}{N_v(N_v+2)}[-2\lambda s_{j,i}^2(N_v+2) + 2\lambda^2 s_{j,i}^2 + \lambda^2\sum_{l=1}^{N_v} s_{j,l}^2]$. The $|\mathbf{\Sigma}_{\mathbf{Y}_i}|$ and the upper bound in Formula (12) can be derived in (37) and (38), respectively.

$$|\mathbf{\Sigma}_{\mathbf{Y}_i}| = (1+\sum_{j=1}^{N_o}\frac{\nu^2\sigma_Z^2}{s_{j,i}^2\sigma_u^2 + T_{j,i}})\prod_{j=1}^{N_o}(s_{j,i}^2\sigma_u^2 + T_{j,i}) \tag{37}$$

$$h(\mathbf{Y}^{N_o}|\mathbf{S}^{N_o},M^{N_o}) \leq \frac{1}{2}\sum_{i=1}^{N_v}\log[(2\pi e)^{N_o}(1+\sum_{j=1}^{N_o}\frac{\nu^2\sigma_Z^2}{s_{j,i}^2\sigma_u^2 + T_{j,i}})\prod_{j=1}^{N_o}(s_{j,i}^2\sigma_u^2 + T_{j,i})] \tag{38}$$

Provably Secure Spread-Spectrum Watermarking Schemes in the Known Message Attack Framework

Jian Cao and Jiwu Huang

School of Information Science and Technology, Sun Yat-Sen University, Guangzhou, China, 510275
phdcaojian@yahoo.cn, isshjw@mail.sysu.edu.cn

Abstract. This paper firstly defines the concept of security classes for spread-spectrum watermarking schemes in a Known Message Attack (KMA) framework. In particular, we define three security classes, namely, by order of increasing security: insecurity, key-security and subspace-security. Then, we present three new spread spectrum watermarking schemes, namely, independent natural watermarking (INW), robust independent natural watermarking (Robust-INW) and independent circular watermarking (ICW). All these new watermarking schemes build on the uniformly distributed random orthogonal matrix which can be estimated by the decoder. And all these schemes (i.e., INW, Robust-INW and ICW) are secure against carriers estimation in the KMA framework.

Keywords: Known message attack, spread spectrum watermarking, watermarking security.

1 Introduction

Watermarking security has become a major concern in the past several years and the basis of cryptanalysis has been cast to watermarking for establishing the concept of watermarking security [5],[8],[1]. Watermarking security refers to *the inability by unauthorized users to have access to the raw watermarking channel* [4] and builds on Kerckhoffs' principle [3]. Kerckhoffs' principle states that all details of a watermarking scheme are publicly known except the secret key of the embedding and decoding processes. *Attacks to security are those aimed at gaining knowledge about the secret key*[8]. In [5], following the Diffie and Hellman methodology, the authors propose a classification of the attacking scenarios. Among them, KMA (known message attack) refers to these attacking scenarios where the attacker both owns several watermarked signals and knows their associated embedded messages. This paper only focuses on the KMA framework unless otherwise stated.

In the case of a spread-spectrum watermarking scheme, the secret key is equivalent to the secret carriers. From an attacker's point of view, the security of a spread-spectrum watermarking scheme is made up of two parts. The first part is

R. Böhme, P.W.L. Fong, and R. Safavi-Naini (Eds.): IH 2010, LNCS 6387, pp. 249–261, 2010.

the embedding subspace spanned by the secret carriers, and the second is the secret carriers. If the attacker can estimate this subspace, he can remove watermark with low distortion. In [5], for instance, the authors present a method for removing the watermark with low distortion by nullifying the watermarked signal's projection in this subspace. By contrast, the secret carriers have more precise information. If the attacker discloses the secret carriers, he has a full access to the watermarking channel and can carry out various malicious attacks, including unauthorized embedding, unauthorized detection and unauthorized removal attack. Classical spread-spectrum modulations, such as additive spread spectrum (SS)[6] and improved spread spectrum (ISS)[7], have already been shown to be insecure against carriers estimation [5]. Recently, two new spread-spectrum modulations were proposed in a watermarked only attack (WOA) framework, which implies that the attacker only owns several watermarked signals. One is called natural watermarking (NW)[11], [10] and is secure against the embedding subspace estimation in the WOA [8] framework. The other is called circular extension of ISS (CW-ISS)[10] ,[9]) and is secure against carriers estimation in the WOA framework. However, as we will see in this paper, both NW and CW-ISS are insecure in the KMA framework. The main flaw of CW-ISS and NW in terms of security is that the watermarked signal's projections in the embedding subspace has a circular distribution (i.e., a distribution invariant under rotations) only under the assumption that the messages are supposed to be independently drawn to a uniform distribution. However, to achieve the security against carriers estimation in a KMA framework, it is necessary that all conditional distributions of the watermarked signal's projections given the messages are circular. Otherwise, there exists a message such that the conditional distribution of the watermarked signal's projections given this message is not circular. It implies that it is possible to estimate the secret carriers when the embedded message is this message.

In this paper, after defining the concept of security classes, we present three new spread spectrum watermarking schemes, namely, independent natural watermarking (INW), robust independent natural watermarking (Robust-INW) and independent circular watermarking (ICW). In this context, *independent* was named after our proposed schemes' ability that independent of the embedded messages, all conditional distributions of the watermarked signal's projections (in the embedding subspace) are circular. All our proposed schemes build on a uniformly distributed random orthogonal matrix that can be estimated by the decoder. This random orthogonal matrix make it possible to randomly use carriers alternating among orthogonal bases of the embedding subspace spanned by the secret carriers. As shown in the paper, ICW, INW and Robust-INW are secure against carriers estimation. Moreover, INW has an interesting property that all conditional distributions of the watermarked signal's projections given the messages are the same as the distribution of the host signal's projections (in the embedding subspace).

The remainder of the paper is organized as follows. Section 2 firstly defines the embedding security classes, then, presents a method for generating a uniformly distributed orthogonal matrix that can be estimated by the decoder without error under the attack-free context. Section 3 illustrates how to use our proposed random orthogonal matrix to design a spread spectrum watermarking scheme called INW. Section 4 illustrates another two SS-based watermarking schemes called ICW and Robust-INW. Finally, Sect. 5 presents experiment results about the security and robustness of the proposed schemes.

2 Embedding Security Classes and Uniformly Distributed Random Orthogonal Matrix

In this section, we firstly define the concept of security classes for spread-spectrum watermarking schemes. Then, we present a method for generating a uniformly distributed orthogonal matrix that under the attack-free context, can be estimated by the decoder without error.

2.1 Notations

In order to embed an N_c-bits message $\mathbf{m} = \{-1, +1\}^{N_c}$ into a host signal $\mathbf{x} \in \mathcal{R}^{N_v}$, we need N_c secret carriers $\{\mathbf{u}_i\}$ to achieve an orthogonal basis of the embedding subspace:

$$\mathbf{u}^H \mathbf{u} = \mathbf{I}_{N_c} \ , \tag{1}$$

where $\mathbf{I}_{N_c}$ is the identity matrix of size $N_c \times N_c$. We obtain the watermarked signal $\mathbf{s}$ as follows:

$$\mathbf{s} = \mathbf{x} + \mathbf{w} \ , \tag{2}$$

where $\mathbf{w}$ denotes the watermark signal. The distortion is assessed by watermark-to-content power ratio (WCR):

$$\text{WCR} = 10 \log \left(\frac{\sigma_w^2}{\sigma_x^2} \right) \ . \tag{3}$$

The robustness attacks are modeled as additive noise:

$$\mathbf{y} = \mathbf{s} + \mathbf{n} \ , \tag{4}$$

where we assume attack noise $\mathbf{n}$ from an uncorrelated white Gaussian random process, i.e., $\mathbf{n} \sim \mathcal{N}(\mathbf{0}, \sigma_n^2 \mathbf{I}_{N_v})$. The attack strength is accessed by means of the signal-to-noise ratio (SNR):

$$\text{SNR} = 10 \log \left(\frac{\sigma_x^2}{\sigma_n^2} \right) \ . \tag{5}$$

Let $\hat{\mathbf{m}}$ denote the decoded message, then we measure decoding performance with bit error rate (BER):

$$\text{BER} = \frac{1}{N_c} \sum_{i=1}^{N_c} E[\mathbf{m}(i) \neq \hat{\mathbf{m}}(i)] \ . \tag{6}$$

The goal of the security attacks is to gain the knowledge about the secret carriers $\{\mathbf{u}_i\}$. Let N_o denote the number of observations, which, of course, are watermarked with the same secret carriers. Because the attacker can model the conditional distribution of the watermarked signals given the keys and the the messages, security attack is actually an optimization problem as follows:

$$\hat{\mathbf{u}} = \arg\max_{\mathbf{u}} P(\mathbf{S}|\mathbf{u}, \mathbf{M}) \ , \tag{7}$$

where $\mathbf{S} = \{\mathbf{S}_1, \mathbf{S}_2, \ldots, \mathbf{S}_{N_o}\}$ denotes N_o watermarked signals and $\mathbf{M}$ their associated messages.

2.2 Embedding Security Classes of SS-Based Watermarking Schemes

Since difficult applications maybe have a very difficult security requirements, we devise accordingly three security classes for the embedding function. Let $\mathcal{U}$ be the set of $N_v \times N_c$ matrices $\mathbf{u}$ with $\mathbf{u}^H\mathbf{u} = \mathbf{I}_{N_c}$.

Definition 1. *(Insecurity): An embedding function is insecure iff (if and only if): $\exists \boldsymbol{M} \in \mathcal{M}$, we have*

$$\forall \boldsymbol{v} \in \mathcal{U}, \boldsymbol{v} \neq \boldsymbol{u}, p(\boldsymbol{S}|\boldsymbol{v}, \boldsymbol{M}) \neq p(\boldsymbol{S}|\boldsymbol{u}, \boldsymbol{M}) \ . \tag{8}$$

An embedding function is then called insecure if there exists a message $\mathbf{M}$ such that there exists an unique key whose associated conditional distribution of watermarked signals given this key matches the distribution of the observations. It implies that there exists a message such that the maximum likelihood estimation of the secret carriers is possible. As we will see in the paper, almost all existing spread-spectrum watermarking schemes are insecure according to this definition.

Definition 2. *(Key-security): An embedding function is key-secure iff (if and only if): $\forall \boldsymbol{M} \in \mathcal{M}$, we have*

$$\forall \boldsymbol{v} \in \mathcal{S}_{\boldsymbol{u}}, p(\boldsymbol{S}|\boldsymbol{v}, \boldsymbol{M}) = p(\boldsymbol{S}|\boldsymbol{u}, \boldsymbol{M}) \ , \tag{9}$$

where $\mathcal{S}_{\mathbf{u}}$ represents the subset of the key space $\mathcal{U}$ which does not modify the probabilistic model of the watermarked signals. Key-security implies that whatever the embedded messages are, it is impossible to estimate the secret carriers. However, it is possible to estimate the invariant subset $\mathcal{S}_{\mathbf{u}}$ and to reduce the uncertainly of the secret carriers. As we will see in the paper, INW, Robust-INW and ICW enable to achieve key-security and the invariant subset is the set of all orthogonal bases of the embedding subspace spanned by the secret carriers $\{\mathbf{u}_i\}$.

Definition 3. *(Subspace-security): An embedding function is subspace-secure iff (if and only if): $\forall \boldsymbol{M} \in \mathcal{M}$, we have*

$$\forall \boldsymbol{v} \in \mathcal{U}, p(\boldsymbol{S}|\boldsymbol{v}, \boldsymbol{M}) = p(\boldsymbol{S}|\boldsymbol{u}, \boldsymbol{M}) \ . \tag{10}$$

By definition, subspace-security implies that there exists no information leakage between the watermarked signals and the secret carriers given the embedded messages.

$$\text{Subspace-security} \Leftrightarrow I(\mathbf{S};\mathbf{u}|\mathbf{M}) = 0 \ . \tag{11}$$

In the WOA framework, the concept of security classes has been already proposed by Cayre [10]. A watermarking scheme, which is secure in the WOA framework, is not necessarily secure in the KMA framework. For example, as we will know, NW and CW-ISS is secure against carriers estimation in the WOA framework but is insecure in the KMA framework.

2.3 Uniformly Distributed Random Orthogonal Matrix of Size $N_c \times N_c$

Let $\mathbf{u}^+$ be a matrix such that $\{\mathbf{u}^+, \mathbf{u}\}$ is an orthogonal matrix. In other words, $\mathbf{u}^+$ is an orthogonal basis of the orthogonal complement of the embedding subspace. This paper assumes that the embedder and decoder share both the secret carriers $\{\mathbf{u}_i\}$ and the matrix $\mathbf{u}^+$. This subsection presents a method for generating a uniformly distributed random orthogonal matrix $\mathbf{Q}$ only depending on both the secret carriers and the projection $\mathbf{x}^H\mathbf{u}^+$. In this context, *uniform* is defined in terms of *Haar measure*, which essentially requires the distribution not change if multiplied by any freely chosen orthogonal matrix. Firstly, we present some background material:

Remark 1. In the case of spread spectrum watermarking schemes, the projection of $\mathbf{x}^H\mathbf{u}^+$ will remain invariant during embedding. In other words, the embedder alters the host only in the embedding subspace to embed a watermark:

$$\mathbf{s}^H\mathbf{u}^+ = \mathbf{x}^H\mathbf{u}^+ \ . \tag{12}$$

Remark 2. Take a $N_c \times N_c$ matrix $\mathbf{A}$ whose elements are independent and identically distributed (i.i.d.) Gaussian variables with zero mean. Let $\mathbf{A}_1, \ldots, \mathbf{A}_{N_c}$ be the columns of $\mathbf{A}$ and $\mathbf{Q}$ be random matrix whose N_c columns are obtained by applying a Gram-Schmidt orthogonalization procedure to $\mathbf{A}_1, \ldots, \mathbf{A}_{N_c}$, then $\mathbf{Q}$ has a uniform distribution on the orthogonal group of size $N_c \times N_c$.

Now, it is time to show how to generate a uniformly distributed random orthogonal matrix only depending on both the secret carriers and the projection $\mathbf{x}^H\mathbf{u}^+$. It is easy to show that projection $\mathbf{x}^H\mathbf{u}^+$ is Gaussian, i.e.,

$$\mathbf{x}^H\mathbf{u}^+ \sim \mathcal{N}(\mathbf{0}, \sigma_x^2\mathbf{I}_{N_v - N_c}) \ . \tag{13}$$

Then, we choose first $N_c \times N_c$ components of the projection $\mathbf{x}^H\mathbf{u}^+$ as the elements of the matrix $\mathbf{A}$, i.e.,

$$\mathbf{A}(i,j) = \mathbf{x}^H\mathbf{u}^+\left((i-1) \times N_c + j\right) \ . \tag{14}$$

According to Remark.2, we can obtain a uniformly distributed random orthogonal matrix $\mathbf{Q}$ by applying a Gram-Schmidt orthogonalization procedure to $\mathbf{A}_1, \ldots, \mathbf{A}_{N_c}$.

For our proposed schemes, namely, INW, Robust-INW and ICW, the decoder needs to estimate the orthogonal matrix used by the embedder. The process of estimating this orthogonal matrix is the same as the process of generating it except replacing the projection $\mathbf{x}^H\mathbf{u}^+$ by the projection $\mathbf{y}^H\mathbf{u}^+$, where $\mathbf{y}$ denotes a potentially degraded version of $\mathbf{s}$. According to Remark.1, it is easy to show that this uniformly distributed random orthogonal matrix $\mathbf{Q}$ used by the embedder, in the attack-free context, can be estimated by the decoder without error.

3 Independent Natural Watermarking

This section presents a method for building a spread spectrum watermarking scheme in such a way that all conditional distributions of $\mathbf{s}^H\mathbf{u}$ given the messages are the same as the distribution of the host signal's projections in the embedding subspace.

3.1 Embedding and Decoding

It is easy to show that the projection $\mathbf{x}^H\mathbf{u}$ is Gaussian, i.e.,

$$p(\mathbf{x}^H\mathbf{u}) = \frac{1}{(\sigma_x\sqrt{2\pi})^{N_c}} \exp\left(-\frac{\rho^2}{2\sigma_x^2}\right) , \tag{15}$$

where $\rho = \sqrt{\mathbf{x}^H\mathbf{u}_1 + \mathbf{x}^H\mathbf{u}_2 + \cdots + \mathbf{x}^H\mathbf{u}_{N_c}}$. Hence, the projection $\mathbf{x}^H\mathbf{u}$ has a distribution invariant under rotations. The goal of INW is to design the embedding in such a way that all conditional distributions of $\mathbf{s}^H\mathbf{u}$ given the messages are the same as the the distribution of the host signal's projections in the embedding subspace, that is to say, $P(\mathbf{s}^H\mathbf{u}|\mathbf{M}) = P(\mathbf{x}^H\mathbf{u})$ for any messages $\mathbf{M}$. The embedding function is given by:

$$\mathbf{s} = \mathbf{x} + \sum_{i=1}^{N_c}\left(\frac{1}{\sqrt{N_c}}\|\mathbf{x}^H\mathbf{v}\|\mathbf{m}(i) - \mathbf{x}^H\mathbf{v}_i\right)\mathbf{v}_i , \tag{16}$$

where $\|\mathbf{x}^H\mathbf{v}\| = \sqrt{\mathbf{x}^H\mathbf{v}_1 + \mathbf{x}^H\mathbf{v}_2 + \cdots + \mathbf{x}^H\mathbf{v}_{N_c}}$ and $\mathbf{v} = \mathbf{u}\mathbf{Q}$. This random orthogonal matrix $\mathbf{Q}$ is generated by the method proposed in Sect.2. Equation (16) states that this embedding rule preserves Euclidean distance between the projection and the origin, i.e., $\|\mathbf{s}^H\mathbf{u}\| = \|\mathbf{x}^H\mathbf{u}\|$. This embedding rule is depicted in Fig.1. We can see in this figure that even when the messages $\mathbf{M}$ are given, the conditional distribution of the watermarked signal's projections in the embedding subspace is still the same as the distribution of the host signal's projections. The decoding rule of INW is given by:

$$\widehat{\mathbf{m}}(i) = \begin{cases} +1 & \text{if} \quad \mathbf{y}^H(\mathbf{u}\hat{\mathbf{Q}})_i > 0 , \\ -1 & \text{if} \quad \mathbf{y}^H(\mathbf{u}\hat{\mathbf{Q}})_i < 0 , \end{cases} \tag{17}$$

where $\hat{\mathbf{Q}}$ denotes the estimation of the orthogonal matrix $\mathbf{Q}$ used by the embedder. The way of estimating this random orthogonal matrix is given in Sect.2. As shown in Sect.2, in the attack-free attack context, this random orthogonal matrix can be estimated by the decoder without error.

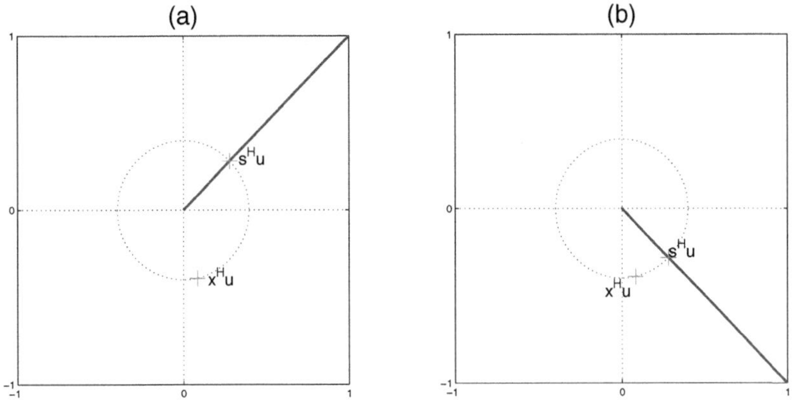

Fig. 1. INW for $\mathbf{m} = (1, 1)$ ($N_c = 2$). (a) is for $\mathbf{Q} = \mathbf{I}_{N_c}$, (b) is for a random chosen orthogonal matrix $\mathbf{Q}$.

3.2 Distortion

The WCR (Watermark-to-Content Ratio) measures the embedding distortion and the first point is to get the variance of the watermark:

$$\sigma_w^2 = \frac{1}{N_v} E[(\mathbf{s} - \mathbf{x})^H (\mathbf{s} - \mathbf{x})] \ . \tag{18}$$

Considering the embedding rule of INW, we have:

$$N_v \sigma_w^2 = 2 N_c \sigma_x^2 \ . \tag{19}$$

Finally, the WCR is as follows:

$$\text{WCR}_{[dB]} = 10 \log_{10} \left(\frac{2N_c}{N_v} \right) \ . \tag{20}$$

3.3 Security

The goal of this section is to prove that all conditional distributions of the projection $\mathbf{s}^H\mathbf{u}$ (given the messages) after INW embedding are the same as the distribution of the projection $\mathbf{x}^H\mathbf{u}$. The embedding formula of INW can be rewritten as:

$$\mathbf{s} = (\mathbf{I}_{N_v} - \mathbf{u}\mathbf{u}^H)\mathbf{x} + \mathbf{u}\mathbf{Q}\mathbf{E} \ . \tag{21}$$

where $\mathbf{E}(i) = \frac{1}{\sqrt{N_c}} \|\mathbf{x}^H\mathbf{u}\| \mathbf{m}(i), 1 \leq i \leq N_c$. It is easy to show that $\mathbf{s}^H\mathbf{u} = \mathbf{Q}\mathbf{E}$.

Theorem 1. *All conditional distributions of $\boldsymbol{s}^H\boldsymbol{u}$ given the messages are the same as the distribution of $\boldsymbol{x}^H\boldsymbol{u}$. That is to say, the random vector $\boldsymbol{QE}$ has a Gaussian distribution with mean $\boldsymbol{0}$ and covariance matrix $\sigma_x^2\boldsymbol{I}_{N_c}$, i.e., $\boldsymbol{QE} \sim \mathcal{N}(\boldsymbol{0}, \sigma_x^2\boldsymbol{I}_{N_c})$.*

Proof. Notice that the random vector $\mathbf{QE} \in \mathcal{R}^{N_c}$ can be expressed by means of its norm and an N_c-dimensional unit vector collinear to $\mathbf{QE}$, that is, we consider a coordinate change:

$$\mathbf{QE} \rightarrow \mathbf{q}.r \ , \tag{22}$$

where $r = \|\mathbf{x}^H\mathbf{u}\|$ and $\mathbf{q} = \frac{1}{\sqrt{N_c}}\mathbf{Qm}$. Here, $\mathbf{m}$ is seen as a column vector. It is clear that the random variables r and $\mathbf{q}$ are mutually independent and $\frac{r}{\sigma_x}$ is distributed according to the chi distribution with N_c degrees of freedom. Meanwhile, we have $\|\mathbf{q}\| = 1$, and for any orthogonal matrix $\mathbf{T}$, we have:

$$\begin{aligned} p(\mathbf{Tq}) &= p(\mathbf{TQ}\frac{1}{\sqrt{N_c}}\mathbf{m}) \\ &= p(\mathbf{q}) \ . \end{aligned} \tag{23}$$

Equation 23 follows from the fact that $\mathbf{Q}$ have a uniformly distribution over the orthogonal group. Equation 23 shows that the random vector $\mathbf{q}$ is uniformly distributed over the surface of the unit-radius hypersphere. Hence, considering the transform $(\mathbf{q}, r) \rightarrow \mathbf{q}.r$, which is a coordinate change, we can deduce that the random vector $\mathbf{QE}$ is Gaussian vector with mean $\mathbf{0}$ and covariance matrix $\sigma_x^2\mathbf{I}_{N_c}$. This completes the proof of this theorem.

4 Independent Circular Watermarking

Independent circular watermarking was named after its ability that independent of the messages, all conditional distribution of $\mathbf{s}^H\mathbf{u}$ (given the messages) are circular and symmetric. It is consistent with the definition of key-security. In particular, all orthogonal bases of the embedded subspace are identical from point of view of the attacker. In other words, if the matrix $\mathbf{Q}$ is an orthogonal matrix of size $N_c \times N_c$, we have $p(\mathbf{S}|\mathbf{u}, \mathbf{M}) = p(\mathbf{S}|\mathbf{uQ}, \mathbf{M})$ for any given messages $\mathbf{M}$. In this case, it is impossible to exactly estimate the secret carriers used to prevent the attacker from reading, writing and removing the embedded message. The uniformly distributed random orthogonal matrix $\mathbf{Q}$ generated by the way proposed in Sect. 2 leaves much room for devising a secure watermarking scheme in the KMA framework. In this paper, we present two implementations of independent circular watermarking. One solution to devise independent circular watermarking is to build on the well-known ISS modulation. For the sake of simplicity, we will refer to this implementation as ICW in the sequel. The other solution to devise independent circular watermarking is to increase the variance of the watermarked signal's projections in the embedding subspace after INW embedding. For the sake of simplicity, we will refer to this implementation as Robust-INW in the sequel.

4.1 ICW and Robust-INW

ICW: ICW is constructed using randomization over the set of all orthogonal bases of the embedding subspace spanned by the secret carriers $\{\mathbf{u}_i\}$. The embedding function of ICW is given by:

$$\mathbf{s} = \mathbf{x} + \sum_{i=1}^{N_c} \left(\alpha \mathbf{m}(i) - \lambda \mathbf{x}^H \mathbf{v}_i\right) \mathbf{v}_i \ . \tag{24}$$

where $\mathbf{v} = \mathbf{uQ}$. ICW requires exactly the same computations as ISS for parameters α and λ.

Robust-INW: The new modulation for Robust-INW is given by:

$$\mathbf{s} = \mathbf{x} + \sum_{i=1}^{N_c} \left(\frac{\lambda}{\sqrt{N_c}} \|\mathbf{x}^H \mathbf{v}\| \mathbf{m}(i) - \mathbf{x}^H \mathbf{v}_i \right) \mathbf{v}_i \ , \tag{25}$$

and the conditional distribution of the projection $\mathbf{s}^H \mathbf{u}$ now follows the distribution $\mathbf{s}^H \mathbf{u} \sim \mathcal{N}(\mathbf{0}, \lambda^2 \sigma_x^2 \mathbf{I}_{N_c})$. If $\lambda = 1$, the embedding rule corresponds to INW, and as the parameter λ increases, the robustness also increases and this leads to Robust-INW. The expression of the WCR becomes:

$$WCR_{\text{dB}} = 10 \log_{10} \left(\frac{(\lambda^2 + 1) N_c}{N_v} \right) \ . \tag{26}$$

The decoding rules of ICW and Robust-INW are the same as the decoding rule of INW. Using ICW and Robust-INW leads also to another important consequence. Since the conditional distribution of the watermarked signal's projections in the embedding subspace is now distinct from distributions of the watermarked signal's projections in other subspaces, it is then possible to estimate the embedding subspace, for example, using principal component analysis (PCA). However, if $N_c > 1$, it is still not possible to estimate the secret carriers because all conditional distributions of the watermarked signal's projections in the embedding subspace are circular.

5 Experiment Results and Analysis

The aim of this section is to assess the security and robustness of various watermarking schemes, namely, NW, CW-ISS, INW, Robust-INW and ICW.

5.1 Security

To assess the security of a spread-spectrum watermarking scheme, we have decided to adopt the following methodology:

1. We generate 5000 observations of watermarked signals using the secret carriers $\{\mathbf{u}_i\}$ and generate the matrix of observations $\mathbf{S}$, where i-th column of $\mathbf{S}$ is i-th observation ($\mathbf{S}_i$).
2. We project these observations into various N_c-dimensional subspaces.

Then, we can obtain the main tuition about the security of a spread-spectrum watermarking scheme by observing how the empirical distribution of the

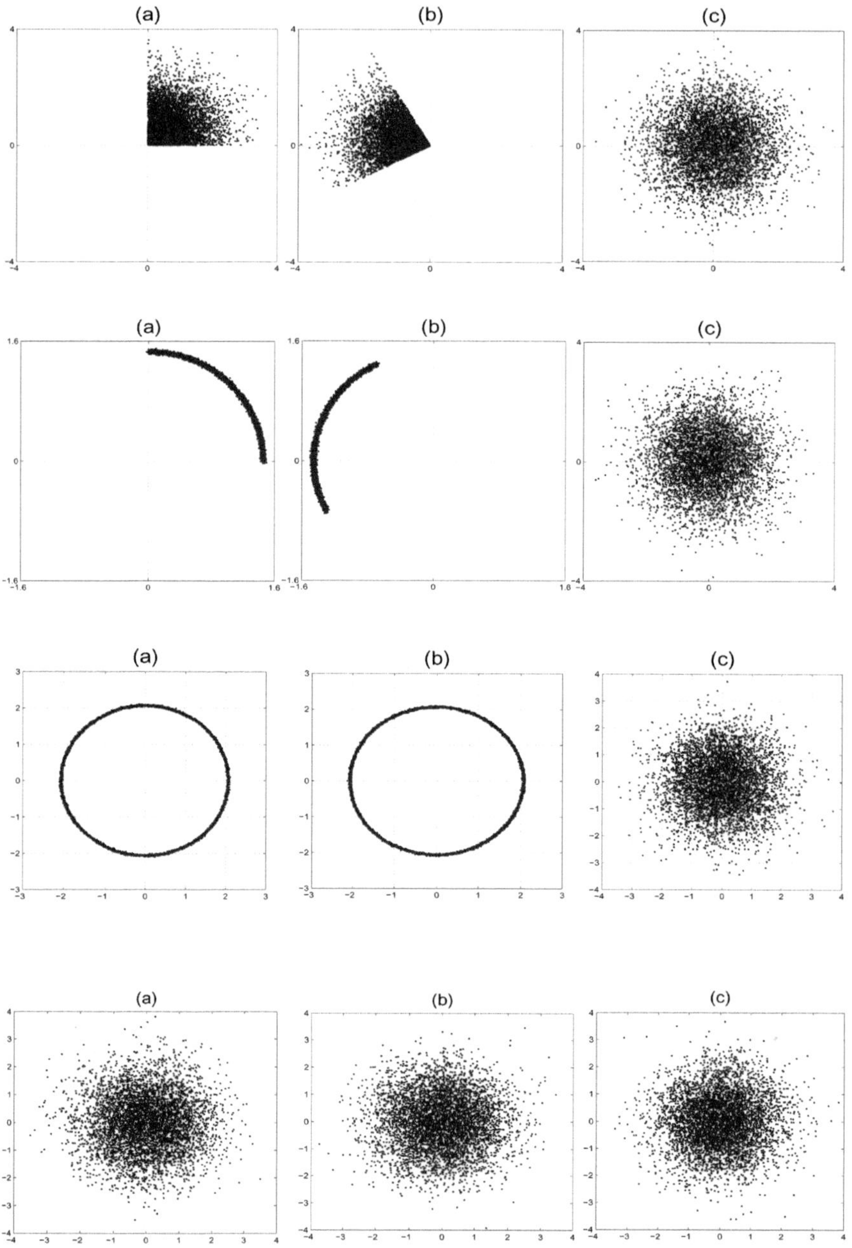

Fig. 2. Empirical distribution of the projection $\mathbf{S}_i^H \mathbf{b}$ for NW, CW-ISS, ICW and INW (from top to bottom). N_o=5000, $N_c = 2$, $N_v = 512$, $\forall i, \mathbf{M}_i = (1, 1)$ and WCR=-21dB. The projections in (a) is for $\mathbf{b} = \mathbf{u}$, in (b) is for $\mathbf{b} = \mathbf{uQ}$ where $\mathbf{Q}$ is randomly chosen orthogonal matrix of size $N_c \times N_c$, and the projections in (c) is for a randomly chosen matrix of size $N_v \times N_c$ with $\mathbf{v}^H \mathbf{v} = \mathbf{I}_{N_c}$.

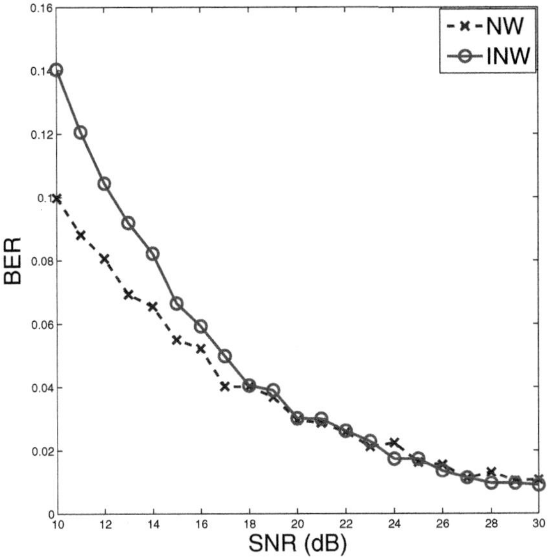

Fig. 3. Comparison of BER for NW and INW. N_v=512, N_c=2.

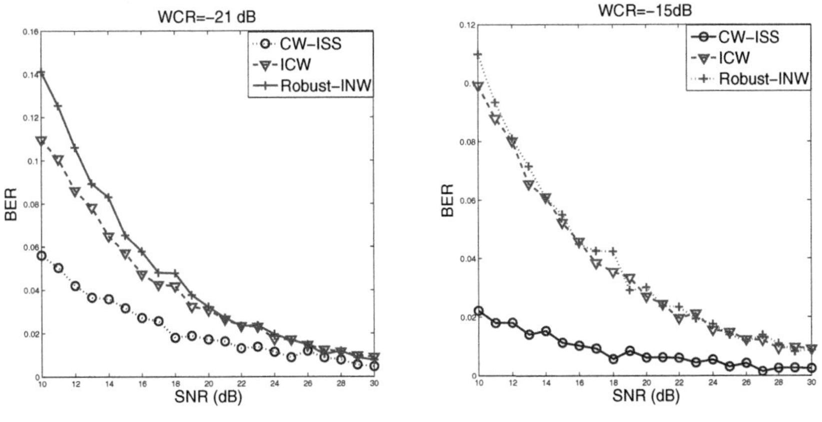

Fig. 4. Comparison of BER for CW-ISS, ICW and Robust-INW. $N_v = 512$, $N_c = 2$.

observation's projections varies with projected subspaces (or more precisely, the basis of subspace). In this paper, we consider two difficult types of subspaces. One is the embedding subspace. The other is the random chosen subspace (hence, approximately orthogonal to the embedding subspace). The conclusion of this security attack is the following:

1. Let **b** denote a random chosen orthogonal basis of the embedding subspace. If the empirical distribution of the observation's projections $\mathbf{S}_i^H \mathbf{b}$ is not the same as the empirical distribution of the observation's projections $\mathbf{S}_i^H \mathbf{u}$, it is possible to estimate the secret carriers.

2. Let $\mathbf{b}$ denote a random chosen orthogonal basis of the embedding subspace and $\mathbf{t}$ an orthogonal basis of random chosen subspace. If the empirical distribution of the observation's projections $\mathbf{S}_i^H\mathbf{b}$ is the same as the empirical distribution of the observation's projections $\mathbf{S}_i^H\mathbf{u}$ but the empirical distribution of the observation's projections $\mathbf{S}_i^H\mathbf{t}$ is not the same as the empirical distribution of the observation's projections $\mathbf{S}_i^H\mathbf{u}$, it is possible to estimate the embedding subspace but it is impossible to estimate the secret carriers.
3. Let $\mathbf{t}$ denote an orthogonal basis of random chosen subspace. If the empirical distribution of the observation's projections $\mathbf{S}_i^H\mathbf{t}$ is the same as the empirical distribution of the observation's projections $\mathbf{S}_i^H\mathbf{u}$, it is impossible to estimate the embedding subspace.

Figure. 2 depicts the empirical distributions of the observation's projections on various subspaces for NW, CW-ISS, ICW and INW when all the embedded messages are same and $\mathbf{M}_i = \{1, 1\}$. We can see that according to conclusion 1, there exists practical algorithm to estimate the secret carriers for NW and CW-ISS. In other words, NW and CW-ISS are insecure in the KMA framework. In the case of ICW, there exists practical algorithm to estimate the embedding subspace.

5.2 Robustness

Figure 3 compares the robustness of INW and NW against AWGN addition for the same hidden channel rate (the same N_c and same N_v) and the same embedding distortion. We can see in this figure that INW has roughly the same robustness as NW for low noise power. Figure 4 compares the robustness of CW-ISS, ICW and Robust-INW against AWGN attack. We believe this figure point out what is the true cost of security for SS-based watermarking techniques.

6 Conclusions

We firstly define three security classes in the Known Message Attack (KMA) framework. The first one is insecurity, which implies that there exists a message such that it is possible to estimate the secret carriers. The second one is key-security, which implies that it is impossible to estimate the secret carriers. the third one is subspace-security, which implies that it is impossible to estimate both the secret carriers and the embedding subspace. Then, we present a method for generating a uniformly distributed orthogonal matrix only depending on both the secret carriers and the host signal's projection in the orthogonal complement of the embedding subspace. This random orthogonal matrix can be estimated by the decoder without error under a attack-free context. Based on this random orthogonal matrix, we present three watermarking schemes, namely, INW, Robust-INW and ICW. All these watermarking schemes are secure against carriers estimation. However, the price to obtain the security against carriers estimation is the relative weak robustness in comparison with NW and CW-ISS.

We would like to propose a practical implementation of INW, ICW and Robust-INW for real-life contents such as images or sounds.

Acknowledgements. The work described in this paper has been supported (in part) by the NSFC (60633030) and the 973 Program (2006CB303104).

References

1. Barni, M., Bartolini, F., Furon, T.: A general framework for robust watermarking security. Signal Processing 83, 2069–2084 (2003)
2. Cachin, C.: An Information-Theoretic Model for Steganography. In: Aucsmith, D. (ed.) IH 1998. LNCS, vol. 1525, pp. 306–318. Springer, Heidelberg (1998)
3. Kerckhoffs, A.: La cryptographie militaire. Journal Des Sciences Militaires IX, 5–38 (1883)
4. Kalker, T.: Considerations on watermarking security. In: Proc. MMSP, Cannes, France, pp. 201–206 (October 2001)
5. Cayre, F., Furon, T., Fontaine, C.: Watermarking security: Theory and practice. IEEE Trans. Signal Process. 53(10), 3976–3987 (2005)
6. Cox, I., Killian, J., Leighton, F., Shamoon, T.: Secure spread spectrum watermarking for multimedia. IEEE Trans. Image Processing 6(12), 1673–1687 (1997)
7. Malvar, H.S., Flôrencio, D.: Improved spread spectrum: a new modulation technique for robust watermarking. IEEE Trans. Signal Process. 53, 898–905 (2003)
8. Comesaña, P., Pérez-Freire, L., Pérez-González, F.: Fundamentals of data hiding security and their application to spread-spectrum analysis. In: Barni, M., Herrera-Joancomartí, J., Katzenbeisser, S., Pérez-González, F. (eds.) IH 2005. LNCS, vol. 3727, pp. 146–160. Springer, Heidelberg (2005)
9. Bas, P., Cayre, F.: Achieving subspace or key security for woa using natural or circular watermarking. In: Proc. ACM Multimedia Security Workshop, Geneva (2006)
10. Cayre, F., Bas, P.: Kerckhoffs-based embedding security classes for WOA Data Hiding. IEEE Trans. Inf. Forensics and Security 3(1), 1–15 (2008)
11. Bas, P., Cayre, F.: Natural watermarking: a secure spread spectrum technique for WOA. In: Camenisch, J.L., Collberg, C.S., Johnson, N.F., Sallee, P. (eds.) IH 2006. LNCS, vol. 4437, pp. 1–14. Springer, Heidelberg (2007)

A New Spread Spectrum Watermarking Scheme to Achieve a Trade-Off between Security and Robustness

Jian Cao, Jiwu Huang, and Jiangqun Ni

School of Information Science and Technology, Sun Yat-Sen University, Guangzhou, China, 510275
phdcaojian@yahoo.cn, isshjw@mail.sysu.edu.cn

Abstract. To achieve a trade-off between robustness and security in the Watermarked Only Attack (WOA) framework, we propose a novel spread spectrum watermarking scheme called natural watermarking-normalized circular watermarking (NW-NCW). Our scheme builds on two spread spectrum watermarking schemes. One is natural watermarking (NW), and has an advantage in security; the other is a novel circular watermarking scheme called normalized circular watermarking (NCW), and has an advantage in robustness. We show that in most typical scenarios, NW-NCW is more suitable for the tradeoff between robustness and security than the existing secure watermarking schemes such as NW and CW-ISS (circular extension of ISS).

Keywords: Circular watermarking, natural watermarking, spread spectrum watermarking, watermarking security, WOA.

1 Introduction

In general, designing a watermarking scheme always involves a trade-off among several conflicting objectives, especially robustness, distortion and security. The notions of security and robustness are very different in essence. According to the definitions proposed in [1], attacks to robustness are those whose target is to increase the probability of error of the watermarking channel, while attacks to security are those whose target is to gain knowledge about the secret key of the embedding and decoding processes. However, the notions of security and robustness are also close in that the attacker can design more powerful robustness attack as long as he can gain enough knowledge about the secret key. Watermarking security builds on Kerckhoffs' principle [2]. It implies that all details of the watermarking scheme are publicly known except the secret key of the embedding and decoding processes. In [4], following the Diffie-Hellman's methodology, the authors proposed a classification of the attacking scenarios. Among them, the Watermarked Only Attack (WOA) framework refers to those attacking scenarios where the attacker only owns several watermarked contents, i.e., he knows neither corresponding original versions nor corresponding hidden messages. This paper only considers the WOA framework unless otherwise stated.

R. Böhme, P.W.L. Fong, and R. Safavi-Naini (Eds.): IH 2010, LNCS 6387, pp. 262–276, 2010.

The security of a spread spectrum watermarking scheme is made up of two layers. The first layer is the embedding subspace spanned by the secret carriers, and the second is the secret carriers. If the attacker can disclose the embedding subspace, he can remove the watermark with low distortion. In [4], for instance, the authors presented a method for removing the watermark with low distortion by nullifying the watermarked signal's projection in the estimated embedding subspace. By contrast, the secret carriers have more precise information. If the attacker can disclose the secret carriers, he has the capacity for full access to the watermarking channel. Hence, to resist unauthorized embedding or extraction attack, ensuring that estimation of the secret carriers is impossible is a basic requirement. Classical spread spectrum watermarking schemes, such as additive spread spectrum (SS)[5] and improved spread spectrum (ISS)[6], have already been shown to be insecure against carriers estimation in [4].

Recently, two spread spectrum embedding functions which are secure against carriers estimation have already been proposed by other authors in [7],[8] and [9]. These two embedding functions have their own advantages and disadvantages. One embedding function is called natural watermarking (NW), and has an advantage in security. For example, there exists a embedding parameter setting such that NW is stego-security, i.e., it is impossible to make the difference between cover and stego contents. Stego-security also implies that it is impossible to estimate the embedding subspace. The disadvantage of NW is that its robustness is relatively poor. The other embedding function is based on the well-known ISS modulation, is called circular extension of ISS (CW-ISS) and has an advantage in robustness. The disadvantage of CW-ISS is that it cannot achieve Stego-security. Actually, it is also impossible to achieve subspace security for CW-ISS. In other words, there exists no the embedding parameters setting such that it is impossible to estimate the embedding subspace. Consequently, neither NW nor CW-ISS is very suitable for the tradeoff between robustness and security. The goal of this paper is to present a new spread spectrum watermarking scheme for the tradeoff between robustness and security. Specifically, we hope that our watermarking scheme has the following three properties. The first one is that it is secure against carriers estimation for any freely chosen embedding parameters. The second one is that there exists a embedding parameters setting such that it is stego-security. The third one is that there exists a embedding parameters setting such that it can achieve roughly the same robustness as CW-ISS. Toward this end, we firstly present a new circular watermarking called normalized circular watermarking (NCW). NCW has an advantage in robustness and an interesting property that CW-ISS does not have, i.e, NCW keeps the watermarked signal's projection (in the embedding subspace) in the same orientation as NW. The property is important to design our new watermarking scheme called natural watermarking-normalized circular watermarking (NW-NCW). As shown in the paper, in most typical scenarios NW-NCW is more suitable for the tradeoff between robustness and security than existing secure watermarking schemes, namely, NW and CW-ISS.

This paper is organized as follows: Section 2 gives a brief description of existing secure spread spectrum watermarking schemes, namely, NW and CW-ISS. Furthermore, Sect. 2 also analyzes the robustness of NW and CW-ISS against the Additive White Gaussian Noise (AWGN) attack, obtaining corresponding expressions of the bit error rate (BER). And in the Sect. 3, we firstly propose a novel circular watermarking called NCW, and then propose our novel watermarking scheme called NW-NCW. The robustness and distortion related to NW-NCW are also presented. Finally, we analyze and compare the performances of various secure SS-based watermarking schemes in Sect. 4.

2 Analysis of Robustness of NW and CW-ISS

This section reviews the principles of NW and CW-ISS and also analyzes the robustness of NW and CW-ISS.

2.1 Spread Spectrum Watermarking

Let $\mathbf{m} \in \{-1, +1\}^{N_c}$ denote an N_c-bits message to be embedded in a host signal $\mathbf{x}$ with N_v coefficients. It is assumed that the coefficients, $\mathbf{x}_i : i = 1, 2, \ldots, N_v$ are independent and identically distributed (i.i.d.) Gaussian variables with zero mean and variance σ_x^2. Further, we need a secret key used to initialize a PRNG (Pseudo-Random Number Generator) in order to provide N_c secret carriers $\{\mathbf{u}_i\}$, and use a Gram-Schmidt procedure to achieve an orthogonal basis of the embedding subspace:

$$\mathbf{u}^T\mathbf{u} = \mathbf{I}_{N_c} \ , \tag{1}$$

where $\mathbf{u}^T$ denotes the transpose of the matrix $\mathbf{u}$, and $\mathbf{I}_{N_c}$ denotes the identity matrix of size $N_c \times N_c$. We obtain the watermarked signal $\mathbf{s}$ as follows:

$$\mathbf{s} = \mathbf{x} + \mathbf{w} \ , \tag{2}$$

where

$$\mathbf{w} = \sum_{i=1}^{N_c} \mu(\mathbf{u}^T\mathbf{x}, \mathbf{m}(i))\mathbf{u}_i \ . \tag{3}$$

where $\mathbf{w}$ denotes the watermark signal. The distortion is assessed by watermark-to-content power ratio (WCR):

$$\mathrm{WCR} = 10 \log \left(\frac{\sigma_w^2}{\sigma_x^2} \right) \ . \tag{4}$$

The robustness attacks are modeled as additive noise:

$$\mathbf{y} = \mathbf{s} + \mathbf{n} \ , \tag{5}$$

where we assume attack noise $\mathbf{n}$ from an uncorrelated white Gaussian random process, i.e., $\mathbf{n} \sim \mathcal{N}(\mathbf{0}, \sigma_n^2 \mathbf{I}_{N_v})$. The attack strength is accessed by means of the Watermark to Noise ratio (WNR):

$$\text{WNR} = 10 \log \left(\frac{\sigma_w^2}{\sigma_n^2} \right) . \tag{6}$$

The decoded message is denoted by $\hat{\mathbf{m}}$. It is estimated from $\mathbf{y}$, a potentially degraded version of $\mathbf{s}$. The decoding rule for spread spectrum watermarking schemes is given by:

$$\widehat{\mathbf{m}}(i) = \begin{cases} +1 & \text{if } \mathbf{y}^T \mathbf{u}_i > 0 , \\ -1 & \text{if } \mathbf{y}^T \mathbf{u}_i < 0 . \end{cases} \tag{7}$$

We measure decoding performance with bit error rate (BER):

$$\text{BER} = \frac{1}{N_c} \sum_{i=1}^{N_c} E[\mathbf{m}(i) \neq \hat{\mathbf{m}}(i)] . \tag{8}$$

2.2 Natural Watermarking and Its Robustness

The embedding function of NW ($\alpha \geq 1$) is given by:

$$\mathbf{s} = \mathbf{x} + \sum_{i=1}^{N_c} \left[\left[-1 + \alpha \mathbf{m}(i) \text{sign} \left(\mathbf{x}^T \mathbf{u}_i \right) \right] \mathbf{x}^T \mathbf{u}_i \right] \mathbf{u}_i . \tag{9}$$

NW presents two interesting properties. One is that NW can achieve stego-security by setting $\alpha = 1$ since in this case the distribution of $\mathbf{x}^H \mathbf{u}$ remains invariant before and after embedding, i.e., $\mathbf{s}^H \mathbf{u} \sim \mathbf{x}^H \mathbf{u}$. The other is that NW is secure against carriers estimation for any freely chosen parameter α since the projection $\mathbf{s}^H \mathbf{u}$ after NW embedding has a distribution invariant under rotations.

When $\alpha = 1$, NW achieves its best security, that is, stego-security. As the embedding parameter α increases, the robustness of NW increases but its security decreases. When $\alpha = \sqrt{\frac{N_v 10^{WCR/10}}{N_c} - 1}$, NW achieves its best robustness. The main disadvantage of NW is the fact that the best robustness that NW can achieve is not very good. Hence, NW is not very suitable for the tradeoff between robustness and security.

Robustness of NW: In [8], the expression of the BER is computed only for $\alpha = 1$. We will deduce the expression of the BER for any chosen embedding parameter α. The i-th bit of the embedded message is estimated by:

$$\hat{\mathbf{m}}(i) = \text{sign}(\mathbf{y}^T \mathbf{u}_i) , \tag{10}$$

where

$$\mathbf{y}^T \mathbf{u}_i = \alpha \mathbf{m}(i) \text{sign}(\mathbf{x}^T \mathbf{u}_i) \mathbf{x}^T \mathbf{u}_i + \mathbf{n}^T \mathbf{u}_i . \tag{11}$$

In particular, let us consider the case when $\mathbf{m}(i) = 1$. Then, an error occurs when $\mathbf{y}^T\mathbf{u}_i < 0$, and therefore the BER for NW is given by:

$$\mathrm{P_e} = \Pr\{\mathbf{y}^T\mathbf{u}_i < 0 | \mathbf{m}(i) = 1\} \tag{12}$$

$$= \int_{-\infty}^{0} g(z)\mathrm{d}z \ , \tag{13}$$

with

$$g(z) = \frac{1}{\sqrt{2\pi(\sigma_n^2 + (\alpha\sigma_x)^2)}} \exp\left(-\frac{z^2}{2(\sigma_n^2 + (\alpha\sigma_x)^2)}\right) \mathrm{erfc}\left(-\frac{\alpha\sigma_x z}{\sqrt{2(\sigma_n^2 + (\alpha\sigma_x)^2)}\sigma_n}\right) .$$

where erfc(.) is the complementary error function. The same BER is obtained under the assumption that $\mathbf{m}(i) = -1$.

2.3 Circular Extension of ISS and Its Robustness

The embedding function of CW-ISS is given by:

$$\mathbf{s} = \mathbf{x} + \sum_{i=1}^{N_c} \left(\sqrt{N_c}\alpha\mathbf{m}(i)\frac{|\mathbf{g}_i|}{\|\mathbf{g}\|} - \lambda\ \mathbf{x}^T\mathbf{u}_i\right)\mathbf{u}_i \ . \tag{14}$$

The vector $\mathbf{g}$ is distributed according to Gaussian variables, $\mathbf{g} \sim \mathcal{N}(\mathbf{0}, \mathbf{I}_{N_c})$, and will be independently drawn at each embedding. When compared with NW, CW-ISS can achieve better robustness (see [7] and [8]). However, the disadvantage of CW-ISS is that there exists no the embedding parameters setting such that it is stego-security or subspace security. Hence, CW-ISS is not applicable for the tradeoff between robustness and security either.

Robustness of CW-ISS: The expression of the BER for CW-ISS is given only for $N_c = 2$ in [8], and no explicit formula is given. In the rest of this subsection, we will show how to deduce the expression of the BER for CW-ISS.

In particular, let us consider the case when $\mathbf{m}(i) = 1$. The projection of the attacked watermarked signal $\mathbf{y}$ on the i-th secret carrier $\mathbf{u}_i$ can be rewritten as:

$$\mathbf{y}^T\mathbf{u}_i = (1-\lambda)\mathbf{x}^T\mathbf{u}_i + \sqrt{N_c}\alpha\frac{|\mathbf{g}_i|}{\|\mathbf{g}\|} + \mathbf{n}^T\mathbf{u}_i \ . \tag{15}$$

It is easy to show that both $\mathbf{x}^T\mathbf{u}_i$ and $\mathbf{n}^T\mathbf{u}_i$ are Gaussian, i.e.,

$$\mathbf{x}^T\mathbf{u}_i \sim \mathcal{N}(0, \sigma_x^2), \quad \mathbf{n}^T\mathbf{u}_i \sim \mathcal{N}(0, \sigma_n^2) \ . \tag{16}$$

Then, $(1-\lambda)\mathbf{x}^T\mathbf{u}_i + \mathbf{n}^T\mathbf{u}_i$ is also Gaussian, i.e.,

$$(1-\lambda)\mathbf{x}^T\mathbf{u}_i + \mathbf{n}^T\mathbf{u}_i \sim \mathcal{N}\left(0, (1-\lambda)^2\sigma_x^2 + \sigma_n^2\right) . \tag{17}$$

Let $\mathbf{d}_i = \frac{\mathbf{g}_i}{\|\mathbf{g}\|}, 1 \le i \le N_c$. In [10], the marginal probability density function of one component $\mathbf{d}_i$ was computed by integrating the probability density function of $\mathbf{d}$ over the surface of an $(N_c - 1)$-dimensional sphere with radius $\sqrt{1-(\mathbf{d}_i)^2}$, resulting in:

$$f(d_i) = \frac{\Gamma(\frac{N_c}{2})}{\sqrt{\pi}\Gamma(\frac{N_c-1}{2})}(1-d_i^2)^{\frac{N_c-3}{2}}, \quad d_i \in [-1, 1]. \tag{18}$$

where $\Gamma(.)$ denotes the complete Gamma function. Then, it is easy to show that the marginal probability density function of random variable $\sqrt{N_c}\alpha|\mathbf{d}_i|$ is

$$f(z) = \begin{cases} \frac{2\Gamma(N_c/2)(N_c\alpha^2 - z^2)^{\frac{N_c-3}{2}}}{\sqrt{\pi}\Gamma((N_c-1)/2)(\sqrt{N_c}\alpha)^{N_c-2}}, \text{if} z \in [0, \alpha), \\ 0, \text{otherwise}, \end{cases} \tag{19}$$

which follows from the fact that if we let $Y = aX$, then $f_Y(y) = \frac{1}{|a|} f_X f_X(\frac{y}{a})$. Applying (15),(17) and (19), the BER can be expressed as follows:

$$\begin{aligned} \mathrm{P_e} &= \Pr\{\mathbf{y}^H \mathbf{u}_i < 0 | \mathbf{m}(i) = 1\} \ , \\ &= \int_0^{\sqrt{N_c}\alpha} h(t) \mathrm{erfc}(\frac{t}{\sqrt{2}\sqrt{(1-\lambda)^2\sigma_x^2 + \sigma_n^2}}) \mathrm{d}t \quad , \end{aligned} \tag{20}$$

with

$$h(t) = \frac{\Gamma(N_c/2)(N_c\alpha^2 - t^2)^{\frac{N_c-3}{2}}}{\sqrt{\pi}\Gamma\left((N_c-1)/2\right)(\sqrt{N_c}\alpha)^{N_c-2}} \ .$$

3 NW-NCW

The robustness of CW-ISS is much better than the robustness of NW in the case of low security, while the robustness of CW-ISS is much worse than the robustness of NW in the case of high security. In other words, CW-ISS is suitable only for low security, and NW is suitable only for high security. As a result, neither NW nor CW-ISS is suitable for the tradeoff between robustness and security. This section shows how to devise a new SS-based watermarking scheme in such a way that it is suitable for any level of security. Before introducing our new watermarking scheme which is suitable for robustness-security tradeoffs, we firstly present a new circular watermarking called NCW.

3.1 NCW

It is easy to show that the projection $\mathbf{x}^T\mathbf{u}$ is Gaussian, that is,

$$\mathbf{x}^T\mathbf{u} \sim \mathcal{N}(0, \sigma_x^2 \mathbf{I}_{N_c}) \ . \tag{21}$$

Note that $\mathbf{x}^T\mathbf{u}$ is indeed isotropically distributed (i.e., with its probability density function invariant under rotations). So random vector $\frac{\mathbf{x}^T\mathbf{u}}{\|\mathbf{x}^T\mathbf{u}\|}$ is uniformly distributed on the surface of the N_c-dimensional sphere of unit radius. The embedding process of normalized CW is as follows:

$$\mathbf{s}^T\mathbf{u}_i = \alpha\mathbf{m}(i)\frac{|\mathbf{x}^T\mathbf{u}_i|}{\|\mathbf{x}^T\mathbf{u}\|} = \alpha\mathbf{m}(i)\frac{\text{sign}(\mathbf{x}^T\mathbf{u}_i)\mathbf{x}^T\mathbf{u}_i}{\|\mathbf{x}^T\mathbf{u}\|} , \tag{22}$$

which means that the embedding function of normalized CW is given by:

$$\mathbf{s} = \mathbf{x} + \sum_{i=1}^{N_c}\left(\alpha\mathbf{m}(i)\frac{\text{sign}(\mathbf{x}^T\mathbf{u}_i)\mathbf{x}^T\mathbf{u}_i}{\|\mathbf{x}^T\mathbf{u}\|} - \mathbf{x}^T\mathbf{u}_i\right)\mathbf{u}_i . \tag{23}$$

where $\|\mathbf{x}^T\mathbf{u}\| = \sqrt{\sum_{i=1}^{N_c}(\mathbf{x}^T\mathbf{u}_i)^2}$. It is easy to show NCW is a special case of circular watermarking. That is, the projection $\mathbf{s}^T\mathbf{u}$ after NCW embedding has a distribution invariant under rotations. In addition, NCW has an property that CW-ISS does not have, that is, the watermarked signal's projection $\mathbf{s}^T\mathbf{u}$(in the embedding space) after NCW embedding is in the same direction as that projection after NW embedding. The property of NCW is important to design our watermarking scheme called NW-NCW.

3.2 NW-NCW

Embedding using NW($\alpha = 1$), we have:

$$\mathbf{s}_{\text{NW}}^T\mathbf{u}_i = \mathbf{m}(i)\text{sign}(\mathbf{x}^T\mathbf{u}_i)\mathbf{x}^T\mathbf{u}_i . \tag{24}$$

Embedding using NCW, we have:

$$\mathbf{s}_{\text{NCW}}^T\mathbf{u}_i = \alpha\mathbf{m}(i)\frac{\text{sign}(\mathbf{x}^T\mathbf{u}_i)\mathbf{x}^T\mathbf{u}_i}{\|\mathbf{x}^T\mathbf{u}\|} . \tag{25}$$

It is easy to see that $\mathbf{s}_{\text{NW}}^T\mathbf{u}$ and $\mathbf{s}_{\text{NCW}}^T\mathbf{u}$ are in the same direction. It is important for us to design a spread spectrum watermarking scheme for the tradeoff between robustness and security in this paper. Now, let us introduce a parameter η. In particular, the projection $\mathbf{s}^T\mathbf{u}$ after NW-NCW is given by:

$$\mathbf{s}^T\mathbf{u} = (1-\eta)\mathbf{s}_{\text{NW}}^T\mathbf{u} + \eta\mathbf{s}_{\text{NCW}}^T\mathbf{u} . \tag{26}$$

where $0 \le \eta \le 1$. The embedding function of NW-NCW is given by:

$$\mathbf{s} = \mathbf{x} + \sum_{i=1}^{N_c}\left(\mathbf{s}^T\mathbf{u}_i - \mathbf{x}^T\mathbf{u}_i\right)\mathbf{u}_i . \tag{27}$$

This embedding rule is depicted in Fig. 1. The parameter η controls the trade-off between security and robustness. When $\eta = 0$, NW-NCW achieves its best security , this is, stego-security. When $\eta = 1$, NW-NCW achieves its best robustness in most typical scenarios (see Fig. 3). By changing the parameter η, the watermarker can choose an appropriate security for a specific application.

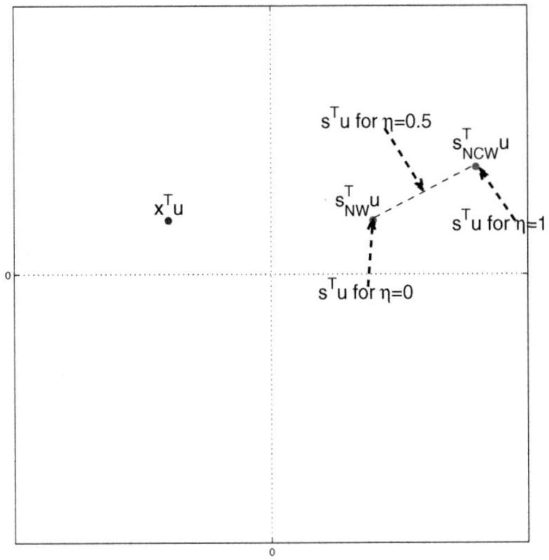

Fig. 1. NW-NCW for $\mathbf{m} = \{1, 1\}$ $(N_c = 2)$

Distortion of NW-NCW: Considering the embedding rule of NW-NCW, we can deduce the watermark signal $\mathbf{w}$ as follows:

$$\mathbf{w} = \sum_{i=1}^{N_c} \left(\left((1-\eta) + \frac{\alpha\eta}{\|\mathbf{x}^T\mathbf{u}\|} \right) \mathbf{m}(i)\mathrm{sign}(\mathbf{x}^T\mathbf{u}_i)\mathbf{x}^T\mathbf{u}_i - \mathbf{x}^T\mathbf{u}_i \right) \mathbf{u}_i \ . \qquad (28)$$

Then, the expectation of $\mathbf{w}^T\mathbf{w}$ is the following:

$$\begin{aligned} E[\mathbf{w}^T\mathbf{w}] &= \sum_{i=1}^{N_c} E[(\mathbf{x}^T\mathbf{u}_i)^2] \\ &+ \sum_{i=1}^{N_c} E\left[-2 \left((1-\eta) + \frac{\alpha\eta}{\|\mathbf{x}^T\mathbf{u}\|} \right) \mathbf{m}(i)\mathrm{sign}(\mathbf{x}^T\mathbf{u}_i)(\mathbf{x}^T\mathbf{u}_i)^2 \right] \\ &+ \sum_{i=1}^{N_c} E\left[\left(\left((1-\eta) + \frac{\alpha\eta}{\|\mathbf{x}^T\mathbf{u}\|} \right) \mathbf{m}(i)\mathrm{sign}(\mathbf{x}^T\mathbf{u}_i)\mathbf{x}^T\mathbf{u}_i \right)^2 \right] \ . \end{aligned} \qquad (29)$$

which follows from $\mathbf{u}^T\mathbf{u} = \mathbf{I}_{N_c}$. It is easy to show that the first and second terms of (29) are:

$$\sum_{i=1}^{N_c} E[(\mathbf{x}^T\mathbf{u}_i)^2] = N_c\sigma_x^2 \ . \qquad (30)$$

$$\sum_{i=1}^{N_c} E[-2 \left((1-\eta) + \frac{\alpha\eta}{\|\mathbf{x}^T\mathbf{u}\|} \right) \mathbf{m}(i)\mathrm{sign}(\mathbf{x}^T\mathbf{u}_i)(\mathbf{x}^T\mathbf{u}_i)^2] = 0 \ . \qquad (31)$$

Then, we compute the third term of (29):

$$\sum_{i=1}^{N_c} E\left[\left(\left((1-\eta)+\frac{\alpha\eta}{\|\mathbf{x}^T\mathbf{u}\|}\right)\mathbf{m}(i)\mathrm{sign}(\mathbf{x}^T\mathbf{u}_i)\mathbf{x}^T\mathbf{u}_i\right)^2\right]$$

$$=\sum_{i=1}^{N_c} E[(1-\eta)^2(\mathbf{x}^T\mathbf{u}_i)^2]+2\alpha\eta(1-\eta)E[\|\mathbf{x}^T\mathbf{u}\|]+(\alpha\eta)^2 \ ,$$

$$=(1-\eta)^2N_c\sigma_x^2+2\alpha\eta(1-\eta)\sigma_x\sqrt{2}\frac{\Gamma\left((Nc+1)/2\right)}{\Gamma(Nc/2)}+(\alpha\eta)^2 \ . \tag{32}$$

where (32) follows from the fact that $\frac{1}{\sigma_x}\|\mathbf{x}^T\mathbf{u}\|$ is distributed according to the chi distribution with N_c degrees of freedom. The WCR is expressed as follows:

$$WCR_{\mathrm{dB}}=10\log_{10}\left(\frac{\left(1+(1-\eta)^2\right)N_c\sigma_x^2+2\alpha\eta(1-\eta)\sigma_x\sqrt{2}\frac{\Gamma((Nc+1)/2)}{\Gamma(Nc/2)}+(\alpha\eta)^2}{N_v\sigma_x^2}\right) \tag{33}$$

We confirm in Fig.2 that there is no difference between this theoretical expression and the practical measurements. Practical WCR tests are made with 10000 host Gaussian signals.

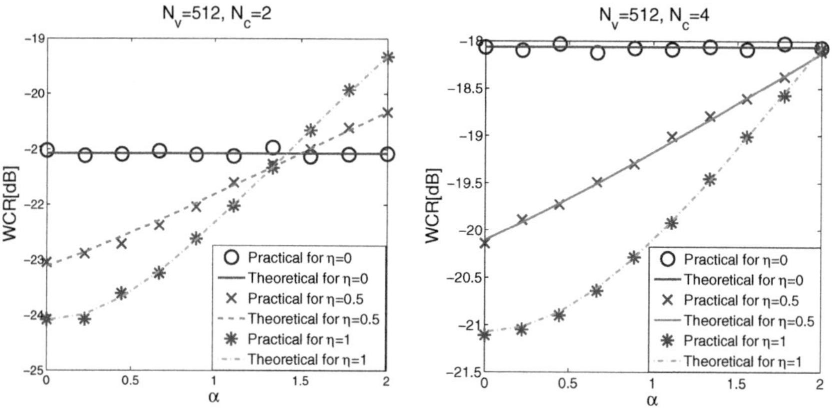

Fig. 2. Comparison between theoretical and practical WCR. $\sigma_x^2 = 1$.

If one wants to specify a target average WCR, the parameter α has the following expression:

$$\alpha=\frac{1}{\eta}\sqrt{Nv\sigma_x^2 10^{WCR/10}+2\sigma_x^2(1-\eta)^2(\frac{\Gamma((N_c+1)/2)}{\Gamma(N_c/2)})^2-(1+(1-\eta)^2)N_c\sigma_x^2}$$
$$-\frac{\sqrt{2}(1-\eta)\Gamma((N_c+1)/2)}{\eta\Gamma(N_c/2)}\sigma_x \ . \tag{34}$$

Robustness of NW-NCW ($0 < \eta < 1$): Decoding is performed by:

$$\hat{\mathbf{m}}(i) = \text{sign}(\mathbf{y}^T \mathbf{u}_i) \ , \tag{35}$$

with

$$\mathbf{y}^T \mathbf{u}_i = (1-\eta)\mathbf{m}(i)\text{sign}(\mathbf{x}^T \mathbf{u}_i)\mathbf{x}^T \mathbf{u}_i + \frac{\alpha\eta}{\|\mathbf{x}^T \mathbf{u}\|}\mathbf{m}(i)\text{sign}\left(\mathbf{x}^T \mathbf{u}_i\right)\mathbf{x}^T \mathbf{u}_i + \mathbf{n}^T \mathbf{u}_i \ . \tag{36}$$

In particular, let us consider the case of $\mathbf{m}(i) = 1$. An error occurs when $\mathbf{y}^T \mathbf{u}_i < 0$. Therefore, the BER is given by:

$$p = \Pr\{\mathbf{y}^T \mathbf{u}_i < 0 | \mathbf{m}(i) = 1\} \ . \tag{37}$$

Given $(1-\eta)\text{sign}(\mathbf{x}^T \mathbf{u}_i)\mathbf{x}^T \mathbf{u}_i = t$, the conditional pdf of $\frac{\alpha\eta}{\|\mathbf{x}^T \mathbf{u}\|}\text{sign}\left(\mathbf{x}^T \mathbf{u}_i\right)\mathbf{x}^T \mathbf{u}_i$ is:

$$f(v) = \frac{2(\alpha\eta)^2 t^{N_c-1}((\alpha\eta)^2 - v^2)^{(N_c-3)/2}}{2^{(N_c-1)/2}\Gamma((N_c-1)/2)(1-\eta)^{N_c}v^{N_c}} \exp(-\frac{t^2((\alpha\eta)^2 - v^2)}{2(1-\eta)^2 v^2}), \quad v \in [0, \alpha\eta] \ . \tag{38}$$

It is easy to show that the random variable $(1-\eta)\text{sign}(\mathbf{x}^T \mathbf{u}_i)\mathbf{x}^T \mathbf{u}_i$ is half-Gaussian:

$$f(t) = \begin{cases} \frac{2}{\sqrt{2\pi}(1-\eta)\sigma_x} \exp(-\frac{t^2}{2(1-\eta)^2\sigma_x^2}), & \text{if} \quad t > 0, \\ 0, \quad \text{otherwise} \ . \end{cases} \tag{39}$$

and the random variable $\mathbf{n}^T \mathbf{u}_i$ is Gaussian:

$$\mathbf{n}^T \mathbf{u}_i \sim \mathcal{N}(0, \sigma_n^2) \ . \tag{40}$$

Finally, the BER is the following:

$$p = \int_0^\infty \frac{2}{\sqrt{2\pi}(1-\eta)\sigma_x} \exp(-\frac{t^2}{2(1-\eta)^2\sigma_x^2}) \int_0^{\alpha\eta} \frac{1}{2} f(v) \text{erfc}(\frac{t+v}{\sqrt{2}\sigma_n}) \mathrm{d}v\mathrm{d}t \ . \tag{41}$$

Robustness of NW-NCW ($\eta = 0$): Remember that NW-NCW ($\eta = 0$) corresponds to NW ($\alpha = 1$). Thereby, the BER of NW-NCW ($\eta = 0$) is:

$$p = \int_{-\infty}^0 g(z) \mathrm{d}z \ , \tag{42}$$

with

$$g(z) = \frac{1}{\sqrt{2\pi(\sigma_{\mathbf{n}}^2 + \sigma_{\mathbf{x}}^2)}} \exp\left(-\frac{z^2}{2(\sigma_{\mathbf{n}}^2 + \sigma_{\mathbf{x}}^2)}\right) \text{erfc}(-\frac{\sigma_{\mathbf{x}} z}{\sqrt{2(\sigma_{\mathbf{n}}^2 + \sigma_{\mathbf{x}}^2)}\sigma_{\mathbf{n}}}) \ .$$

Robustness of NW-NCW ($\eta = 1$): NW-NCW ($\eta = 1$) corresponds to NCW. In this case, the correlation of the attacked watermarked signal $\mathbf{y}$ with the i-th carrier is:

$$\mathbf{y}^T\mathbf{u}_i = \frac{\alpha\eta}{\|\mathbf{x}^T\mathbf{u}\|}\mathbf{m}(i)\text{sign}\left(\mathbf{x}^T\mathbf{u}_i\right)\mathbf{x}^T\mathbf{u}_i + \mathbf{n}^T\mathbf{u}_i \ . \tag{43}$$

In particular, let us consider the case when $\mathbf{m}(i) = 1$. An error occurs when $\mathbf{y}^T\mathbf{u}_i < 0$. Therefore, the BER is given by:

$$p = \Pr\{\mathbf{y}^T\mathbf{u}_i < 0|\mathbf{m}(i) = 1\} \ . \tag{44}$$

It is clear that the random variable $\frac{\text{sign}\left(\mathbf{x}^T\mathbf{u}_i\right)\mathbf{x}^T\mathbf{u}_i}{\|\mathbf{x}^T\mathbf{u}\|}$ has the same distribution as the random variable $\frac{|\mathbf{g}_i|}{\|\mathbf{g}\|}$. Hence, the probability density function of the random variable $\alpha\frac{\text{sign}\left(\mathbf{x}^T\mathbf{u}_i\right)\mathbf{x}^T\mathbf{u}_i}{\|\mathbf{x}^T\mathbf{u}\|}$ is:

$$f(t) = \frac{2\Gamma(N_c/2)(\alpha^2 - t^2)^{\frac{N_c-3}{2}}}{\sqrt{\pi}\Gamma\left((N_c - 1)/2\right)(\alpha)^{N_c-2}}, \quad t \in [0, \alpha] \ . \tag{45}$$

Now, it is easy to show that the BER is:

$$p = \frac{1}{2}\int_0^\alpha f(t)\text{erfc}(\frac{t}{\sqrt{2}\sigma_n})dt \ . \tag{46}$$

Fig.3 shows no difference between this theoretical expression and the practical measurements, where practical tests are made with 10000 host Gaussian signals.

Security of NW-NCW: Figure 4 shows the distribution of the projections $\mathbf{x}^T\mathbf{u}$ and the distribution of the projections $\mathbf{s}^T\mathbf{u}$ after NW-NCW embedding for $N_c = 2$, where security tests are made with 10000 host Gaussian signals. As we can see, when $\eta = 0$, the distribution of the projections $\mathbf{s}^T\mathbf{u}$ is the same as the distribution of the host signal's projections in the embedding subspace , and

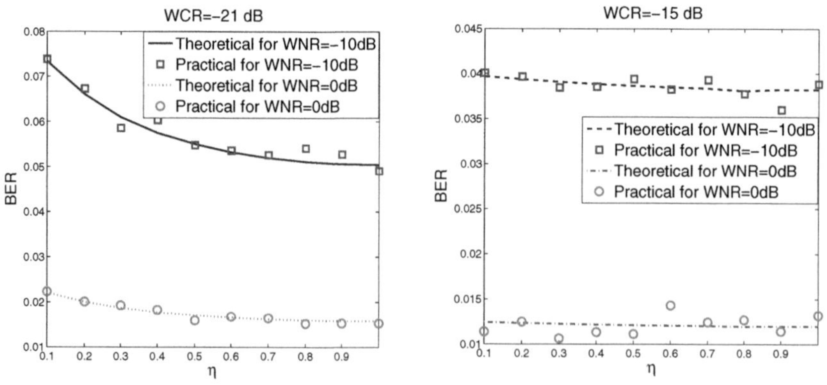

Fig. 3. Comparison of theoretical and practical BER. $N_v = 512, N_c = 2$.

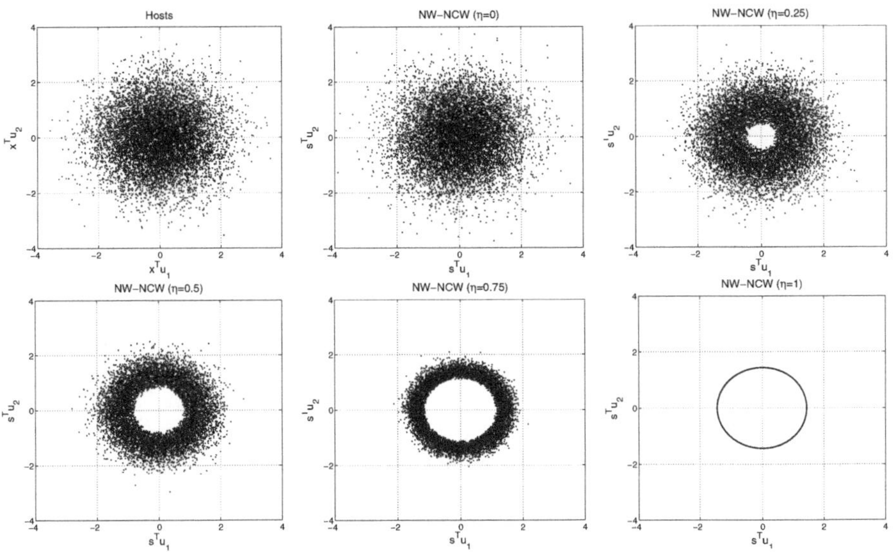

Fig. 4. The distribution of the projections $\mathbf{x}^T\mathbf{u}$(top-left) and the distributions of the projections $\mathbf{s}^T\mathbf{u}$ after NW-NCW embedding for various η. $N_v = 512$, $N_c = 2$ and WCR=-21dB.

for the other parameter, η, the distribution of the projections $\mathbf{s}^T\mathbf{u}$ is circular (i.e., invariant under rotations) but is not the same as the distribution of the projections $\mathbf{x}^T\mathbf{u}$. These results imply that when $\eta = 0$, NW-NCW achieves stego-security, and when $\eta \neq 0$, NW-NCW does not achieve stego-security but is still secure against carriers estimation.

4 Simulation Results and Analysis

In this section, we will access how the robustness and security of NW-NCW vary with the parameter η, and compare the performance of NW, CW-ISS and NW-NCW from the point of the view of the tradeoff.

4.1 Security-Robustness Tradeoff Analysis for NW-NCW

In the last section, we have shown that in most typical scenarios, the robustness of NW-NCW increases with the parameter η (see Fig. 3). Next, we will analyze how the security of NW-NCW varies with the parameter η. The security of NW-NCW, since it is circular watermarking for any freely chosen embedding parameters, is equivalent to the difficulty of estimating the embedding subspace spanned by the secret carriers $\{\mathbf{u}_i\}$. To access the difficulty of subspace estimation, we decide to adopt the methodology called PCA (principal component analysis), and we resort to the chordal distance [11] to measure the distance between the embedding subspace $\mathbf{u}$ and the estimated subspace $\hat{\mathbf{u}}$. PCA involves

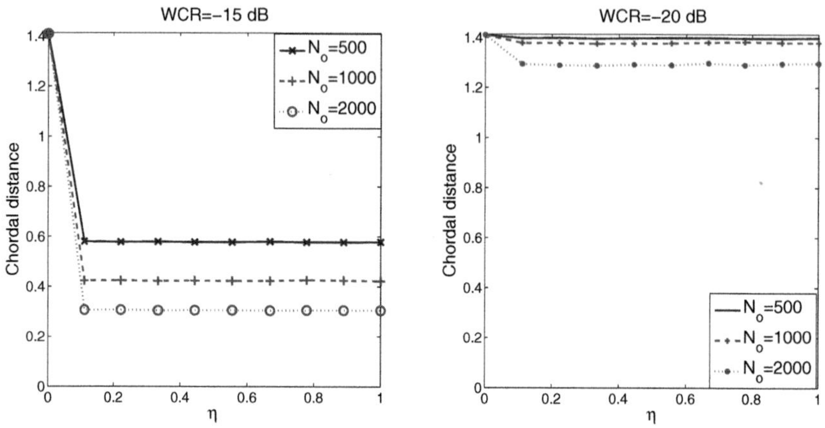

Fig. 5. Chordal distance between the estimated subspace and the original one for NW-NCW. $N_c = 2$, $N_v = 512$.

the calculation of the eigenvalue decomposition of the covariance matrix of watermarked contents $\mathbf{S}$ (taken column-wise). In this context, the estimated base of the embedding subspace $\hat{\mathbf{u}}$ is the set of N_c eigenvectors of the covariance matrix $\mathbf{S}$ corresponding to N_c biggest eigenvalues. The chordal distance is defined as $d_c(\mathbf{u}, \hat{\mathbf{u}}) = \frac{1}{\sqrt{2}}\|\mathbf{u}\mathbf{u}^T - \hat{\mathbf{u}}\hat{\mathbf{u}}^T\|_F$, where $\|.\|_F$ denotes the Frobenius norm for matrices. The chordal distance achieves it maximum, $\sqrt{N_c}$, when the two subspaces are perfectly orthogonal, and it equals 0 when both matrices generate the same subspace. The chordal distance close to $\sqrt{N_c}$ illustrates the fact that the subspace estimation is impossible. Figure 5 presents the chordal distance between the estimated subspace and the real one after NW-NCW embedding for 100 experiments. As expected, NW-NCW is secure against the embedding subspace estimation only in the case of $\eta = 0$.

4.2 Comparison of NW, CW-ISS and NW-NCW

We only consider the case when the attacker obtains enough observations. In this context, there exist only two security requirements, that is, ensuring it is impossible to estimate the embedding subspace (subspace-security) or ensuring it is impossible to estimate the secret carriers (key-security).

Subspace-security: Remember that NW is a special case of NW-NCW. Hence, NW-NCW can achieve this security requirement. In this case, it is easy to show the robustness of NW-NCW is the same as the robustness of NW. However, CW-ISS cannot achieve this security requirement.

Key-security: We choose the parameter $\eta = 1$ for NW-NCW . This is because the robustness of NW-NCW is best when $\eta = 1$ in most typical scenarios. Figure 6 compares the robustness of the various proposed watermarking schemes to AWGN addition for the same hidden channel rate (the same N_c and the same

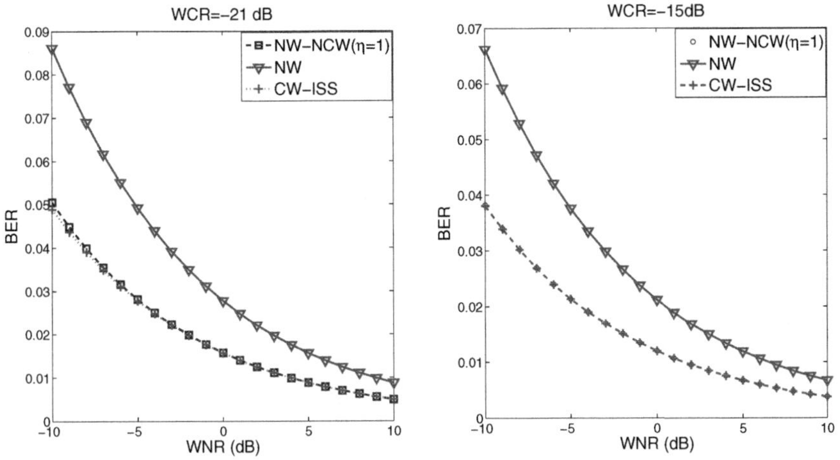

Fig. 6. Comparison of BER for NW, CW-ISS and NW-NCW ($\eta = 1$). $N_c = 2, N_v = 512$.

N_v). Formulas for NW and CW-ISS are given in Sect.2 and NW-NCW is given in Sect.3. We can see in this figure that NW-NCW achieves roughly the same robustness as CW-ISS and has a better robustness than NW.

5 Conclusion and Future Works

This paper has presented a spread spectrum watermarking scheme called NW-NCW to provide a framework for the tradeoff between robustness and security. NW-NCW presents three interesting properties. The first one is that NW-NCW can achieve stego-security, which implies that it is impossible to make the difference between the cover and stego contents. Stego-security also implies that it is impossible to estimate both the secret carriers and the embedding subspace. The second one is that NW-NCW can achieve roughly the same robustness as CW-ISS and has a better robustness than NW. The third one is that for any freely chosen embedding parameters, since it is circular watermarking, NW-NCW is secure against carriers estimation. Furthermore, in order to design the NW-NCW, another new circular watermarking called NCW is also proposed in this paper. Finally, in order to compare the robustness of various presented watermarking schemes, the expressions of BER for NW, CW-ISS and NW-NCW are also computed. Our future works will concentrate on tradeoff analysis for NW-NCW. In particular, we will compute a close expression for the security of NW-NCW from an information-theoretic point of view by means of the equivocation about the secret carriers. In the case of NW-NCW, the equivocation measures the difficulty of estimating the embedding subspace.

Acknowledgements. The authors appreciate the supports received from NSFC (60970145, 60633030), and 973 Program (2006CB303104).

References

1. Comesaña, P., Pérez-Freire, L., Pérez-González, F.: Fundamentals of data hiding security and their application to spread-spectrum analysis. In: Barni, M., Herrera-Joancomartí, J., Katzenbeisser, S., Pérez-González, F. (eds.) IH 2005. LNCS, vol. 3727, pp. 146–160. Springer, Heidelberg (2005)
2. Kerckhoffs, A.: La cryptographie militaire. Journal des Sciences militaires IX, 5–38 (1883)
3. Kalker, T.: Considerations on watermarking security. In: Proc. MMSP, Cannes, France, pp. 201–206 (October 2001)
4. Cayre, F., Furon, T., Fontaine, C.: Watermarking security: Theory and practice. IEEE Trans. Signal Process. 53(10), 3976–3987 (2005)
5. Cox, I., Killian, J., Leighton, F., Shamoon, T.: Secure spread spectrum watermarking for multimedia. IEEE Trans. Image Processing 6(12), 1673–1687 (1997)
6. Malvar, H.S., Flôrencio, D.: Improved spread spectrum: a new modulation technique for robust watermarking. IEEE Trans. Signal Process. 53, 898–905 (2003)
7. Bas, P., Cayre, F.: Achieving subspace or key security for woa using natural or circular watermarking. In: Proc. ACM Multimedia Security Workshop, Geneva (2006)
8. Cayre, F., Bas, P.: Kerckhoffs-based embedding security classes for WOA Data Hiding. IEEE Trans. Inf. Forensics and Security 3(1), 1–15 (2008)
9. Bas, P., Cayre, F.: Natural watermarking: a secure spread spectrum technique for WOA. In: Camenisch, J.L., Collberg, C.S., Johnson, N.F., Sallee, P. (eds.) IH 2006. LNCS, vol. 4437, pp. 1–14. Springer, Heidelberg (2007)
10. Pérez-Freire, L., Pérez-González, F.: Spread spectrum watermarking security. IEEE Trans. Inf. Forensice and Security 4(1), 2–24 (2009)
11. Conway, J.H., Hardin, R.H., Sloane, N.J.A.: Packing lines, plaries, etc.: Packings in grassmanian spaces. Experimental Mathematics 5(2), 139–159 (1996)

Author Index